AI for Cybersecurity: Robust models for Authentication, Threat and Anomaly Detection

AI for Cybersecurity: Robust models for Authentication, Threat and Anomaly Detection

Editors

Francesco Bergadano
Giorgio Giacinto

MDPI • Basel • Beijing • Wuhan • Barcelona • Belgrade • Manchester • Tokyo • Cluj • Tianjin

MDPI

Editors

Francesco Bergadano
University of Torino
Torino, Italy

Giorgio Giacinto
University of Cagliari
Cagliari, Italy

Editorial Office
MDPI
St. Alban-Anlage 66
4052 Basel, Switzerland

This is a reprint of articles from the Special Issue published online in the open access journal *Algorithms* (ISSN 1999-4893) (available at: https://www.mdpi.com/journal/algorithms/special_issues/AI_Cybersecurity_Model).

For citation purposes, cite each article independently as indicated on the article page online and as indicated below:

LastName, A.A.; LastName, B.B.; LastName, C.C. Article Title. *Journal Name* **Year**, *Volume Number*, Page Range.

ISBN 978-3-0365-8264-1 (Hbk)
ISBN 978-3-0365-8265-8 (PDF)

Contents

algorithms

MDPI

Editorial

Special Issue "AI for Cybersecurity: Robust Models for Authentication, Threat and Anomaly Detection"

Francesco Bergadano [1,*] and Giorgio Giacinto [2]

[1] Department of Computer Science, University of Torino, Via Pessinetto 12, 10149 Torino, Italy
[2] Department of Electrical and Electronic Engineering, University of Cagliari, Piazza d'Armi, 09123 Cagliari, Italy; giacinto@unica.it
[*] Correspondence: francesco.bergadano@unito.it

1. Introduction

Cybersecurity models include provisions for legitimate user and agent authentication, as well as algorithms for detecting external threats, such as intruders and malicious software. In particular, we can define a continuum of cybersecurity measures ranging from user identification to risk-based and multilevel authentication, complex application and network monitoring, and anomaly detection. We refer to this as the "anomaly detection continuum".

Machine learning and other artificial intelligence technologies can provide powerful tools for addressing such issues, but the robustness of the obtained models is often ignored or underestimated. On the one hand, AI-based algorithms can be replicated by malicious opponents, and attacks can be devised so that they will not be detected (evasion attacks). On the other hand, data and system contexts can be modified by attackers to influence the countermeasures obtained from machine learning and render them ineffective (active data poisoning).

This Special Issue presents ten papers [1–10] that can be grouped under five main topics.

2. Cyber Physical Systems (CPSs) [1–3]

AI techniques are particularly needed for the security of CPSs. This is due to the high number and large variety of devices that cannot be manually controlled and monitored. Security automation is also needed in this context because of the deployment of the target infrastructure, which is often remote and difficult to access physically. The first paper [1] reviews existing studies and datasets for anomaly detection in CPSs. In [2], the authors propose a new approach for multi-vector attack detection in the IoT domain, using machine learning algorithms and providing an experimental evaluation. In article [3], classifiers obtained via machine learning were applied to the security monitoring of smart grids, and an adaptive deep learning algorithm is proposed and evaluated with the NSL-KDD dataset.

3. Intrusion Detection [4,5]

Intrusion detection is traditionally a common target of AI applications in the context of cybersecurity because machine learning can provide a means to train models that distinguish normal traffic from malicious attacks. The fourth paper [4] studies such issues in the particular context of cooperative intelligent transportation systems, proposing algorithms and an intrusion detection architecture evaluated on the NGSIM dataset. The fifth paper [5] is devoted to network intrusion detection and addresses the problems of high false negative rates and low predictability for minority classes.

4. Malware Analysis [6]

Malware detection, analysis, and response can be partly automated with artificial intelligence. The number and variety of malware attacks make this a necessity, as manual inspection, as well as ad hoc countermeasures, would be impossible. In [6], the authors

Citation: Bergadano, F.; Giacinto, G. Special Issue "AI for Cybersecurity: Robust Models for Authentication, Threat and Anomaly Detection". *Algorithms* **2023**, *16*, 327. https://doi.org/10.3390/a16070327

Received: 29 June 2023
Accepted: 30 June 2023
Published: 7 July 2023

compare different ensemble learning methods that have been proposed in this context: Random Forests, XGBoost, CatBoost, GBM, and LightGBM. Experiments were performed on different datasets, finding that tree-based ensemble learning algorithms can achieve good performance with limited variability.

5. Access Control [7,8]

As stated above, access control can be viewed as another point in the anomaly detection continuum. Again, distinguishing a legitimate user from impostors can be automated through machine learning. The seventh paper [7] addresses this in the context of face recognition systems (FRSs) and proposes a practical white box adversarial attack algorithm. The method is evaluated with the CASIA WebFace and the LFW datasets. In [8], the authors used the legitimate user's iris image, combined with a secret key, to generate a public key and subsequently use such data to limit access to protected resources.

6. Threat Intelligence [9,10]

Not only do we want to recognize and block attacks as they occur—we also need to observe external data and the overall network context to predict relevant events and new attack patterns, addressing the so-called threat intelligence landscape. In [9], the authors used two well-known threat databases (CVE and MITRE) and proposed a technique to link and correlate these two sources. The tenth paper [10] used formal ontologies to monitor new threats and identify the corresponding risks in an automated way.

7. Conclusions

In conclusion, we observed that AI is increasingly being used in cybersecurity, with three main directions of current research: (1) new areas of cybersecurity are addressed, such as CPS security and threat intelligence; (2) more stable and consistent results are being presented, sometimes with surprising accuracy and effectiveness; and (3) the presence of an AI-aware adversary is recognized and analyzed, producing more robust and reliable solutions.

Author Contributions: Special issue editorial by both authors. All authors have read and agreed to the published version of the manuscript.

Funding: This research received no external funding.

Conflicts of Interest: The authors declare no conflict of interest.

References

1. Tushkanova, O.; Levshun, D.; Branitskiy, A.; Fedorchenko, E.; Novikova, E.; Kotenko, I. Detection of Cyberattacks and Anomalies in Cyber-Physical Systems: Approaches, Data Sources, Evaluation. *Algorithms* **2023**, *16*, 85. [CrossRef]
2. Lysenko, S.; Bobrovnikova, K.; Kharchenko, V.; Savenko, O. IoT Multi-Vector Cyberattack Detection Based on Machine Learning Algorithms: Traffic Features Analysis, Experiments, and Efficiency. *Algorithms* **2022**, *15*, 239. [CrossRef]
3. Li, X.J.; Ma, M.; Sun, Y. An Adaptive Deep Learning Neural Network Model to Enhance Machine-Learning-Based Classifiers for Intrusion Detection in Smart Grids. *Algorithms* **2023**, *16*, 288. [CrossRef]
4. Almalki, S.A.; Abdel-Rahim, A.; Sheldon, F.T. Adaptive IDS for Cooperative Intelligent Transportation Systems Using Deep Belief Networks. *Algorithms* **2022**, *15*, 251. [CrossRef]
5. Mijalkovic, J.; Spognardi, A. Reducing the False Negative Rate in Deep Learning Based Network Intrusion Detection Systems. *Algorithms* **2022**, *15*, 258. [CrossRef]
6. Louk, M.H.L.; Tama, B.A. Tree-Based Classifier Ensembles for PE Malware Analysis: A Performance Revisit. *Algorithms* **2022**, *15*, 332. [CrossRef]
7. Lang, D.; Chen, D.; Huang, J.; Li, S. A Momentum-Based Local Face Adversarial Example Generation Algorithm. *Algorithms* **2022**, *15*, 465. [CrossRef]
8. Matveev, I.; Safonov, I. From Iris Image to Embedded Code: System of Methods. *Algorithms* **2023**, *16*, 87. [CrossRef]

9. Grigorescu, O.; Nica, A.; Dascalu, M.; Rughinis, R. CVE2ATT&CK: BERT-Based Mapping of CVEs to MITRE ATT&CK Techniques. *Algorithms* **2022**, *15*, 314.
10. Shaked, A.; Margalit, O. Sustainable Risk Identification Using Formal Ontologies. *Algorithms* **2022**, *15*, 316. [CrossRef]

algorithms

MDPI

Article

Detection of Cyberattacks and Anomalies in Cyber-Physical Systems: Approaches, Data Sources, Evaluation

Olga Tushkanova [1,†], Diana Levshun [1,†], Alexander Branitskiy [1,†], Elena Fedorchenko [1,2,*,†], Evgenia Novikova [1,†] and Igor Kotenko [1,2,†]

[1]   Computer Security Problems Laboratory, St. Petersburg Federal Research Center of the Russian Academy of Sciences, 199178 Saint-Petersburg, Russia
[2]   Department of Computer Science and Engineering, Saint-Petersburg Electrotechnical University ETU "LETI", 197022 Saint-Petersburg, Russia
*   Correspondence: doynikova@comsec.spb.ru
†   These authors contributed equally to this work.

Abstract: Cyberattacks on cyber-physical systems (CPS) can lead to severe consequences, and therefore it is extremely important to detect them at early stages. However, there are several challenges to be solved in this area; they include an ability of the security system to detect previously unknown attacks. This problem could be solved with the system behaviour analysis methods and unsupervised or semi-supervised machine learning techniques. The efficiency of the attack detection system strongly depends on the datasets used to train the machine learning models. As real-world data from CPS systems are mostly not available due to the security requirements of cyber-physical objects, there are several attempts to create such datasets; however, their completeness and validity are questionable. This paper reviews existing approaches to attack and anomaly detection in CPS, with a particular focus on datasets and evaluation metrics used to assess the efficiency of the proposed solutions. The analysis revealed that only two of the three selected datasets are suitable for solving intrusion detection tasks as soon as they are generated using real test beds; in addition, only one of the selected datasets contains both network and sensor data, making it preferable for intrusion detection. Moreover, there are different approaches to evaluate the efficiency of the machine learning techniques, that require more analysis and research. Thus, in future research, the authors aim to develop an approach to anomaly detection for CPS using the selected datasets and to conduct experiments to select the performance metrics.

Keywords: anomaly detection; attack detection; cyber-physical system; machine learning; datasets; evaluation metrics

Citation: Tushkanova, O.; Levshun, D.; Branitskiy, A.; Fedorchenko, E.; Novikova, E.; Kotenko, I. Detection of Cyberattacks and Anomalies in Cyber-Physical Systems: Approaches, Data Sources, Evaluation. *Algorithms* 2023, 16, 85. https://doi.org/10.3390/a16020085

Academic Editors: Francesco Bergadano and Giorgio Giacinto

Received: 21 December 2022
Revised: 28 January 2023
Accepted: 30 January 2023
Published: 3 February 2023

1. Introduction

Cybersecurity risks are highly relevant nowadays. It is almost impossible to completely exclude security risks for modern information systems, including cyber-physical systems (CPS) and Internet of Things (IoT). Thus, it is essential to continuously detect cyberattacks and anomalies to monitor security risks and provide security awareness.

Cyberattacks against cyber-physical systems can lead to severe impacts on physical, environmental, as well as economical safety of the population [1]. For example, the attack on the Colonial Pipeline disrupted fuel supply on the US East Coast in 2021 [2], and the attack on the Venezuelan hydroelectric power plant led to a nationwide blackout in 2019 [3]. In 2022, Germany's internal fuel distribution system was disrupted by a cyberattack [4]. Thus, it is extremely important to detect such attacks at early stages.

There are several challenges in this area, and one of the most critical challenges is the detection of the previously unknown attacks. Another challenge relates to the availability of the datasets used to train analytical models, as the performance of the attack detection strongly depends on the quality of the training datasets. The first challenge relates to the fact that machine learning models are usually trained on datasets with known attack patterns, and as a result, they are unable to detect previously unseen attacks. One of the possible solutions is to use anomaly detection techniques based on the analysis of the cyber-physical entities' behaviour [5–7]. However, such approaches require high-quality datasets to model normal behaviour or apply unsupervised or semi-supervised machine learning techniques. The lack of datasets close to the real world is explained by the fact that organizations do not want to share data, as they can include confidential data. There are attempts to generate such datasets using cyber-physical or software test beds, but the completeness and validity of such generated datasets are questionable. The last challenge relates to the validation of the attack and anomaly detection models. The analysis of the research papers has shown that different researchers use different approaches to calculate performance metrics that complicate the comparison of the models.

In this paper, the authors review existing approaches to attack and anomaly detection, outline the most commonly used datasets, and evaluate the applicability of the selected datasets in the anomaly detection task. We also revealed that researchers use different approaches to calculate performance metrics to evaluate machine learning models. These metrics consider the fact that the anomalies in CPS have a certain duration, and malicious activity may result in a delayed response of the system process; however, the variety of used metrics makes the comparison of the obtained experimental results complicated.

Thus, the *contribution* of the research is as follows:

- analysis of the approaches to anomaly detection for the cyber-physical systems;
- analysis of the selected datasets, namely, ToN_IoT [8], SWAT [9], and HAI [10] containing normal and anomaly related data for the cyber-physical systems, and selection of the dataset for the experiments;
- overview of the metrics used to evaluate the anomaly and attack detection models.

The paper is organized as follows. Section 2 provides the results of the review of the approaches to anomaly and attack detection for cyber-physical systems. Section 3 analyzes the datasets used for the attack and anomaly detection that contain the data from the cyber-physical systems. Section 4 researches the metrics for the evaluation of the attack and anomaly detection models. The paper ends with a conclusion.

2. Approaches to the Anomaly and Attack Detection for the Cyber-Physical Systems

Anomaly detection is the process of identifying anomalous events that do not match the expected behaviour of the system. This allows the detection of new and hidden attacks. Currently, anomaly detection approaches are often implemented using machine learning, such as shallow (or traditional) learning and deep learning [7,11,12]. In this case, the profile of normal behaviour can be built using many data sources.

Anomaly and attack detection in CPS based on shallow learning methods uses algorithms such as support vector machine (SVM) [13], Bayesian classification [14], k-nearest neighbor (kNN) [15], Random Forest (RF) [16,17], Isolation Forest [18], XGBoost [19], and artificial neural networks (ANN) [20,21]. They are based on training intelligent models to profile the normal behaviour of a cyber-physical system, and then inconsistent observations are identified as anomalies. For example, Elnour et al. [18] propose an attack detection framework based on dual isolation forest (DIF). Two isolated forest models are trained independently using normalized raw data and a preprocessed version of the data using principal component analysis (PCA). The principle of the approach is to detect and separate anomalies using the concept of isolation after analyzing the data in the original and PCA-transformed representations. Mokhtari et al. [16] and Park and Lee [17] explore such supervised learning algorithms for anomaly detection as k-nearest neighbours, decision tree classifier, and random forest. In both studies, the random forest shows the best detection result.

The analysis of related works has shown that the research focus has now shifted towards the use of deep neural networks to detect anomalies in technological processes. A number of authors compare classical and deep learning approaches to anomaly detection.

So Inoue et al. [22] compare one-class SVM with radial basis function kernel deep and dense neural network with a layer of long short-term memory (LSTM), and the experiments have shown that the deep learning model is characterized by a lower rate of false positive alarms. Gaifulina and Kotenko [23] experimentally compare several models of deep neural networks for anomaly detection. Shalyga et al. [24] propose several methods to improve the quality of anomaly detection, including exponentially weighted smoothing to reduce the false positive rate, individual error weight for each feature, non-overlapping prediction windows, etc. The authors also propose their own anomaly detection model based on a multilayer perceptron (MLP).

Traditional machine learning methods tend to be inefficient when processing large-scale data and unevenly distributed samples. Deep learning models are more productive when analyzing such data. Researchers often use autoencoders (AE) [5,6], recurrent neural networks [25], convolutional neural networks (CNN) [26–28], and generative adversarial networks (GAN) [29,30] as deep neural networks for anomaly detection in CPS. Often, the approaches propose a hybrid use of neural network data. For example, Xie et al. [25] and Wu et al. [31] use CNN for data dimensionality reduction and gated recurrent units (GRU) for data prediction. GRU is one of the types of recurrent networks, as well as LSTM. Bian X. [32] also uses GRU for anomaly detection. The main idea of the anomaly detection method is to predict the value of the next moment and determine if an anomaly occurs due to a deviation between the predicted value and the actual value.

The autoencoder is trained on normal data, and then the incoming events are reconstructed based on the normal model. Exceeding the reconstruction error threshold indicates an anomaly. Such an approach is used in the APAD (Autoencoder-based Payload Anomaly Detection) model by Kim et al. [5]. Wang et al. [6] propose an approach to anomaly detection using a composite model. The proposed model consists of three components: the encoder and decoder used to reconstruct the error, and the LSTM classifier, which takes the encoder output as input and makes predictions. To detect an anomaly, both model outputs, i.e., reconstruction error and prediction value, are considered together to calculate the anomaly score. The authors also compare the change ratio of each attribute during the current period and the previous one, and those attributes that have changed more are considered anomalous.

Generative adversarial networks can be used to investigate the distribution of normal data for recognizing anomalies from unknown data. The generator creates new data instances, and the discriminator evaluates them for authenticity. In the MAD-GAN (Multivariate Anomaly Detection with GAN) approach by Li et al. [29], both generator and discriminator components are represented by LSTM. The discriminator is trained to distinguish anomalies from normal data, and the anomaly score is computed as a combination of the discrimination output and reconstruction error produced by the generator component. A similar approach is proposed by Neshenko et al. [30]. The building blocks for the proposed GAN are the recurrent neural network and convolutional neural network. The authors also extended the anomaly detection approach by incorporating a module that attributes potentially attacked sensors. This task is solved by the application of various techniques starting with feature importance evaluation and finishing with KernelShap [33] and LIME [34] techniques that are model agnostic methods.

We should also mention approaches to anomaly detection using graph probabilistic models, such as Bayesian networks (BN) and Markov models. For example, Lin et al. [35] propose TABOR (Time Automata and Bayesian netwORk). Time Automata simulate the operation of the sensors and actuators, and the Bayesian network (BN) models the dependencies among random variables from the sensors and the actuators. This approach allows for the detection of timing anomalies, anomalies of sensor, and actuator value range, as well as a violation in their dependencies. Another popular way to represent normal behaviour is the hidden Markov model (HMM). Sukhostat L. [36] uses hierarchical HMM to detect anomalies in sensor values.

Application of the proposed techniques requires high-quality datasets that allow proper modelling of the CPS system functioning. Depending on the technique, it is required to have only normal data; some techniques require having both samples with normal and abnormal behaviour.

The first group of datasets is the data containing the indicators of the sensors of the cyber-physical system in the form of logs. The analysis of the research papers showed that currently, the most commonly used CPS dataset is SWAT dataset [9]. It is used in [5,6,18,24,25,29,30,35,36]. This dataset contains records from sensors, actuators, control programmable logic controllers (PLCs), and network traffic. Another new dataset for anomaly detection is HAI [10], which is used in research [16,17,32]. The dataset contains the parameters of sensors for an industrial power generation system using steam turbines and pumped storage power plants. To detect anomalies in IoT devices, the authors in the papers [19,20,31] use the TON_IoT dataset [8]. The ToN_IoT dataset includes telemetry from heterogeneous IoT and Industrial Internet of Things (IIoT) sensors.

Another group of datasets that are often used to detect anomalies and attacks in CPS are represented by network traffic datasets. They include such datasets such as NSL-KDD [37], CICIDS2017 [38] and UNSW-NB15 [39], and are used in the following research papers [23,27,31,40]. However, these datasets are represented mainly by network data that could be given in form of the PCAP (Packet Capture) files or labelled network flows. Section 3 discusses datasets in detail.

We should note that differences in the experimental conditions affect the possibility of comparing the results of anomaly detection. For example, Elnour et al. [18] exclude the stabilization time from the SWaT dataset. The way metrics are calculated can also vary, and research papers do not always provide a way to calculate these metrics. In general, the above machine learning methods show high anomaly detection results and can be used in further developments. A promising area of research and development is the creation of hybrid machine learning models for anomaly detection. In particular, combined networks with RNNs are used to capture temporal relationships [6,29], and combined networks with CNNs are applicable for context analysis (e.g., packet order and content) [25,30].

3. Datasets for the Attack and Anomaly Detection

An essential challenge of anomaly detection research is generating or finding a suitable dataset for the experiments. The authors analyzed existing datasets to select the dataset for further research.

The authors specified the following requirements of the dataset based on the research goal of anomaly detection in cyber-physical systems:

R1: the dataset should be gathered from the cyber-physical system;
R2: the dataset should contain event logs;
R3: the dataset should contain anomalies;
R4: the dataset should be labelled (what is normal and what is abnormal);
R5: the dataset should be close to real data (i.e., data from the real or semi-real system).

Currently, there are a lot of datasets available for various purposes and systems; they represent the functioning of the computer networks and cyber-physical systems, including the Internet of Things, Industrial Internet of Things, and Industrial Control Systems (ICS), such as SCADA (Supervisory Control And Data Acquisition) system [41].

Alsaedi et al. [8] present the comparative analysis of the available datasets for security purposes. Thus, there are datasets containing computer network traffic that was generated for attack detection purposes: KDDCUP99, NSL-KDD [37], UNSW-NB15 [39], and CICIDS2017 [38]. Such datasets do not contain sensors' data that is specific to CPS. Moreover, they do not include CPS network traffic, both normal and abnormal.

There are also datasets generated for cyber-physical systems security research purposes. Choi et al. [42] provide a comparison of the existing datasets generated for ICSs security research based on attack paths. Lemay and Fernandez [43] generate the SCADA network datasets (Modbus dataset) for intrusion detection research. The SCADA network datasets by Rodofile et al. [44] contain attacks on the S7 protocol. These datasets are SCADA specific and contain a limited set of protocol specific attacks.

There are also multiple datasets for IoT and IIoT. Suthaharan et al. [45] propose the labelled wireless sensor network dataset (LWSNDR). It contains homogeneous data collected from a humidity-temperature sensor. The sensor is deployed in single-hop and multi-hop wireless sensor networks (WSNs). The dataset does not contain attack scenarios, but does contain anomalies introduced by the author using a hot water kettle. Sivanathan et al. [46] propose the datasets gathered from a smart home testbed. It contains network traffic characteristics of IoT devices. The dataset is generated for the IoT devices classification. The dataset does not contain attack scenarios.

There are also multiple network-based IoT datasets [37–39,46–48]. These datasets do not consider sensor data; thus, they do not allow for the detection of the attacks that manipulate sensors' data.

The datasets that are suitable considering the set requirements, i.e., that contain labelled sensors and network data, are as follows: TON_IoT [8], SWaT [9], and HAI [10]. The authors conducted a more detailed analysis of these datasets.

3.1. TON_IoT Dataset Analysis

The TON_IoT dataset is created by the Intelligent Security Group of the UNSW Canberra, Australia, and positioned by its authors as realistic telemetry datasets of IoT and IIoT sensors. It contains data from seven IoT devices, namely, a smart fridge, GPS tracker, smart sense motion light, remotely activated garage door, Modbus device, smart thermostat, and weather monitoring system. All the data were generated using a testbed of Industry 4.0/Industrial IoT networks developed by the authors. The data include several normal and cyber-attack events, namely, scanning, DoS, DDoS, ransomware, backdoor, data injection, cross-site scripting, password cracking attacks, and man-in-the-middle. The TON_IoT dataset incorporates the ground truth indicating normal and attack classes for binary classification, and the feature indicating the classes of attacks for multi-classification problems. Statistics on class balance for device samples from the TON_IoT dataset are presented in Table 1.

Table 1. The statistics on the TON_IoT dataset class balance by devices.

IoT Device	Normal	Attack	Total	Class Balance, %
Fridge	35,000	24,944	59,944	58/42
Garage Door	35,000	24,587	59,587	59/41
GPS Tracker	35,000	23,960	58,960	59/41
Modbus	35,000	16,106	51,106	68/32
Motion Light	35,000	24,488	59,488	59/41
Thermostat	35,000	17,774	52,774	66/34
Weather	35,000	24,260	59,260	59/41

Alsaedi et al. and Moustafa [8] also tried several popular machine learning methods to show that the TON_IoT dataset may be used to train classifiers for intrusion detection purposes. To justify the results and ensure that attacks are indeed identifiable, we have tried to follow the course of the authors' experiment with binary classification. It should be mentioned that the authors reported very high accuracy for the majority of the investigated methods (more than 0.8 for the F-measure in most cases). As we tried to follow the authors, at first we applied the same preprocessing procedures, namely, transformed categorical features with two unique values into binary ones, applied the min-max scaling technique to numeric features, and randomly split data into train and test subsamples in 80% to 20% stratified proportion.

It should be noted that during data preprocessing, we found several artefacts in the data. For example, *'temp_condition'* feature for the fridge contains values 'high', 'low', 'low', 'high', 'low', 'high' values, and *'sphone_signal'* for fridge contains 'true', 'false', '0', '1' values. As there are no special notes about that in the paper or the dataset description, we supposed that those were inaccuracies in the data and fixed them.

Figure 1 shows the correlation between features for different devices, both with each other and with the anomaly behaviour label. We can note a high correlation between the features of the dataset for a fridge, garage door, GPS tracker, and motion light. At the same time, the correlation value between these features and the label is low. The correlation of features for Modbus, thermostat, and weather is close to zero.

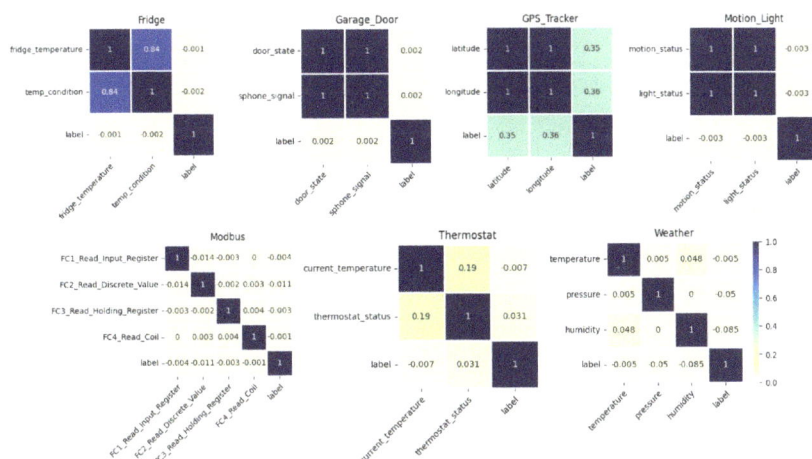

Figure 1. IoT device feature and label correlation.

We applied the same machine learning models to those mentioned in the original paper, namely, Logistic Regression (LR), Linear Discriminant Analysis (LDA), k-Nearest Neighbour (kNN), Classification and Regression Trees (CART), Random Forest (RF), Naïve Bayes (NB), Support Vector Machine (SVM), and Long Short-Term Memory (LSTM), with the hyperparameters that authors specified, and also tried to tune those hyperparameters using 4-fold cross-validation.

We did not manage to reach the reported accuracy for most of the datasets, in either case. Table 2 shows the best values for F-measure that we received for classifiers trained on 80% of the data for each device calculated on the remaining 20% of the data.

Table 2. The F-measure values for the best hyperparameters of the model trained on the TON_IoT dataset calculated for the test subsample.

IoT Device	LR	LDA	kNN	RF	CART	NB	SVM	LSTM
Fridge	0	0	0.37	0.02	0	0	0	0
Garage Door	0.58	0	0.56	0	0	0	0	0
GPS Tracker	0.51	0.43	0.95	0.95	0.93	0.43	0.81	0.85
Modbus	0	0	0.87	0.97	0.97	0	0	0
Motion Light	0	0	0.50	0	0	0	0	0
Thermostat	0	0	0.26	0.31	0.33	0	0	0
Weather	0.10	0.10	0.95	0.98	0.97	0.53	0.58	0.61

The best F-measure values were obtained for the GPS Tracker dataset. We assume that this is due to the strongest correlation between features and anomaly class labels in this device dataset in comparison to the other device datasets. For other datasets, correlations are close to zero, that is, very weak. The strong correlations between features and the weak correlations between features and anomaly class labels for fridge, garage door, and motion light may explain the low F-measure values for these datasets.

Further investigation of the data showed that anomaly class labels relate only to data and time of device events; although, according to the authors' experiment design, date and time are not taken into account. Figure 2 shows an example distribution of anomaly class labels in time for the temperature feature of the smart fridge.

Figure 2. Normal and attack events for temperature feature of smart fridge.

Conclusions. We analyzed the obtained results considering the set requirements. Requirements R1, R2, and R4 are satisfied; requirement R3 is partially satisfied, as soon as the dataset contains attack scenarios. However, the performed experiments showed that these attacks do not affect IoT telemetry. The requirement R5 is not satisfied. The analysis demonstrated that there is no connection between the data in the network dataset and the data in the sensor's dataset. Moreover, the sensors do not follow any normal behaviour scenario and the obtained accuracy results are rather low. Thus, this data set is not suitable for the goals of further research.

3.2. SWaT Dataset Analysis

The Secure Water Treatment (SWaT) dataset [9] is generated by the Singapore University of Technology and Design (SUTD). The researchers deployed a six-stage SWaT testbed simulating a real-world industrial water treatment plant. The collected dataset contains both normal and attack traffic. It should be noticed that the deployed plant was run non-stop for eleven days: during the first seven it operated without any attacks, while during the remaining days, cyber and physical attacks were conducted against the plant. The collected dataset contains both the data from sensors and actuators of the plant (25 sensors and 26 actuators) and network traffic. Currently, there are several versions of this dataset; the researchers regularly update it by organizing cybersecurity events using it, thus, generating new data with different attack types.

We conducted a series of experiments with different machine learning models for anomaly detection using the SWaT dataset 2015 to evaluate this dataset and check its compliance with the criteria proposed above. The dataset incorporates three CSV files with anomaly (or attack) and normal data: "Attack_v0.csv", "Normal_ v0.csv", and "Normal_v1.csv". The attacks were performed on different technological processes, and Table 3 shows the number of abnormal records for different technological processes. It should be also noted that some network attacks do not impact the readings from physical sensors.

Table 3. Distribution of the attacks per processes in SWaT dataset

Record Type	Number of Impacted Processes	Impacted Processes	Number of Samples
Normal	0	0	399,157
Attack	1	P1	4053
	1	P2	1809
	1	P3	37,860
	1	P4	1700
	1	P5	1044
	2	P3, P4	1691
	2	P1, P3	1445
	2	P3, P6	697
	2	P4, P5	463

Experiment 1. For this experiment series, we tried both time and random train-test splits on the "Attack_v0.csv" dataset containing 449,919 rows in total, including 395,298 normal records and 54,621 anomaly records that correspond to attacks, meaning that the contamination rate is 0.138 for this subsample. For the time train-test split mode, the training sample was incorporated all rows before 2 January 2016 , while the testing sample contained rows after 2 January 2016 (inclusively). Due to uneven distribution of anomalies across time, the class balance for train and test subsamples was different: the train subsample included 344,436 normal instances and 51,483 attack instances meaning that the contamination rate was equal to 0.149; the test subsample included 50,862 normal instances and 3138 attack instances with a contamination rate of 0.062. The results of the experiment for the train-test split mode and different anomaly detection machine learning models are provided in Table 4. The best results were obtained for the K Nearest Neighbors method (KNN) with *F*1-measure 0.784, AUC-ROC 0.935, and AUC-PRC 0.739 on the test subsample.

Table 4. The results of Experiment 1 for the time split mode for the SWaT dataset.

Optimal Threshold	Train Data						Test Data					
	P	R	FPR	F1	AUC-ROC	AUC-PRC	P	R	FPR	F1	AUC-ROC	AUC-PRC
Sklearn												
ocSVM	0.300	0.795	0.205	0.436	0.723	0.087	0.355	0.193	0.807	0.250	0.654	0.051
isoF	0.045	0.240	0.760	0.076	0.868	0.072	0.065	0.839	0.161	0.120	0.567	0.051
PYOD												
ECOD	0.806	0.668	0.331	0.731	0.879	0.772	0.310	0.270	0.730	0.289	0.791	0.240
COPOD	0.879	0.662	0.338	0.755	0.878	0.791	0.497	0.268	0.732	0.348	0.796	0.236
KNN	0.252	0.008	0.993	0.015	0.204	0.087	0.819	0.752	0.248	0.784	0.935	0.739
Deep-SVDD	0.803	0.011	0.989	0.022	0.633	0.187	0.965	0.079	0.921	0.147	0.566	0.143
VAE	0.729	0.745	0.255	0.737	0.892	0.666	0.364	0.493	0.507	0.419	0.785	0.201
AutoEnc	0.721	0.753	0.247	0.737	0.894	0.672	0.305	0.460	0.540	0.367	0.793	0.205
AnoGAN	0.896	0.653	0.347	0.756	0.875	0.777	0.422	0.212	0.788	0.282	0.695	0.182

For the random train-test split mode, we used 80% to 20% ratio so the train subsample contained 316,238 normal instances and 43,697 attack instances, while the test subsample contained 79,060 normal instances and 10,924 attack instances with a contamination rate of 0.138 for both. The results of experiment 1 for the random train-test split mode and different anomaly detection machine learning models are provided in Table 5. It can be seen that rather close results were obtained for the ECOD (F1-measure 0.743, AUC-ROC 0.878, and AUC-PRC 0.758 on the testing sample), COPOD (F1-measure 0.744, AUC-ROC 0.874, and AUC-PRC 0.768 on the testing sample), VAE (F1-measure 0.766, AUC-ROC 0.892, and AUC-PRC 0.661 on the testing sample), AutoEnc (F1-measure 0.767, AUC-ROC 0.892, and AUC-PRC 0.660 on the testing sample), and AnoGAN (F1-measure 0.750, AUC-ROC 0.864, and AUC-PRC 0.753 on the testing sample).

Table 5. The results of Experiment 1 for the random split mode for the SWaT dataset.

Optimal Threshold	Train Data						Test Data					
	P	R	FPR	F1	AUC-ROC	AUC-PRC	P	R	FPR	F1	AUC-ROC	AUC-PRC
					Sklearn							
ocSVM	0.211	0.017	0.983	0.031	0.813	0.072	0.237	0.019	0.981	0.036	0.811	0.073
isoF	0.209	0.861	0.139	0.336	0.859	0.07	0.210	0.862	0.138	0.338	0.86	0.069
					PYOD							
ECOD	0.928	0.615	0.385	0.740	0.876	0.757	0.934	0.617	0.383	0.743	0.878	0.758
COPOD	0.942	0.610	0.390	0.741	0.873	0.769	0.946	0.613	0.387	0.744	0.874	0.768
KNN	0.121	1.000	0.000	0.217	0.227	0.085	0.121	0.999	0.000	0.217	0.232	0.085
Deep-SVDD	0.191	0.675	0.325	0.298	0.583	0.150	0.191	0.672	0.329	0.297	0.585	0.153
VAE	0.853	0.689	0.311	0.763	0.89	0.653	0.861	0.690	0.310	0.766	0.892	0.661
AutoEnc	0.853	0.690	0.310	0.763	0.89	0.652	0.860	0.691	0.309	0.767	0.892	0.660
AnoGAN	0.989	0.604	0.396	0.750	0.862	0.750	0.989	0.605	0.395	0.750	0.864	0.753

Experiment 2. For this experiment series, we used the data from "Attack_v0.csv" and "Normal_v0.csv" files to form train, test, and validation subsamples. The train and test subsamples incorporated all instances before 2 January 2016, with 672,989 normal instances and 41,186 attack instances for train and 168,247 normal instances and 10,297 attack instances for test (contamination is equal to 0.061 for both) after 80% to 20% stratified train test split. Meanwhile, the validation sample consisted of all instances after 2 January 2016 (inclusively), with 50,862 normal instances and 3138 attack instances and contamination of 0.062. The results of experiment 2 for different anomaly detection machine learning models are provided in Tables 6 and 7. It can be seen that rather close results are obtained for the ECOD (F1-measure 0.718, AUC-ROC 0.864, and AUC-PRC 0.530 on the testing sample), COPOD (F1-measure 0.729, AUC-ROC 0.867, and AUC-PRC 0.563 on the testing sample), VAE (F1-measure 0.732, AUC-ROC 0.896, and AUC-PRC 0.505 on the testing sample), AutoEnc (F1-measure 0.732, AUC-ROC 0.896, and AUC-PRC 0.505 on the testing sample), and AnoGAN (F1-measure 0.746, AUC-ROC 0.851, and AUC-PRC 0.555 on the testing sample).

Table 6. The results of Experiment 2 for the SWaT dataset (for train and test data).

Optimal threshold	Train data						Test data					
	P	R	FPR	F1	AUC-ROC	AUC-PRC	P	R	FPR	F1	AUC-ROC	AUC-PRC
					Sklearn							
ocSVM	0.00	0.00	0.00	0.0	0.00	0.00	0.891	0.617	0.383	0.729	0.211	0.180
isoF	0.00	0.00	0.00	0.00	0.00	0.00	0.805	0.623	0.377	0.702	0.862	0.032
					PYOD							
ECOD	0.862	0.623	0.377	0.724	0.865	0.540	0.856	0.619	0.381	0.718	0.864	0.530
COPOD	0.897	0.621	0.379	0.734	0.868	0.575	0.890	0.617	0.383	0.729	0.867	0.563
KNN	0.058	1.000	0.000	0.109	0.209	0.040	0.058	0.999	0.000	0.109	0.213	0.041
DeepSVDD	0.067	0.832	0.168	0.124	0.490	0.054	0.067	0.826	0.174	0.124	0.489	0.055
VAE	0.772	0.696	0.304	0.732	0.896	0.509	0.770	0.696	0.304	0.732	0.896	0.505
AutoEnc	0.772	0.696	0.304	0.732	0.896	0.509	0.770	0.696	0.304	0.732	0.896	0.505
AnoGAN	0.899	0.644	0.356	0.751	0.854	0.568	0.893	0.641	0.359	0.746	0.851	0.555

Experiment 3. The data from "Attack_v0.csv", "Normal_v0.csv", and "Normal_v1.csv" files together were used to train algorithms in novelty detection or unsupervised mode in this experiment series. The data contain 1,441,719 instances in total, including 1,387,098 normal instances and 54,621 attack instances with contamination of 0.039. To train algorithms, all instances from "Normal_v0.csv" and "Normal_v1.csv" files (except stabilization period of 3 hours) were used, while all instances from "Attack_v0.csv" file were used for testing. The train sample included 972,000 normal instances and no attack instances. The test sample included 395,298 normal instances and 54,621 attack instances, that is, contamination is

equal to 0.139. The results of experiment 3 for the novelty detection mode and different anomaly detection machine learning models are provided in Table 8. It can be seen that the results are rather close for different models with a rather low false positive rate on the testing sample.

Table 7. The results of Experiment 2 for the SWaT dataset (for validation data).

	Validation Data				
	ACC	P	R	FPR	F1
Sklearn					
ocSVM	0.942	0.000	0.000	1.000	0.000
isoF	0.935	0.022	0.003	0.997	0.005
PYOD					
ECOD	0.942	0.466	0.011	0.989	0.021
COPOD	0.942	0.000	0.000	1.000	0.000
kNN	0.058	0.058	1.000	0.000	0.110
DeepSVDD	0.601	0.045	0.293	0.707	0.079
VAE	0.933	0.000	0.000	1.000	0.000
AutoEncoder	0.933	0.000	0.000	1.000	0.000
AnoGan	0.943	0.672	0.043	0.957	0.081

Table 8. The results of Experiment 3 for the SWaT dataset.

	Train Data	Test Data						
Optimal Threshold	ACC	ACC	P	R	FPR	F1	AUC-ROC	AUC-ROC
Sklearn								
ocSVM	0.990	0.936	0.998	0.585	0.415	0.738	0.808	0.082
isoF	0.960	0.777	0.124	0.932	0.068	0.219	0.833	0.072
PYOD								
ECOD	0.900	0.833	0.981	0.598	0.402	0.743	0.858	0.758
COPOD	0.960	0.919	0.948	0.619	0.381	0.749	0.855	0.756
KNN	0.960	0.127	0.987	0.636	0.364	0.774	0.816	0.727
DeepSVDD	0.960	0.766	0.991	0.646	0.354	0.783	0.838	0.732
VAE	0.960	0.410	0.991	0.633	0.368	0.772	0.820	0.732
AutoEnc	0.960	0.410	0.991	0.633	0.368	0.772	0.820	0.732

Conclusions. We analyzed the obtained results considering the dataset requirements listed above. All specified requirements are satisfied for this dataset. It is generated using physical devices and components, and this impacts the efficiency of the network attacks; not all network attacks result in changes in the readings of the sensors. Thus, we consider that this dataset is a realistic one. The preliminary results of the analysis of the sensors data are in conformance with the results obtained by other researchers [6,29,30,35,36]. Interestingly, all considered papers do not analyze network and sensor data together, and we believe that joint analysis of such data could significantly enhance the performance of the analysis models targeted to detect anomalies and network attacks.

3.3. HAI Dataset Analysis

The dataset describes the parameters of an industrial control system testbed with an embedded simulator. The testbed comprises four elements: a boiler, turbine, water-treatment component, and a hardware-in-the-loop (HIL) simulator. The HIL simulation implements a simulation of the thermal power and pumped-storage hydropower genera-tion.

When forming the dataset, several different attack scenarios were used, aimed at three types of devices: the Emerson Ovation, GE Mark-VIe, and Siemens S7-1500.

During the attack, the attacker operates with four types of variables: set points, process variables, control variables, and control parameters. The set of certain values of these variables in a given period of time determines one of two behaviours of the system: anomalous or normal. When the system is operating normally, the values of the process variables change within a predefined range. To this end, the operator adjusts the set point values, which allows for achieving stable and predictable results in the behaviour of the sensors, and the entire system as a whole.

This dataset has three versions: HAI 20.07, HAI 21.03, and HAI 22.04. Statistical information about each of them is given in Table 9.

Figure 3 shows 10 features which keep the highest correlation value with the class label for files test1.csv within HAI 20.07, HAI 21.03, and HAI 22.04.

Table 10 contains the values of F-measure (F1) and accuracy (ACC) in percentages for 5 classifiers: decision tree (DT), KNN, random forest, logistic regression, and neural network (NN).

Conclusions. We analyzed the obtained results considering the set requirements. The requirements R1, R2, R3, and R4 are satisfied. The requirement R5 is also satisfied; however, considering the existence of the simulated part of the test bed, the quality of the dataset depends on the quality of the simulated part of the test bed. The preliminary experimental results are in line with the results obtained in other research papers. Thus, this dataset is consistent and suitable for the intrusion detection task.

Table 9. Statistical data on the HAI dataset class balance by version.

File	Normal	Attack	Total	Class Balance, %	Features
hai-20.07/train1.csv.gz	309,600	0	309,600	100/0	59
hai-20.07/train2.csv.gz	240,424	776	241,200	99.7/0.3	59
hai-20.07/test1.csv.gz	280,062	11,538	291,600	96/4	59
hai-20.07/test2.csv.gz	147,011	5989	51,106	96.1/3.9	59
hai-21.03/train1.csv.gz	216,001	0	216,001	100/0	79
hai-21.03/train2.csv.gz	226,801	0	226,801	100/0	79
hai-21.03/train3.csv.gz	478,801	0	478,801	100/0	79
hai-21.03/test1.csv.gz	42,572	629	43,201	98.5/1.5	79
hai-21.03/test2.csv.gz	115,352	3449	118,801	97.1/2.9	79
hai-21.03/test3.csv.gz	106,466	1535	108,001	98.6/1.4	79
hai-21.03/test4.csv.gz	38,444	1157	39,601	97.1/2.9	79
hai-21.03/test5.csv.gz	90,224	2177	92,401	97.6/2.4	79
hai-22.04/train1.csv	93,601	0	93,601	100/0	86
hai-22.04/train2.csv	201,600	0	201,600	100/0	86
hai-22.04/train3.csv	126,000	0	126,000	100/0	86
hai-22.04/train4.csv	86,401	0	86,401	100/0	86
hai-22.04/train5.csv	237,600	0	237,600	100/0	86
hai-22.04/train6.csv	259,200	0	259,200	100/0	86
hai-22.04/test1.csv	85,515	885	86,400	99/1	86
hai-22.04/test2.csv	79,919	2881	82,800	96.5/3.5	86
hai-22.04/test3.csv	58,559	3841	62,400	93.8/6.2	86
hai-22.04/test4.csv	125,177	4423	129,600	96.6/3.4	86

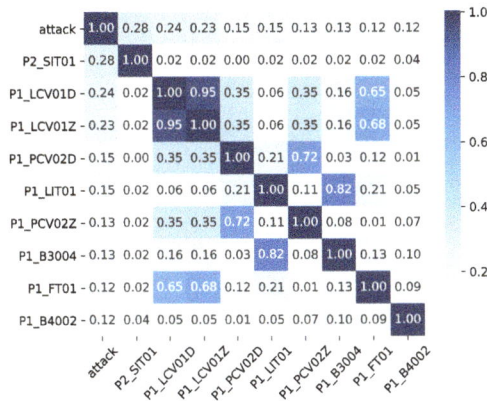

(a) HAI 20.07 dataset.

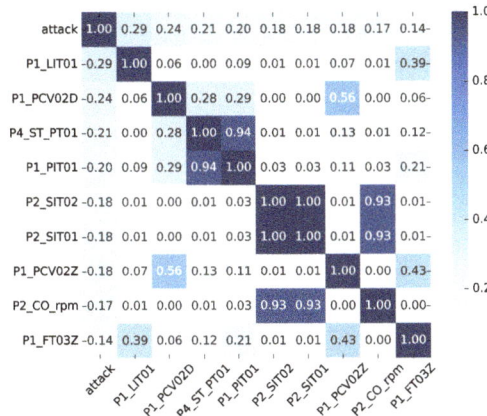

	attack	P2_SIT01	P1_LCV01D	P1_LCV01Z	P1_PCV02D	P1_LIT01	P1_PCV02Z	P1_B3004	P1_FT01	P1_B4002
attack	1.00	0.28	0.24	0.23	0.15	0.15	0.13	0.13	0.12	0.12
P2_SIT01	0.28	1.00	0.02	0.02	0.00	0.02	0.02	0.02	0.02	0.04
P1_LCV01D	0.24	0.02	1.00	0.95	0.35	0.06	0.35	0.16	0.65	0.05
P1_LCV01Z	0.23	0.02	0.95	1.00	0.35	0.06	0.35	0.16	0.68	0.05
P1_PCV02D	0.15	0.00	0.35	0.35	1.00	0.21	0.72	0.03	0.12	0.01
P1_LIT01	0.15	0.02	0.06	0.06	0.21	1.00	0.11	0.82	0.21	0.05
P1_PCV02Z	0.13	0.02	0.35	0.35	0.72	0.11	1.00	0.08	0.01	0.07
P1_B3004	0.13	0.02	0.16	0.16	0.03	0.82	0.08	1.00	0.13	0.10
P1_FT01	0.12	0.02	0.65	0.68	0.12	0.21	0.01	0.13	1.00	0.09
P1_B4002	0.12	0.04	0.05	0.05	0.01	0.05	0.07	0.10	0.09	1.00

(b) HAI 21.03 dataset.

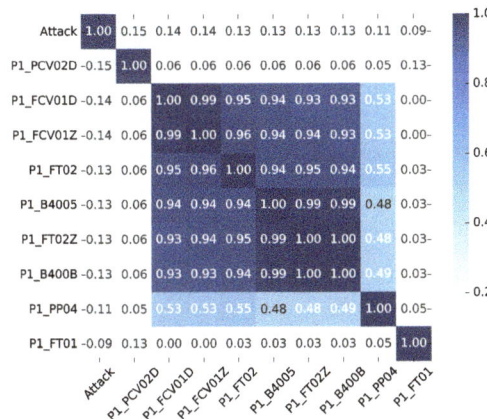

	attack	P1_LIT01	P1_PCV02D	P4_ST_PT01	P1_PIT01	P2_SIT02	P2_SIT01	P1_PCV02Z	P2_CO_rpm	P1_FT03Z
attack	1.00	0.29	0.24	0.21	0.20	0.18	0.18	0.18	0.17	0.14
P1_LIT01	-0.29	1.00	0.06	0.00	0.09	0.01	0.01	0.07	0.01	0.39
P1_PCV02D	-0.24	0.06	1.00	0.28	0.29	0.00	0.00	0.56	0.00	0.06
P4_ST_PT01	-0.21	0.00	0.28	1.00	0.94	0.01	0.01	0.13	0.01	0.12
P1_PIT01	-0.20	0.09	0.29	0.94	1.00	0.03	0.03	0.11	0.03	0.21
P2_SIT02	-0.18	0.01	0.00	0.01	0.03	1.00	1.00	0.01	0.93	0.01
P2_SIT01	-0.18	0.01	0.00	0.01	0.03	1.00	1.00	0.01	0.93	0.01
P1_PCV02Z	-0.18	0.07	0.56	0.13	0.11	0.01	0.01	1.00	0.00	0.43
P2_CO_rpm	-0.17	0.01	0.00	0.01	0.03	0.93	0.93	0.00	1.00	0.00
P1_FT03Z	-0.14	0.39	0.06	0.12	0.21	0.01	0.01	0.43	0.00	1.00

(c) HAI 22.04 dataset.

	Attack	P1_PCV02D	P1_FCV01D	P1_FCV01Z	P1_FT02	P1_B4005	P1_FT02Z	P1_B400B	P1_PP04	P1_FT01
Attack	1.00	0.15	0.14	0.14	0.13	0.13	0.13	0.13	0.11	0.09
P1_PCV02D	-0.15	1.00	0.06	0.06	0.06	0.06	0.06	0.06	0.05	0.13
P1_FCV01D	-0.14	0.06	1.00	0.99	0.95	0.94	0.93	0.93	0.53	0.00
P1_FCV01Z	-0.14	0.06	0.99	1.00	0.96	0.94	0.94	0.93	0.53	0.00
P1_FT02	-0.13	0.06	0.95	0.96	1.00	0.94	0.95	0.94	0.55	0.03
P1_B4005	-0.13	0.06	0.94	0.94	0.94	1.00	0.99	0.99	0.48	0.03
P1_FT02Z	-0.13	0.06	0.93	0.94	0.95	0.99	1.00	1.00	0.48	0.03
P1_B400B	-0.13	0.06	0.93	0.93	0.94	0.99	1.00	1.00	0.49	0.03
P1_PP04	-0.11	0.05	0.53	0.53	0.55	0.48	0.48	0.49	1.00	0.05
P1_FT01	-0.09	0.13	0.00	0.00	0.03	0.03	0.03	0.03	0.05	1.00

Figure 3. Features with the highest correlation value with class label.

Table 10. Result of evaluating classifiers on HAI dataset.

File	DT		KNN		RF		LR		NN	
	F1, %	ACC, %	F1, %	ACC, %	F1, %	ACC, %	F1, %	ACC, %	F1, %	ACC, %
hai-20.07/test1.csv.gz	99.00	99.85	86.42	98.28	99.67	99.95	80.88	97.73	96.70	99.50
hai-20.07/test2.csv.gz	99.48	99.92	94.93	99.30	99.76	99.96	97.29	99.60	99.30	99.90
hai-21.03/test1.csv.gz	98.61	99.91	93.39	99.59	99.31	99.95	90.10	99.43	49.57	98.29
hai-21.03/test2.csv.gz	97.60	99.73	89.89	98.99	99.52	99.95	74.99	98.03	88.52	98.81
hai-21.03/test3.csv.gz	99.38	99.96	99.61	99.61	99.38	99.96	90.96	99.53	72.23	98.76
hai-21.03/test4.csv.gz	99.45	99.94	95.65	99.52	99.67	99.96	99.22	99.91	49.25	97.05
hai-21.03/test5.csv.gz	98.26	99.84	92.72	99.38	99.30	99.94	81.30	98.63	49.39	97.59
hai-22.04/test1.csv	98.66	99.95	90.87	99.67	99.11	99.97	78.76	99.39	49.75	98.99
hai-22.04/test2.csv	97.85	99.70	89.80	98.73	99.39	99.92	72.80	97.46	49.07	96.34
hai-22.04/test3.csv	98.79	99.73	94.45	98.83	99.64	99.92	90.00	98.03	94.30	98.65
hai-22.04/test4.csv	98.38	99.78	88.26	98.65	99.45	99.93	62.88	97.02	49.12	96.53

4. Performance Metrics for Anomaly and Attack Detection

Finally, in this section, we describe performance metrics used for anomaly and attack detection. Precision, recall, and F-measure are the most used evaluation metrics. There is no specialized metric to measure the performance of anomaly detection methods. The listed metrics are classic for machine learning methods, on which most anomaly detection methods are based. However, we discovered that there are different approaches to calculating them [28,49,50]. This section reviews proposed approaches.

Let us denote the time series signal observed from K sensors during time T as

$$X = \{x_1, \ldots, x_T\}, x_t \in \mathbb{R}^N.$$

The normalized signal is divided into a number of time windows:

$$W = \{w_1, \ldots, w_{T-h+\tau}\},$$
$$w_t = \{x_t, \ldots, x_{t+h-\tau}\},$$

where h—window size, τ—step length.

The purpose of the time series anomaly detection method is to predict the binary label of the presence of an anomaly $(\hat{y}_t)$, either for individual X instances or for time windows W. The labels are obtained by comparing the anomaly estimates A with a threshold δ. For the specific instances:

$$\hat{y}_t = \begin{cases} 1, & \text{if } A(x_t) > \delta, \\ 0, & \text{otherwise.} \end{cases}$$

For all windows in the test dataset:

$$\hat{y}_t = \begin{cases} 1, & \text{if } A(w_t) > \delta, \\ 0, & \text{otherwise.} \end{cases}$$

A set of test data may contain several sequences (segments) of anomalies within a certain period of time. Let us denote S as a set of M segments of anomalies:

$$S = \{S_1, \ldots, S_M\},$$
$$S_m = \{x_{t^{ms}}, \ldots, x_{t^{me}}\},$$

where t^{ms} and t^{me} are the S_m starting and ending time, accordingly.

Below, several approaches to calculate the performance metrics of anomaly detection are described.

Point-wise calculation approach. The calculation of the performance metrics is implemented using separate records within the dataset [28,49]. The calculation of precision (P), recall (R), and F-measure ($F1$) is implemented using all points within the dataset:

$$P = \frac{TP}{TP + FP}, \qquad R = \frac{TP}{TP + FN}, \qquad F1 = 2 \times \frac{P \times R}{P + R},$$

where

- TP—correctly detected anomaly ($y_t = 1, \hat{y}_t = 1$);
- FP—false detected anomaly ($y_t = 0, \hat{y}_t = 1$);
- TN—correctly assigned norm ($y_t = 0, \hat{y}_t = 0$);
- FN—false assigned norm ($y_t = 1, \hat{y}_t = 0$).

Point-adjusted (PA) calculation approach . The calculation of the performance metrics is implemented using the corrected labels. If at least one observation of an anomalous segment is detected correctly, all other observations of the segment are also considered to be correctly detected, even if they were not detected [28,49]. Observations outside the true anomaly segment are processed as usual. It can be specified as follows:

$$\hat{y}_t^{pa} = \begin{cases} 1, & \text{if } A(x_t) > \delta \text{ or } \exists A(x_{t'} > \delta), x_t, x_t' \in S_m, \\ 0, & \text{otherwise.} \end{cases}$$

The metrics are calculated considering the corrected labels in the dataset:

$$P_{pa} = \frac{TP_{pa}}{TP_{pa} + FP_{pa}}, \qquad R_{pa} = \frac{TP_{pa}}{TP_{pa} + FN_{pa}}, \qquad F1_{pa} = 2 \times \frac{P_{pa} \times R_{pa}}{P_{pa} + R_{pa}},$$

This idea is represented in Figure 4.

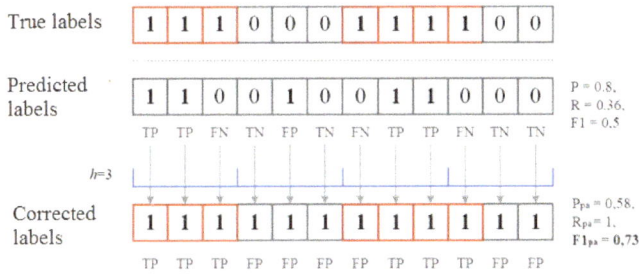

Figure 4. True, corrected, and predicted labels in case of the PA approach to metrics calculation.

Revised point-adjusted (RPA, event-wise) calculation approach. The calculation of metrics is implemented using time windows of records [50]. If any point at the anomaly window is labelled as anomalous, then one true positive result is fixed. If the anomalies were not labelled, then one false negative result is fixed. Any predicted anomalies outside the anomaly windows are considered false positives. This can be specified as follows:

$$P_{rpa} = \frac{TP_{rpa}}{TP_{rpa} + FP_{rpa}}, \qquad R_{pa} = \frac{TP_{rpa}}{TP_{rpa} + FN_{rpa}}, \qquad F1_{pa} = 2 \times \frac{P_{rpa} \times R_{rpa}}{P_{rpa} + R_{rpa}},$$

where

- TP_{rpa}—any part of the predicted anomaly sequence intersects with a sequence that actually has an anomaly;
- FN_{rpa}—if no sequence that is predicted to be anomalous intersects with a real anomalous sequence;
- FP_{rpa}—all predicted anomalous sequences that do not intersect with any really anomalous sequence.

This idea is represented in Figure 5.

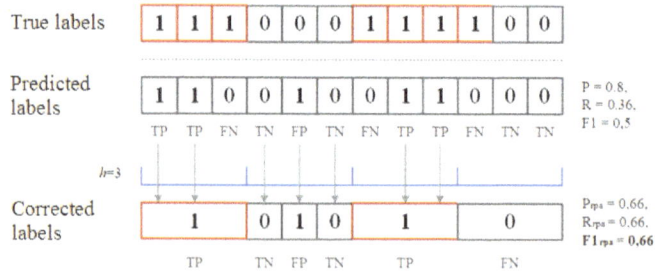

Figure 5. True, predicted, and corrected labels in case of the RPA approach to the metrics calculation.

Another metric is the composite $F1$ score [50]. For this metric, precision is considered as P (by the number of points), and recall is calculated as R_{rpa} (by the number of segments):

$$F1_c = 2 \times \frac{P \times R_{rpa}}{P + R_{rpa}}.$$

This idea is represented in Figure 6.

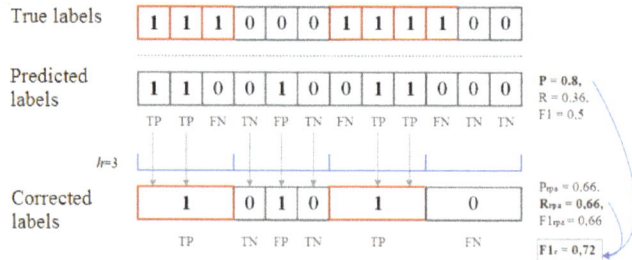

Figure 6. True, predicted, and corrected labels in case of the composite $F1$ score approach to the metrics calculation.

Conclusions. There are various approaches to the calculation of metrics for performance evaluation of the machine learning models. In addition to the classical way of calculating through TP, TN, FN, and FP, researchers present options with adjusted indicators. This is aimed at improving the quality of anomaly detection in a large amount of data, or at reducing the number of false positives. In this case, the choice of metrics strongly depends on the detection problem being solved. To select the appropriate approach to calculation, additional experiments are required: the authors plan to implement and compare all the described metrics in future experiments with anomaly detection methods.

5. Conclusions

In the paper, the authors considered existing approaches in the anomaly detection area, existing datasets that can be used for the experiments, and existing performance metrics. The analysis of the related works showed that the research focus has shifted to the application of deep neural networks to anomaly detection in technological processes; however, there are still solutions based on classical anomaly detection techniques. The application of machine learning techniques requires high-quality datasets. High-quality datasets are datasets that are relevant to the subject domain, meaningful, and reliable. We formulated five requirements for the datasets that consider these properties and evaluated three different datasets that are currently proposed for testing and evaluation of cybersecurity applications. The selected datasets are SWaT, HAI, and TON_IoT. Our experiments revealed that TON_IoT is not suitable for the intrusion detection task, as we did not discover any relations between sensor data and network data. We consider that SWaT and HAI datasets are more relevant for cybersecurity tasks, primarily due to the fact that they were generated using real physical test beds. The SWaT dataset contains both network and sensor data; this makes it preferable for intrusion detection, as authors believe that joint analysis of the network and sensor data could benefit the early detection of the attacks a lot, including multi-step attacks.

Another interesting finding relates to the performance evaluation of the machine learning techniques proposed to detect anomalies. These techniques consider the specificity of the anomalies of the CPS systems—their duration and the delayed response of the system. Although these features could significantly enhance the evaluation process of the proposed cybersecurity solutions, they require more analysis and research.

Finally, in future research, the authors plan to develop an approach to anomaly detection in cyber-physical systems that will provide accurate and explainable results, and will conduct experiments to select the performance metrics.

Author Contributions: Conceptualization, E.N., E.F. and I.K.; methodology, E.N. and E.F.; software, O.T., D.L. and A.B.; validation, E.N., E.F., O.T. and I.K.; investigation, E.F., E.N., O.T., D.L. and A.B.; writing—original draft preparation, E.F.; writing—review and editing, E.N., O.T., D.L. and A.B.; visualization, D.L. and A.B.; supervision, I.K., E.N. and E.F.; funding acquisition, I.K. All authors have read and agreed to the published version of the manuscript.

Funding: This research is being supported by the grant of RSF #21-71-20078 in SPC RAS.

Data Availability Statement: Not applicable

Conflicts of Interest: The authors declare no conflict of interest.

References

1. Levshun, D.; Chechulin, A.; Kotenko, I. Design of Secure Microcontroller-Based Systems: Application to Mobile Robots for Perimeter Monitoring. *Sensors* **2021**, *21*, 8451. [CrossRef]
2. Turton, W.; Mehrotra, K. Hackers Breached Colonial Pipeline Using Compromised Password. 4 June 2021. Available online: https://www.bloomberg.com/news/articles/2021-06-04/hackers-breached-colonial-pipeline-using-compromised-password (accessed on 20 December 2022). .
3. Jones, S. Venezuela Blackout: What Caused It and What Happens Next. *The Guardian* 13 March 2019. Available online: https://www.theguardian.com/world/2019/mar/13/venezuela-blackout-what-caused-it-and-what-happens-next (accessed on 20 December 2022).
4. Graham, R. Cyberattack Hits Germany's Domestic Fuel Distribution System. 1 February, 2022. Available online: https://www.bloomberg.com/news/articles/2022-02-01/mabanaft-hit-by-cyberattack-that-disrupts-german-fuel-deliveries (accessed on 20 December 2022).
5. Kim, S.; Jo, W.; Shon, T. APAD: Autoencoder-based payload anomaly detection for industrial IoE. *Appl. Soft Comput.* **2020**, *88*, 106017. [CrossRef]
6. Wang, C.; Wang, B.; Liu, H.; Qu, H. Anomaly Detection for Industrial Control System Based on Autoencoder Neural Network. *Wirel. Commun. Mob. Comput.* **2020**, *2020*, 8897926:1–8897926:10. [CrossRef]
7. Kotenko, I.; Gaifulina, D.; Zelichenok, I. Systematic Literature Review of Security Event Correlation Methods. *IEEE Access* **2022**, *10*, 43387–43420. [CrossRef]
8. Alsaedi, A.; Moustafa, N.; Tari, Z.; Mahmood, A.; Anwar, A. TON_IoT telemetry dataset: A new generation dataset of IoT and IIoT for data-driven intrusion detection systems. *IEEE Access* **2020**, *8*, 165130–165150. [CrossRef]
9. Goh, J.; Adepu, S.; Junejo, K.N.; Mathur, A. A dataset to support research in the design of secure water treatment systems. In Proceedings of the Critical Information Infrastructures Security: 11th International Conference, CRITIS 2016, Paris, France, 10–12 October 2016; Revised Selected Papers 11; Springer: New York, NY, USA, 2017; pp. 88–99.
10. Shin, H.K.; Lee, W.; Yun, J.H.; Kim, H. HAI 1.0: HIL-based augmented ICS security dataset. In Proceedings of the 13th USENIX Conference on Cyber Security Experimentation and Test, Boston, MA, USA, 10 August 2020; p. 1.
11. Meleshko, A.; Shulepov, A.; Desnitsky, V.; Novikova, E.; Kotenko, I. Visualization Assisted Approach to Anomaly and Attack Detection in Water Treatment Systems. *Water* **2022**, *14*, 2342. [CrossRef]
12. Shulepov, A.; Novikova, E.; Murenin, I. Approach to Anomaly Detection in Cyber-Physical Object Behavior. In *Intelligent Distributed Computing XIV*; Camacho, D., Rosaci, D., Sarné, G.M.L., Versaci, M., Eds.; Springer International Publishing: Cham, Switzerland, 2022; pp. 417–426.
13. Khan, A.A.; Beg, O.A.; Alamaniotis, M.; Ahmed, S. Intelligent anomaly identification in cyber-physical inverter-based systems. *Electr. Power Syst. Res.* **2021**, *193*, 107024. [CrossRef]

14. Parto, M.; Saldana, C.; Kurfess, T. Real-time outlier detection and Bayesian classification using incremental computations for efficient and scalable stream analytics for IoT for manufacturing. *Procedia Manuf.* **2020**, *48*, 968–979. [CrossRef]
15. Mohammadi Rouzbahani, H.; Karimipour, H.; Rahimnejad, A.; Dehghantanha, A.; Srivastava, G. Anomaly detection in cyber-physical systems using machine learning. In *Handbook of Big Data Privacy*; Springer: New York, NY, USA, 2020; pp. 219–235.
16. Mokhtari, S.; Abbaspour, A.; Yen, K.K.; Sargolzaei, A. A machine learning approach for anomaly detection in industrial control systems based on measurement data. *Electronics* **2021**, *10*, 407. [CrossRef]
17. Park, S.; Lee, K. Improved Mitigation of Cyber Threats in IIoT for Smart Cities: A New-Era Approach and Scheme. *Sensors* **2021**, *21*, 1976. [CrossRef] [PubMed]
18. Elnour, M.; Meskin, N.; Khan, K.; Jain, R. A Dual-Isolation-Forests-Based Attack Detection Framework for Industrial Control Systems. *IEEE Access* **2020**, *8*, 36639–36651. [CrossRef]
19. Gad, A.R.; Haggag, M.; Nashat, A.A.; Barakat, T.M. A Distributed Intrusion Detection System using Machine Learning for IoT based on ToN-IoT Dataset. *Int. J. Adv. Comput. Sci. Appl.* **2022**, *13*, 548–563. [CrossRef]
20. Kumar, P.; Tripathi, R.; Gupta, G.P. P2IDF: A privacy-preserving based intrusion detection framework for software defined Internet of Things-fog (SDIoT-Fog). In Proceedings of the Adjunct 2021 International Conference on Distributed Computing and Networking, Nara, Japan, 5–8 January 2021; pp. 37–42.
21. Huč, A.; Šalej, J.; Trebar, M. Analysis of machine learning algorithms for anomaly detection on edge devices. *Sensors* **2021**, *21*, 4946. [CrossRef] [PubMed]
22. Inoue, J.; Yamagata, Y.; Chen, Y.; Poskitt, C.M.; Sun, J. Anomaly Detection for a Water Treatment System Using Unsupervised Machine Learning. In Proceedings of the 2017 IEEE International Conference on Data Mining Workshops (ICDMW), Orleans, LA, USA, 18–21 November 2017; pp. 1058–1065. [CrossRef]
23. Gaifulina, D.; Kotenko, I. Selection of deep neural network models for IoT anomaly detection experiments. In Proceedings of the 2021 29th Euromicro International Conference on Parallel, Distributed and Network-Based Processing (PDP), Valladolid, Spain, 10–21 March 2021; IEEE: Hoboken, NJ, USA, 2021; pp. 260–265.
24. Shalyga, D.; Filonov, P.; Lavrentyev, A. Anomaly Detection for Water Treatment System based on Neural Network with Automatic Architecture Optimization. *arXiv* **2018**, arXiv:1807.07282.
25. Xie, X.; Wang, B.; Wan, T.; Tang, W. Multivariate abnormal detection for industrial control systems using 1D CNN and GRU. *IEEE Access* **2020**, *8*, 88348–88359. [CrossRef]
26. Nagarajan, S.M.; Deverajan, G.G.; Bashir, A.K.; Mahapatra, R.P.; Al-Numay, M.S. IADF-CPS: Intelligent Anomaly Detection Framework towards Cyber Physical Systems. *Comput. Commun.* **2022**, *188*, 81–89. [CrossRef]
27. Fan, Y.; Li, Y.; Zhan, M.; Cui, H.; Zhang, Y. IoTDefender: A Federated Transfer Learning Intrusion Detection Framework for 5G IoT. In Proceedings of the 2020 IEEE 14th International Conference on Big Data Science and Engineering (BigDataSE), Guangzhou, China, 29 December 2020–1 January 2021; pp. 88–95. [CrossRef]
28. Audibert, J.; Michiardi, P.; Guyard, F.; Marti, S.; Zuluaga, M.A. USAD: UnSupervised Anomaly Detection on Multivariate Time Series. In Proceedings of the KDD'20, 26th ACM SIGKDD International Conference on Knowledge Discovery & Data Mining, Virtual Event, CA, USA, 6–10 July 2020; Association for Computing Machinery: New York, NY, USA, 2020; pp. 3395–3404. [CrossRef]
29. Li, D.; Chen, D.; Shi, L.; Jin, B.; Goh, J.; Ng, S.K. MAD-GAN: Multivariate Anomaly Detection for Time Series Data with Generative Adversarial Networks. In Proceedings of the International Conference on Artificial Neural Networks, Munich, Germany, 17–19 September 2019.
30. Neshenko, N.; Bou-Harb, E.; Furht, B. A behavioral-based forensic investigation approach for analyzing attacks on water plants using GANs. *Forensic Sci. Int. Digit. Investig.* **2021**, *37*, 301198. [CrossRef]
31. Wu, P.; Moustafa, N.; Yang, S.; Guo, H. Densely connected residual network for attack recognition. In Proceedings of the 2020 IEEE 19th International Conference on Trust, Security and Privacy in Computing and Communications (TrustCom), Guangzhou, China, 29 December 2020–1 January 2021; IEEE: Hoboken, NJ, USA, 2020; pp. 233–242.
32. Bian, X. Detecting Anomalies in Time-Series Data using Unsupervised Learning and Analysis on Infrequent Signatures. *J. IKEEE* **2020**, *24*, 1011–1016.
33. Lundberg, S.M.; Lee, S.I. A Unified Approach to Interpreting Model Predictions. In Proceedings of the NIPS'17, 31st International Conference on Neural Information Processing Systems, Long Beach, CA, USA, 4–9 December 2017; Curran Associates Inc.: Red Hook, NY, USA, 2017; pp. 4768–4777.
34. Ribeiro, M.T.; Singh, S.; Guestrin, C. "Why Should I Trust You?": Explaining the Predictions of Any Classifier. In Proceedings of the KDD'16, 22nd ACM SIGKDD International Conference on Knowledge Discovery and Data Mining, San Francisco, CA, USA, 13–17 August 2016; Association for Computing Machinery: New York, NY, USA, 2016; pp. 1135–1144. [CrossRef]
35. Lin, Q.; Adepu, S.; Verwer, S.; Mathur, A. TABOR: A Graphical Model-Based Approach for Anomaly Detection in Industrial Control Systems. In Proceedings of the ASIACCS'18, 2018 on ACM Asia Conference on Computer and Communications Security, Incheon, Republic of Korea, 4–8 June 2018; Association for Computing Machinery: New York, NY, USA, 2018; pp. 525–536. [CrossRef]
36. Sukhostat, L. Anomaly Detection in Industrial Control System Based on the Hierarchical Hidden Markov Model. In *Cybersecurity for Critical Infrastructure Protection via Reflection of Industrial Control Systems*; IOS Press: Amsterdam, The Netherlands, 2022; pp. 48–55.

37. Tavallaee, M.; Bagheri, E.; Lu, W.; Ghorbani, A.A. A detailed analysis of the KDD CUP 99 data set. In Proceedings of the 2009 IEEE Symposium on Computational Intelligence for Security and Defense Applications, Ottawa, ON, Canada, 8–10 July 2009; pp. 1–6. [CrossRef]
38. Sharafaldin, I.; Habibi Lashkari, A.; Ghorbani, A. Toward Generating a New Intrusion Detection Dataset and Intrusion Traffic Characterization. In Proceedings of the 4th International Conference on Information Systems Security and Privacy (ICISSP 2018), Funchal, Portugal, 22–24 January 2018; pp. 108–116. [CrossRef]
39. Moustafa, N.; Slay, J. UNSW-NB15: A comprehensive data set for network intrusion detection systems (UNSW-NB15 network data set). In Proceedings of the 2015 Military Communications and Information Systems Conference (MilCIS), Canberra, Australia, 10–12 November 2015; IEEE: Hoboken, NJ, USA, 2015; pp. 1–6.
40. Qin, Y.; Kondo, M. Federated Learning-Based Network Intrusion Detection with a Feature Selection Approach. In Proceedings of the 2021 International Conference on Electrical, Communication, and Computer Engineering (ICECCE), Kuala Lumpur, Malaysia, 12–13 June 2021; pp. 1–6. [CrossRef]
41. Murenin, I.; Doynikova, E.; Kotenko, I. Towards Security Decision Support for large-scale Heterogeneous Distributed Information Systems. In Proceedings of the 2021 14th International Conference on Security of Information and Networks (SIN), Edinburgh, UK, 15–17 December 2021; Volume 1, pp. 1–8. [CrossRef]
42. Choi, S.; Yun, J.H.; Kim, S.K. A Comparison of ICS Datasets for Security Research Based on Attack Paths. In Proceedings of the CRITIS, Kaunas, Lithuania, 24–26 September 2018.
43. Lemay, A.; Fernandez, J.M. Providing SCADA Network Data Sets for Intrusion Detection Research. In Proceedings of the 9th Workshop on Cyber Security Experimentation and Test (CSET 16), Austin, TX, USA, 8 August 2016; USENIX Association: Austin, TX, USA, 2016.
44. Rodofile, N.R.; Schmidt, T.; Sherry, S.T.; Djamaludin, C.; Radke, K.; Foo, E. Process Control Cyber-Attacks and Labelled Datasets on S7Comm Critical Infrastructure. In *Information Security and Privacy*; Pieprzyk, J., Suriadi, S., Eds.; Springer International Publishing: Cham, Switzerland, 2017; pp. 452–459.
45. Suthaharan, S.; Alzahrani, M.; Rajasegarar, S.; Leckie, C.; Palaniswami, M. Labelled data collection for anomaly detection in wireless sensor networks. In Proceedings of the 2010 Sixth International Conference on Intelligent Sensors, Sensor Networks and Information Processing, Brisbane, Australia, 7–10 December 2010; pp. 269–274. [CrossRef]
46. Sivanathan, A.; Gharakheili, H.H.; Loi, F.; Radford, A.; Wijenayake, C.; Vishwanath, A.; Sivaraman, V. Classifying IoT Devices in Smart Environments Using Network Traffic Characteristics. *IEEE Trans. Mob. Comput.* **2019**, *18*, 1745–1759. [CrossRef]
47. Koroniotis, N.; Moustafa, N.; Sitnikova, E.; Turnbull, B.P. Towards the Development of Realistic Botnet Dataset in the Internet of Things for Network Forensic Analytics: Bot-IoT Dataset. *Future Gener. Comput. Syst.* **2019**, *100*, 779–796. [CrossRef]
48. Hamza, A.; Gharakheili, H.H.; Benson, T.A.; Sivaraman, V. Detecting Volumetric Attacks on IoT Devices via SDN-Based Monitoring of MUD Activity. In Proceedings of the 2019 ACM Symposium on SDN Research, San Jose, CA, USA, 3–4 April 2019.
49. Xu, H.; Chen, W.; Zhao, N.; Li, Z.; Bu, J.; Li, Z.; Liu, Y.; Zhao, Y.; Pei, D.; Feng, Y.; et al. Unsupervised Anomaly Detection via Variational Auto-Encoder for Seasonal KPIs in Web Applications. In Proceedings of the WWW'18, 2018 World Wide Web Conference, Lyon, France, 23–27 April 2018; International World Wide Web Conferences Steering Committee: Geneva, Switzerland, 2018; pp. 187–196. [CrossRef]
50. Hundman, K.; Constantinou, V.; Laporte, C.; Colwell, I.; Soderstrom, T. Detecting Spacecraft Anomalies Using LSTMs and Nonparametric Dynamic Thresholding. In Proceedings of the KDD'18, 24th ACM SIGKDD International Conference on Knowledge Discovery & Data Mining, London, UK, 19–23 August 2018; Association for Computing Machinery: New York, NY, USA, 2018; pp. 387–395. [CrossRef]

algorithms

MDPI

Article

IoT Multi-Vector Cyberattack Detection Based on Machine Learning Algorithms: Traffic Features Analysis, Experiments, and Efficiency

Sergii Lysenko [1,*], Kira Bobrovnikova [1], Vyacheslav Kharchenko [2,*] and Oleg Savenko [1]

[1] Computer Engineering and Information Systems Department, Khmelnytskyi National University, 29016 Khmelnytskyi, Ukraine; bobrovnikova.kira@gmail.com (K.B.); savenko_oleg_st@ukr.net (O.S.)
[2] Department of Computer Systems, Networks and Cybersecurity, National Aerospace University "KhAI", 61001 Kharkiv, Ukraine
* Correspondence: sirogyk@ukr.net (S.L.); v.kharchenko@csn.khai.edu (V.K.); Tel.: +380-68-772-81-79 (S.L.); +380-67-915-19-89 (V.K.)

Abstract: Cybersecurity is a common Internet of Things security challenge. The lack of security in IoT devices has led to a great number of devices being compromised, with threats from both inside and outside the IoT infrastructure. Attacks on the IoT infrastructure result in device hacking, data theft, financial loss, instability, or even physical damage to devices. This requires the development of new approaches to ensure high-security levels in IoT infrastructure. To solve this problem, we propose a new approach for IoT cyberattack detection based on machine learning algorithms. The core of the method involves network traffic analyses that IoT devices generate during communication. The proposed approach deals with the set of network traffic features that may indicate the presence of cyberattacks in the IoT infrastructure and compromised IoT devices. Based on the obtained features for each IoT device, the feature vectors are formed. To conclude the possible attack presence, machine learning algorithms were employed. We assessed the complexity and time of machine learning algorithm implementation considering multi-vector cyberattacks on IoT infrastructure. Experiments were conducted to approve the method's efficiency. The results demonstrated that the network traffic feature-based approach allows the detection of multi-vector cyberattacks with high efficiency.

Keywords: Internet of Things; cybersecurity; cyber threats; malware detection; machine learning; network traffic

Citation: Lysenko, S.; Bobrovnikova, K.; Kharchenko, V.; Savenko, O. IoT Multi-Vector Cyberattack Detection Based on Machine Learning Algorithms: Traffic Features Analysis, Experiments, and Efficiency. *Algorithms* **2022**, *15*, 239. https://doi.org/10.3390/a15070239

Academic Editors: Francesco Bergadano and Giorgio Giacinto

Received: 19 June 2022
Accepted: 9 July 2022
Published: 12 July 2022

Publisher's Note: MDPI stays neutral with regard to jurisdictional claims in published maps and institutional affiliations.

1. Introduction

1.1. Motivation

The Internet of Things is a concept that aggregates many technologies and physical objects—devices that exchange data and interact over the internet, as well as big data that generate these devices. Internet of Things devices have various purposes and complexities, from wearable things or technology to intelligent devices in smart homes and critical infrastructure. The Internet of Things was designed to make many areas of human life more comfortable and safer. However, the Internet of Things not only brings increased comfort but also new challenges and problems related to cybersecurity [1,2].

Security issues surrounding the Internet of Things infrastructure are determined by the specific features of an environment. One possible feature involved in building an IoT infrastructure is an IoT system of groups of identical or similar technical characteristic devices. If a specified device has a vulnerability, such homogeneity multiplies its impact [3–5].

Important issues include security issues with protocols used in the internet infrastructure, the use of unsafe network services, such as Telnet and SSH, and vulnerabilities in routers and open ports. With the ability to monitor and collect data on the IoT, even specialized compromised IoT devices with limited resources can be used to leverage critical infrastructure systems, such as database servers. Vulnerability in the IoT device communication protocol can spread to other devices that use the vulnerable protocol in the IoT infrastructure [6].

Thus, vulnerabilities in the protocols used in the IoT network can have devastating effects on the entire IoT infrastructure. The criticalities of these effects depend on the environments in which the compromised IoT devices operate.

Moreover, in some cases, the deployment conditions of IoT devices make it difficult or impossible to reconfigure or upgrade IoT devices. Often, IoT devices cannot be upgraded due to the discontinuation of device support from the manufacturer. This leads to the possibility of new vulnerabilities and threats to the IoT device in the future, as the current security mechanisms of device deployment may be out of date. Technical support and management of IoT smart devices are important cybersecurity issues in the long run. Another specific problem surrounding IoT cybersecurity is the fact that the internal operation of a smart device and the data streams generated by the device may be unknown to the user. The situation is complicated by the constant availability of IoT devices on a network and the ignorance of users (i.e., concerning potential cybersecurity risks). It may lead to the use of dangerous settings on IoT devices (default), direct network connections of internet devices to the internet, the use of obsolete or unreliable devices, and weak passwords.

One important IoT cybersecurity risk is that the functionality of smart devices can be changed by the device manufacturer without the consent or knowledge of the user (by updating the device firmware). It creates a new vulnerability that can allow the smart device to partially change the functionality or perform undesirable actions on the user's device, such as collecting sensitive user data without the user's knowledge.

However, the risks are not limited to data confidentiality. Attacks on IoT infrastructure can not only target compromised devices to steal sensitive data or cause financial losses but also disrupt or damage IoT devices physically. Compromised IoT devices can even lead to the injuries or deaths of people who depend on these devices or work with them.

Thus, non-compliance with basic security requirements (for both manufacturers and the users of smart devices) is the main cause of IoT cybersecurity problems. Common causes of security breaches in IoT infrastructure due to manufacturers are vulnerabilities in the IoT device software, lack of support for automatic updates, lack of firmware updates, and dangerous update mechanisms. This situation is often caused by manufacturers attempting to launch new smart devices as soon as possible. Vulnerabilities in software and web applications can lead to the theft of sensitive information or the spread of malicious firmware updates. Another common problem is unsafe authentication methods provided by the device manufacturers. The above weaknesses of the current IoT state of affairs, as well as the heterogeneity of the IoT environment, make IoT devices more vulnerable than computers and servers on conventional networks. Vulnerable components of IoT can be IoT devices, device software, and communication channels of the IoT infrastructure. The main threats in IoT infrastructure are distributed denial of service (DDoS), disclosure of confidential information, falsification, spoofing, and elevation of privilege. These threats are commonly used by cybercriminals as entry points, followed by other criminal activities: infecting devices with malicious software, stealing sensitive data, or blocking network connections.

Mentioned factors contribute to the high probability of compromising IoT devices, the spread of malicious software, and various multi-vector cyberattacks on IoT infrastructure (MVIA). At the same time, compromised IoT devices can be used as sources of attacks both inside and outside the IoT infrastructure.

The next subsection presents a brief analysis of the modern ideas and methods addressed to solve the problem of IoT malware detection by analyzing the advantages and disadvantages.

1.2. Objectives and Contribution

The main objectives of the work were to study the possibility of a multi-vector cyberattack detection in the IoT infrastructure based on a flow analysis and a deeper traffic analysis that takes into account IoT protocol features. This research aims to improve detection efficiency via various machine learning algorithm usages. The proposed approach deals with the set of network traffic features that may indicate the presence of cyberattacks in the IoT infrastructure and compromised IoT devices.

Thus, the novelty of this work involves the approach used for IoT multi-vector cyberattack detection, which involves a flow-based features analysis. It enables decreased detection time and is scalable. On the other hand, if the flow-based feature analysis was unable to conclude the attack presence, a deep analysis of network traffic with the use of MQTT-based, DNS-based, and HTTP-based features analysis was employed.

This paper is organized as follows. Section 2 presents the state-of-the-art. Section 3 describes the machine learning algorithms for cyberattack detection. Section 4 discusses the stages of the proposed IoT multi-vector cyberattack detection technique based on machine learning algorithms with the traffic features analysis. Section 5 proposes the experiments and the efficiency of the proposed approach. Finally, we present our conclusions and future research.

2. The State-of-the-Art

The scientific community is focusing on the increasing problems concerning cybersecurity today. Solutions devoted to cyberattack detection against Internet of Things infrastructure are widely presented [7,8]. Quite possibly, the most encouraging approaches for IoT cyberattack detection are based on machine learning algorithms (MLA) [9–13].

To solve the cyberattack detection problem, the authors of [14] proposed an approach that executes the IoT malware traffic analysis. It is based on the usage of multilevel artificial intelligence and involves neural networks and binary visualization. In addition, the approach proposes efficiency improvement via learning from the misclassification approach, which includes three main stages, is designed to collect the network traffic, perform the binary visualization to store the collected network traffic in ASCII, convert it to 2D images, and process/analyze the obtained binary image. An analysis of the binary images is executed using the TensorFlow tool, an end-to-end open-source platform designed to use machine learning for different problem solutions. It can find and classify patterns automatically. The main advantage of the tool is the ability to organize the system retraining as well as the possibility to make the image recognition. The approach proposes the use of the algorithm to perform the visualization of the collected traffic characteristics as an image (in the form of tiles using the Binvis tool). The TensorFlow machine tool can make predictions. The use of graphic tiles allows the determination of the tile combination on which the image is based. It is able to detect needed objects regardless of the location within the obtained image. The provided method can perform the IoT device protection on the gateway level, bypassing the IoT environment constraints.

The authors of [15] presented a survey on the experimental studies with a detailed analysis of a set of machine learning algorithms. The article included comparative data concerning the algorithm detection efficiency of anomalous behavior in IoT networks. Experimental results have shown that the best efficiency concerning used datasets is produced by the random forest algorithm. Nevertheless, all investigated machine learning algorithms demonstrated to be very close to random forest algorithm and detection efficiency results; sometimes the choice of an appropriate algorithm depends on the nature of the analyzed data.

Article [16] is devoted to machine learning classifiers involved in the botnet traffic analysis in the IoT environment. Nine IoT devices were employed for dataset construction, consisting of several botnet attack types. To evaluate the efficiency of the proposed approach, true positive, true negative, false positive, false negative, F1-score accuracy, precision, and recall were used. The experimental results of the research demonstrated that the random forest algorithm produced the best results while the support vector machine produced the lowest results. The main disadvantage of the approach is the strong need for data analysis of all features in processed datasets.

The IoT cyberattack detection approach for the IoT network is presented in [17]. It is based on the use of intelligent technologies. The produced intelligent system operates with a set of network features. The approach aims to reduce the feature number via its ranking with the usage of the correlation coefficient, random forest algorithm, and the gain ratio. The base for the experimental research involves three feature sets, where using the proposed algorithm is to be combined to obtain an optimized feature set. The means of data processing the authors used were K-nearest neighbor, random forest, and XGBoost machine learning algorithms. All experiments were based on the usage of NSL-KDD, BoT-IoT, and DS2OS datasets. The investigation of the detection efficiency of the proposed system was executed. For this purpose, the metrics of accuracy, detection rate, F1-score, and precision were evaluated.

An approach for IoT attack detection based on the usage of cloud technologies and software-defined networks (SDNs) is presented in [18]. It employs a decentralized two-layer SDN and is able to perform attack mitigation in the wireless IoT infrastructure. To execute the network traffic control for each subnet domain, the predefined local domain controller of the specified domain was employed. The core of the approach is a special controller connected to a local controller and it is placed in the cloud environment. The approach also involves some special local controllers to perform the traffic collection from the investigated domains to perform the feature extraction, and, as a result, to find out the facts of the DDoS attack presence in the domain. The attack detection process is based on the analysis of 155 features, collected via the SPAN function of the Cisco switch. The obtained feature values were evaluated by detection modules placed within all defined local controllers to detect DDoS attacks. The approach used an extreme learning machine (ELM) as a decision-maker for attack detection. The feed-forward neural network with semi-supervised learning was used. The main advantage of ELM implementation is the training time reduction as it performs the random selection of the initial parameters. As a result, usage of ELM decreases the detection time. An attack mitigation module is also presented on each local controller. There is the possibility to organize the data exchange between each local controller, as well as with the universal controller. The proposed attack mitigation technique involves a set of attack mitigation scenarios able to perform in the wireless internet environment for different fixed devices.

The authors of [19] propose an intrusion detection system for IoT infrastructures. It is based on deep learning (DL-IDS). The approach for the IoT infrastructure intrusion detection involves the network traffic analysis; the data normalization procedure (to avoid the uncertainties in the obtained dataset); the data similarity evaluation on the usage of the Minkowski distance (to take into account the missing values, to eliminate possible redundancy, and to remove from the dataset the redundant and duplicate data); the replacement of the missing feature values in the obtained dataset (taking into account the evaluated values of the nearest neighbor on the basis of the K-nearest neighbor in the Euclidean distance to produce the average values for proceed data (to not take into account the classification results based on the data obtained from the more frequent entries); the traffic feature selection procedure on the basis of the spider monkey optimization algorithm usage (the set of features that are able to indicate the intrusion into the IoT infrastructure); and the exact intrusion detection procedure based on the stacked-deep polynomial network for the incoming data classification to mark it as normal or abnormal. The proposed approach is able to detect intrusions concerning the IoT environment (a remote-to-local attack, a DDoS attack, a probing attack, a user-to-root attack, etc.).

The study [20] provides research devoted to the usage of machine learning algorithms for anomaly detection in the Internet of Things infrastructures. To do this, the authors investigated the effectiveness and the main aspects of the usage of several single algorithms or their combinations for detection. The efficiency of the anomaly detection involved performance metrics, such as false positives, false negatives, specificity, sensitivity, and overall accuracy. The experimental part of the study is based on the Nemenya and Friedman tests that made it possible to perform a statistical analysis of the classifiers' differences. Another aspect of the research was the evaluation of the classifiers' response time. For this purpose, specific IoT infrastructure (as part of the implemented IDS) was employed. As a result of the conducted experiments, the authors of the study concluded that the most acceptable classification accuracy and the time of response were provided by the classification trees, regression trees, and extreme gradient boosting.

An approach for cyberattack detection as an AD-IoT system is presented in [21]. The proposed system is designed for the smart city infrastructure and is based on the random forest machine learning algorithm. The system aims to detect the compromised IoT devices that are placed in the distributed fog nodes. The division of normal and malicious behaviors of IoT devices is executed on the basis of monitoring and analyzing the fog nodes' network traffic. Such analysis is performed to verify whether the fog level attacks are detected and to inform the cloud security services concerning the evaluated results. The presented approach demonstrates sufficient detection efficiency and applies to the smart city infrastructure.

An approach for DDoS attack detection is presented in [22]. It is based on the hybrid optimization algorithms of Metaheuristic lion and Firefly. It was designed to perform data collecting, data preprocessing for noise removing, and filling missing data. The feature extraction was performed by employing recursive feature elimination (RFE). An important item of the proposed technique is the possibility of detecting low-rate attacks using the hybrid ML-F optimization algorithm. For the attack classification, a random forest classifier was used.

The article [23] introduces an IDS, which is based on the technique that uses an ensemble-based voting classifier. This approach uses multiple classifiers as a base learner. The final prediction is formed via producing the classifier's vote for the traditional classifier predictions. As the mean of the efficiency evaluation of the presented approach, a set of IoT devices with the usage of different sensors (garage door, light motion, GPS sensor, fridge sensor, thermostat, modbus, and weather) were employed. Multi-class attacks, such as XSS, Ransomeware, scanning injection, DDoS, and backdoor, were involved in the technique efficiency verification. The efficiency of the presented method was compared with the set of new intrusion detection approaches provided by scientists. The comparison was constructed on the basis of the accuracy, precision, recall, and F-score metrics. Furthermore, a set of machine learning algorithms, such as decision tree, naive Bayes, random forest, and K-nearest neighbors were involved in the comparison procedure. The experimental results demonstrated that the proposed approach has a high detection efficiency.

The authors of [24] propose a detection method for DoS/DDoS attacks against the IoT using machine learning. The approach aims to detect and apply the mitigation scenarios in the situation of DoS/DDoS attacks. To do this, the approach employs a multiclass classifier ("Looking back"). In addition, the ability of the technique to detect "malicious" packets makes it possible to apply mitigation measures against attacks that employ specific packet types.

The approach in [25] provides a botnet detection system for IoT devices. It is based on the algorithm named local–global best bat, which is used for neural networks and is able to process the botnet's feature sets to distinguish malicious and benign network traffic. As an experimental part of the study, the botnets Mirai and Gafgyt were used to infect several commercial IoT devices. In addition, to classify 10 botnet classes, the proposed algorithm was used. It was designed to tune the neural network hyperparameters and optimize the weight. The authors made the efficiency comparison of the provided algorithm with other approaches. The experimental results demonstrated that the proposed botnet detection approach accuracy was up to 90%, while BA-NN was 85.5%, and PSO-NN was 85.2%.

The authors of [26] proposed a taxonomy of intrusions detection systems that utilizes the data objects as the dimensions to summarize and classify machine learning- and deep learning-based IDS. The survey clarifies the concept of IDSs. Moreover, machine learning-based algorithms, metrics, and benchmark datasets frequently used in IDSs were introduced. IDSs applied to various data sources, i.e., logs, sessions, packets, and flow, were analyzed. The proposed taxonomic system was presented as a baseline and key IDS issues with using machine learning and deep learning algorithms. Moreover, future developments and challenges of IDS were discussed.

The authors of [27] introduced a probabilistic-driven ensemble (PDE)-based approach. This approach operates with several classification algorithms, wherein the effectiveness of these algorithms has been improved by applying a probabilistic criterion. Thus, the proposed approach allows maximizing the possibility of detecting intrusion events, regardless of the operational scenario, using several evaluation models. This makes it possible to distinguish ordinary events from related events to all classes of attacks. Experiments performed by using real-world data show that the proposed ensemble approach has better capability in detecting intrusion events (concerning known solutions).

The authors of [28] presented machine learning-based IDS. The feature reduction approach has two components: (1) Auto-encoder as a deep learning instance for dimensionality reduction; and (2) principal component analysis. The resulting set of low-dimensional features from both approaches was used to build different classifiers, i.e., Bayesian network, random forest, linear discriminant analysis, and quadratic discriminant analysis for designing IDS. The obtained experimental findings show better performance in terms of detection rate, false alarm rate, accuracy, and F-measure for binary and multi-class classification. This approach is able to reduce the feature dimensions of the CICIDS2017 dataset from 81 to 10, with high accuracy in both multi-class and binary classifications.

The objective of [29] was to apply various approaches for handling imbalanced datasets to design an effective IDS from the CIDDS-001 dataset. The effectiveness of sampling methods based on CIDDS-001 was studied and experimentally evaluated via random forest, deep neural networks, variational autoencoder, voting, and stacking machine learning classifiers. The developed system makes it possible to detect attacks with high accuracy when processing an unbalanced distribution of classes using a smaller number of samples. It makes it possible to apply the proposed system to data classification problems if it is necessary to merge data in real-time.

In [30], the authors were devoted to solving cybersecurity problems, such as the difficulty in distinguishing illegitimate activities from legitimate ones due to their high degrees of heterogeneity and similar characteristics. To solve this problem, a local feature engineering approach was proposed. This approach is based on the adoption of a data pre-processing strategy that allows reducing the number of network event patterns, increasing their characterization. The main distinguishing feature of the approach is that it operates locally in the feature space of each single network event, allowing to introduce new features and discretizing their values. The experimental results showed that the proposed approach improves the performance of known solutions.

The results of the machine learning algorithm efficiency analysis for detecting cyberattacks in the Internet of Things infrastructure are presented in Table 1.

The analysis of related works allows concluding that most studies had good detection accuracy; nevertheless, the main disadvantage of the investigated works is that they do not cover most features that may indicate the attack presence.

The analysis shows that the known approaches for detecting IoT cyberattacks demonstrate high-efficiency levels. Nevertheless, there are limitations—the inability to detect and respond to unknown attacks (zero-day attacks), the low efficiency of detection of multi-vector attacks; a high level of false positives, a significant response time that is unacceptable for real-time systems, and the need for significant amounts of computing resources. Another important aspect is the need to select a minimum and sufficient set of informative network traffic features that are able to indicate the presence of cyberattacks in the IoT infrastructure.

Table 1. Machine learning algorithm (MLA) efficiency for cyberattack detection in the Internet of Things infrastructure.

Authors	Goal	MLA	Data Set	Result
Shire, R.; Shiaeles, S.; Bendiab, K.; Ghita B.; Kolokotronis, N. [14]	malware detection, zero-day malware classification	Convolutional Neural Network and binary visualization	Real network environments	Accuracy of 91.32%, Precision of 91.67%, Recall of 91.03%
Elmrabit, N.; Zhou, F.; Li, F.; Zhou H. [15]	anomaly detection, attack detection	Logistic Regression, Decision Tree, Adaptive boosting, KNN, Random Forest, Naive Bayes, Gated Recurrent Units, Simple Recurrent Neural Network, Convolutional Neural Network and Long short-Term Memory, Convolutional Neural Network, Long short-Term Memory, Deep Neural Network	UNSW-NB15, CICIDS-2017, ICS Cyberattack	Performance about 99.9% using Random Forest (CICIDS-2017)
Bagui, X. Wang; Bagui, S. [16]	intrusion detection	Logistic regression, SVM, random forest	UCI Machine Learning Repository	Accuracy of about 99%
Kumar, P.; Gupta, G.P.; Tripathi, R. [17]	cyber-attack detection against IoT networks	K-nearest neighbor, random forest, XGBoost	DS2OS, NSL-KDD, BoT-IoT	Accuracy up to 99%, detection 90–100%
Ravi N.; Shalinie S.M. [18]	DDoS attacks detection and attacks mitigation	ELM, semi-supervised extreme learning machines	UNB-ISCX	Accuracy of about 96.28%
Otoum, Y.; Liu, D.; Nayak A. [19]	DoS, user-to-root (U2R), remote-to-local (R2L) detection, probe, intrusions	Stacked-deep polynomial network	NSL-KDD	Accuracy up to 99.02%, Precision up to 99.4%, recall up to 98.3%, F1-score up to 98.8%
Verma, A.; Ranga, V. [20]	Survey on machine learning algorithms for DoS attacks detection	AdaBoost, extremely randomized trees, multilayer perceptron, classification and regression trees, random forest, gradient boosted machine, extreme gradient boosting	UNSW-NB15, NSL-KDD, CIDDS-001	Regression trees, classification trees, and EG boosting show the best results—accuracy up to 96.7%, specificity up to 96.2%, sensitivity up to 97.3%
Alrashdi, I.; Alqazzaz, A.; Aloufi, E.; Alharthi, R.; Zohdy, M.; Ming, H. [21]	Detection of DDoS attacks	Bat Algorithm	N-BaIoT	Accuracy up to 90%
Krishna, E.S.; Thangavelu, A. [22]	Detection of the DDoS attacks	Random Forest	NSL-KDD, NBaIoT	Accuracy up to 99.98%, precision up to 99.87%, recall up to 100%, and F-score up to 99.73%
Mihoub, A.; Fredj, O.B.; Cheikhrouhou, O.; Derhab, A.; Krichen, M. [23]	Investigation of DoS/DDoS attacks detection for IoT based on ML algorithms	Looking-back-enabled random forest	IoT-Bot	Accuracy up to 99.81%

Table 1. *Cont.*

Authors	Goal	MLA	Data Set	Result
Khan, M.A.; Khan Khattk, M.A.; Latif, S.; Shah, A.A.; Ur Rehman, M.; Boulila, W.; Ahmad, J. [24]	intrusion detection	Combined decision tree, naive Bayes, random forest, and K-Nearest Neighbors using a voting-based technique	TON IoT	Accuracy up to 88%, Precision up to 90%, Recall up to 88%, F-score of 88% for DT-RF-NB based on binary classification with a combined IoT dataset
Alharbi, A.; Alosaimi, W.; Alyami, H.; Rauf, H.T. [25]	detection of DDoS attacks	Bat algorithm	N-BaIoT	Accuracy up to 90%
Saia, R.; Carta, S.; Recupero, D.R. [27]	intrusion events detection	Multilayer perceptron, decision tree, adaptive boosting, gradient boosting, random forests	NSL-KDD	Better performance compared to single classifiers in terms of specificity, without significant degradation in other aspects, since there is little degradation in terms of mean F-score, but a positive mean AUC (compared to competitor approaches), demonstrates the effectiveness of the approach
Abdulhammed, R.; Musafer, H.; Alessa, A.; Faezipour, M.; Abuzneid, A. [28]	developing the features dimensionality reduction approaches for machine learning-based IDS	Bayesian network, random forest, linear discriminant analysis, quadratic discriminant analysis	CICIDS2017	Reducing the feature dimensions of a dataset from 81 to 10, with high accuracy of 99.6% in both multi-class and binary classification
Abdulhammed, R.; Faezipour, M.; Abuzneid, A.; AbuMallouh, A. [29]	applying various approaches for handling imbalanced datasets to design effective IDS	Random forest, deep neural networks, variational autoencoder, voting, stacking	CIDDS-001	Attacks detection with up to 99.99% accuracy
Carta, S.; Podda, A.S.; Recupero, D.R.; Saia, R. [30]	solving such cybersecurity problems, as the difficulty of distinguishing illegitimate activities from legitimate ones	Random forests, decision tree, gradient boosting, adaptive boosting, multilayer perceptron	NSL-KDD, CICIDS2017, UNSW-NB15	Improving the performance of the state-of-the-art canonical solutions

To summarize, there is a strong need to evolve new methods for cyberattack detection in the IoT infrastructure. To do this, we are to eliminate technique drawbacks and increase the detection efficiency of detecting known and unknown cyberattacks in the IoT infrastructure.

3. Machine Learning Algorithms for Cyberattack Detection

The current study has involved five MLAs for IoT multi-vector cyberattack detections, as they were mostly used in (recent) research for efficient object classification [15–17,20,22,30]; we relied on our own experience in MLA use for cyberattack detection [11]:

1. Decision tree (DT) [31,32];
2. Random forest (RF) [33–38];
3. K-Nearest Neighbor (KNN) [39];
4. Extreme Gradient Boosting (XGBoost) [40];
5. Support Vector Machine (SVM) [41–43].

4. IoT Multi-Vector Cyberattack Detection Based on Machine Learning Algorithms

4.1. Detection Steps

The approach for IoT cyberattack detection includes the following steps (Figure 1):

1. Traffic obtaining;
2. Grouping packets by type, source device, and time. Packets from each device are grouped by type and by N records, according to the last connection time;
3. Feature extraction;
4. Feature classification based on the machine learning algorithm;
5. Result producing.

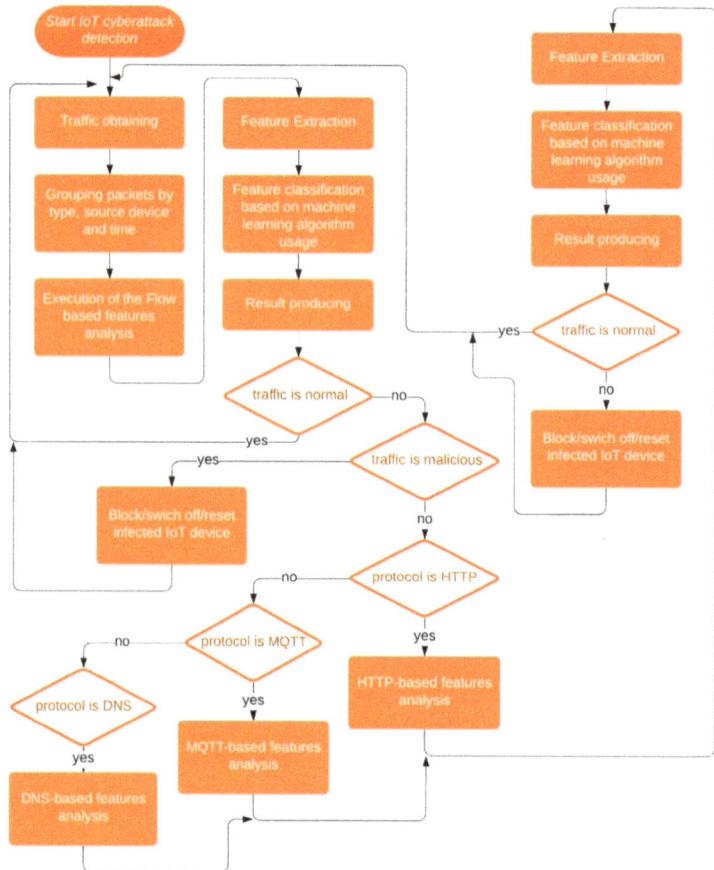

Figure 1. IoT cyberattack detection scheme.

4.2. Features Description

An important task is to speed up the detection of attack traffic. Early detection of attack traffic provides an opportunity to increase the security of the Internet of Things infrastructure, as it prevents the further spread of malicious software compromising not yet infected devices in the IoT infrastructure. Therefore, to speed up the detection of cyberattacks in the infrastructure, four types of features are involved:

- Flow-based features;
- MQTT-based features;
- DNS-based features;
- HTTP-based features.

Using only flow-based features (Table 2) makes it possible to speed up the detection of attacks on the network by faster extraction of features from streams and their analyses. In the case of suspicious traffic behavior that cannot be unambiguously classified as an attack, an in-depth traffic analysis is applied with the MQTT-based (Table 3), DNS-based (Table 4), and HTTP-based (Table 5) feature extractions.

Table 2. Flow-based features.

#	Features Designation	Value Description
1	f_1	Protocol type
2	f_2	Source IP address
3	f_3	Destination IP address
4	f_4	Source port
5	f_5	Destination port
6	f_6	Last connection time
7	f_7	Transaction bytes from f2 to f3
8	f_8	Transaction bytes from f3 to f2
9	f_9	Mean packet size transmitted by f2
10	f_{10}	Mean packet size transmitted by f3
11	f_{11}	Source bits per second
12	f_{12}	TTL value, f2 to f3
13	f_{13}	TTL value, f3 to f2
14	f_{14}	Interpacket interval
15	f_{15}	Bandwidth
16	f_{16}	Packet jitter

Table 3. MQTT-based features.

#	Features Designation	Value Description
1	f_{18}	The amount of connections to f_3 in N gathered records according to f_6
2	f_{19}	The amount of connections of f_2 in N gathered records according to f_6
3	f_{20}	The amount of connections of f_2 and f_5 in N gathered records according to f_6
4	f_{21}	The amount of connections to f_3 and f_4 in N gathered records according to f_6
5	f_{22}	The amount of connections of f_2 and f_3 in N gathered records according to f_6

Table 4. HTTP-based features.

#	Features Designation	Value Description
1	f_{48}	HTTP request method (GET, POST, HEAD)
2	f_{49}	HOST header value
3	f_{50}	Length of the HOST header value
4	f_{51}	URL in the request
5	f_{52}	Length of URL
6	f_{53}	HTTP pipelining depth
7	f_{54}	Uncompressed size of the transferred data from the client

Table 4. *Cont.*

#	Features Designation	Value Description
8	f_{55}	Uncompressed size of the transferred data from the server
9	f_{56}	Percentage of f_{48} with the same f_{49} in N records according to f_6
10	f_{57}	Percentage of the f_{49} the with same the f_{51} in N records according to f_6
11	f_{58}	Percentage of f_{48} with the same f_{51} in N records according to f_6

Table 5. DNS-based features.

#	Features Designation	Value Description
1	f_{23}	Requested domain name
2	f_{24}	Value specifying the request type
3	f_{25}	Length of f_{23}
4	f_{26}	Amount of unique characters in f_{23}
5	f_{27}	Entropy of f_{23}
6	f_{28}	TTL-period, mode (the value that appears most often in a set of data), in N records according to f_6
7	f_{29}	TTL-period, median (the numerical value separating the higher half of a data sample from the lower half), in N records according to f_6
8	f_{30}	TTL-period, average value, in N records according to f_6
9	f_{31}	Amount of A-records corresponding to f_{23} in the incoming DNS messages (the feature is used if value $f_{31} > 1$), in N records according to f_6
10	f_{32}	Amount of IP addresses concerned with f_{23} (feature is used if value $f_{31} = 1$), in N records according to f_6
11	f_{33}	Average distance between the IP addresses concerned with f_{23} (feature is used if value $f_{31} = 1$), in N records according to f_6
12	f_{34}	Average distance between the IP addresses in the set of A-records for f_{23} in the incoming DNS message (feature is used if value $f_{31} > 1$), in N records according to f_6
13	f_{35}	Amount of unique IP addresses in sets of A-records corresponding to f_{23} in the DNS messages (feature is used if value $f_{31} > 1$), in N records according to f_6
14	f_{36}	Average distance between unique IP addresses in sets of A-records corresponding to f_{23} in the DNS messages (feature is used if value $f_{31} > 1$), in N records according to f_6
15	f_{37}	Domain name amounts that share IP addresses corresponding to $f_{23,}$ in N records according to f_6
16	f_{38}	Sign of the usage of uncommon types of DNS records, or DNS records that are not commonly used by a typical client (e.g., TXT are most often used for tunneling (excluding mail servers), KEY, or NULL)
17	f_{39}	The entropy of the DNS records, which are contained in the DNS messages (CNAME, TXT, NS, MX, KEY, NULL, etc.)
18	f_{40}	Maximum size of the DNS messages about f_{23}, in N records according to f_6
19	f_{41}	Sign of success of DNS query ($f_{41} = 0$ if DNS query failed, and $f_{41} = 1$ if DNS query was successful)
20	f_{42}	Answer length
21	f_{43}	Mean class value in N records according to f_6
22	f_{44}	Mean type value in N records according to f_6
23	f_{45}	Amount of f_2 and f_{23} in N records according to f_6
24	f_{46}	Amount of f_{23} to the same f_2 in N records according to f_6
25	f_{47}	Percentage of the domain in N records according to f_6

This section presents the involvement of four feature types for multi-vector cyberattack detection in the IoT infrastructure. The features based on flow analysis enable the possibility of speeding up attack detections through faster analyses and make the detection algorithm scalable, allowing us to analyze high-bandwidth IoT traffic. On the other hand, the features based on deep packet analyses enable us to improve the accuracy of detection in cases where the use of a sign based on flow analysis does not provide an unambiguous answer about the presence of a cyberattack (and also allows detecting the multi-vector attacks).

5. Experiments

5.1. Evaluation Setting

To conduct the experiments, a Wi-Fi network of IoT devices was created. A Raspberry Pi 3 was configured as a middlebox, which acted as a Wi-Fi access point. To simulate DoS attacks as a source of malicious traffic, a computer system with a virtual Kali Linux was used. As a victim of DoS attacks, Raspberry Pi 2 with an installed Apache web server was used. All devices were connected to create a Wi-Fi network access point.

Three IoT devices (router, thermostat, camcorder) were also connected to the Wi-Fi network. To obtain normal traffic, a simulation of user interactions with the devices of the created IoT network was performed. To do this, actions such as transmitting video from the camera and installing software updates on connected IoT devices were performed. To obtain malicious traffic, a simulation of performing the most common classes of DoS attacks was executed.

An HTTP GET flood attack was simulated with the Goldeneye tool [44]; TCP SYN and UDP flood were simulated with Kali Linux hping3 utility [45]. The iodine utility was used to perform DNS tunneling attacks [46].

Malicious/benign traffic was collected at the Wi-Fi access point. The IoT traffic collection was executed via the Zeek tool [47]. It gives capacities to the network intrusion detection systems (IDS) and empowers security operation centers (SOC). The Zeek tool was used as a network traffic analyzer with an in-built classification engine.

In the collected DoS traffic samples, the source IP addresses and MAC addresses were substituted for the IP addresses and MAC addresses of the devices of the created IoT network. The time of sending malicious packets was modified so that the total collected IoT traffic replicated the activity of the attacking and normal activity devices.

Thus, the execution of DoS attacks of different types by each IoT device was simulated.

5.2. Dataset Description

To hold the experiments, the traffic generated by Mirai, Gafgyt, Dark Nexus botnets, UCI Machine Learning Repository, DS2OS, Bot-IoT, N-BaIoT, CIDDS, UNSW-NB15, and NSL-KDD traffic datasets [48–54] were used.

The DS2OS dataset contains traces gathered from the application layer of the IoT environment from devices such as movement sensors, light controllers, thermometers, batteries, thermostats, smart doors, etc. This dataset can be used to assess anomaly-based attack detection algorithms.

The UNSW-NB15 dataset contains data on nine types of attacks, such as Fuzzers, Analysis, Backdoors, DoS, Exploits, Generic, Reconnaissance, Shellcode, and Worms. A total of 49 features were extracted to describe these types of attacks.

The N_BaIoT dataset offers real-world IoT traffic data collected from nine IoT devices infected by Mirai and BASHLITE. Malicious data are divided into 10 attacks as well as harmless data (with 115 different features).

The Kitsune Network Attack Dataset contains nine network capture datasets in total that relate to different types of attack traffic against the IoT Infrastructure.

The BoT-IoT dataset was created by deploying a realistic IoT infrastructure network environment and it includes legitimate IoT network traffic as well as various types of attacks. The BoT-IoT includes DDoS and DoS for different protocols, OS scan, service scan, data exfiltration, and keylogging attacks.

The CIDDS and NSL-KDD datasets are built on network intrusion data describing "bad" connections, which are called intrusions (or attacks) and "good" connections (legitimate connections). These databases describe a wide range of intrusions and take into account user behavior scenarios.

Furthermore, experiments dealt with the set of traffic features presented in the above-mentioned datasets for three IoT devices: router, thermostat, and camcorder that were infected by Mirai, Gafgyt, and Dark Nexus botnets. The set of traffic features corresponds to four types of attacks (TCP, UDP, HTTP GET, and DNS tunneling).

As each dataset contains different samples and features, the preprocessing and feature selection processes were executed via each file type analysis and their parsing into the needed presentation for the next preprocessing. Such files as .csv, .pcap, Argus files, Zeek files, and .txt were processed.

Mirai is well-known malware that is able to infect an IoT device and turn such a smart device into a remotely-controlled network of bots—a botnet. The main negative impact of Mirai is the ability to launch massive DDoS attacks, as well as the ability to scan the internet for IoT smart devices based on the ARC processor. Such vulnerability as the usage of a stripped-down Linux version makes it possible to perform the logging into the device and execute malicious actions. In addition, the Mirai botnet uses a great amount of hijacked IoT devices to increase its spread and it is very dangerous for its mutating [55].

Gafgyt is a botnet that uses the vulnerabilities of IoT devices. It employs infected devices for large-scale (DDoS) attack execution. Moreover, Gafgyt uses known vulnerabilities (e.g., CVE-2017-17215, CVE-2018-10561) to implement the downloading of the next-stage payloads to compromised devices. New versions of the Gafgyt botnet include Mirai-based components to perform DDoS attacks; HTTP flooding to send a great number of HTTP requests to server targets to overwhelm them; UDP flooding to send special UDP packets to server victims to exhaust them; TCP flood attacks; STD attacks to send a random string to a specified IP address [56].

Dark Nexus is an IoT botnet that launches DDoS attacks. It was designed to launch credential stuffing attacks against different kinds of IoT devices (video recorders; DLink, Dasan Zhone, ASUS routers, thermal cameras, etc.) [57].

5.3. Training and Testing

The proposed approach involves five ML algorithms (decision tree, random forest, K-nearest neighbor, extreme gradient boosting, and support vector machine) to compare their detection possibilities. All algorithms were trained and tested using the dataset with training and testing percentages of 75% and 25%.

The BotGRABBER framework uses the scikit-learn library–an open-source platform for MLA in Python [58]. The configuration of each used MLA relies on the appropriate set of algorithm parameters. The optimal used values of algorithm parameters are presented in Tables 6–10 [59–63].

Table 6. Decision tree algorithm parameters [59].

Parameter	Value	Description
criterion	gini	The function to measure the quality of a split.
splitter	best	The strategy used to choose the split at each node.
max_depth	None	The maximum depth of the tree.
min_samples_split	3	The minimum number of samples required to split an internal node.
min_samples_leaf	1	The minimum number of samples required to be at a leaf node.
min_weight_fraction_leaf	0.0	The minimum weighted fraction of the sum total of weights (of all the input samples) required to be at a leaf node.
max_features	auto	The number of features to consider when looking for the best split.
random_state	RandomState instance	Controls the randomness of the estimator.
class_weight	balanced	Weights associated with classes.
ccp_alpha	0.0	Complexity parameter used for minimal cost complexity pruning.

Table 7. Random forest algorithm parameters [60].

Parameter	Value	Description
n_estimators	100	The number of trees in the forest.
criterion	gini	The function to measure the quality of a split.
max_depth	None	The maximum depth of the tree.
min_samples_split	2	The minimum number of samples required to split an internal node.

Table 7. *Cont.*

Parameter	Value	Description
min_samples_lea	1	The minimum number of samples required to be at a leaf node.
min_weight_fraction_leaf	0.0	The minimum weighted fraction of the sum total of weights.
max_features	$\log^2$	The number of features to consider when looking for the best split.
class_weight	balanced	Weights associated with classes.
ccp_alpha	0.0	Complexity parameter used for minimal cost complexity pruning.

Table 8. K-Nearest Neighbor algorithm parameters [61].

Parameter	Value	Description
n_neighbors	5	Number of neighbors.
weights	distance	Weight function used in prediction.
algorithm	kd_tree	The algorithm used to compute the nearest neighbors.
leaf_size	30	Leaf size passed to KDTree.
p	2	Power parameter for the Minkowski metric.
metric	str	The distance metric to use for the tree.
metric_params	dict	The number of parallel jobs to run for the neighbors' search.

Table 9. Extreme gradient boosting algorithm parameters [62].

Parameter	Value	Description
loss	exponential	The loss function to be optimized.
learning_rate	0.1	Learning rate shrinks the contribution of each tree.
n_estimators	100	The number of boosting stages to perform.
subsample	1.0	The fraction of samples to be used for fitting the individual base learners.
criterion	squared_error	The function to measure the quality of a split.
min_samples_split	2	The minimum number of samples required to split an internal node.
min_weight_fraction_leaf	0.0	The minimum weighted fraction of the sum total of weights (of all the input samples) required to be at a leaf node.
max_depth	3	The maximum depth of the individual regression estimator.
random_state	RandomState instance	Controls the random seed given to each tree estimator at each boosting iteration.
max_features	None	The number of features to consider when looking for the best split.
max_leaf_nodes	None	Grow trees with max_leaf_nodes in the best-first fashion.
validation_fraction	0.1	The proportion of training data to set aside as the validation set for early stopping.
n_iter_no_change	None	The decision as to whether early stopping will be used to terminate training when the validation score does not improve.
tol	1×10^3	Tolerance for the early stopping.
ccp_alpha	0.0	Complexity parameter used for minimal cost complexity pruning.

Table 10. Support vector machine parameters [63].

Parameter	Value	Description
C	1.0	Regularization parameter.
kernel	rbf	Specifies the kernel type to be used in the algorithm.
gamma	auto	Kernel coefficient.
tol	1×10^3	Tolerance for stopping criterion.
cache_size	100	Specify the size of the kernel cache (in MB).
max_iter	−1	Hard limit on iterations (no limit).
random_state	RandomState instance	Controls the pseudo-random number generation to shuffle the data for probability estimates.

5.4. Implementation Platform

To perform the feature extraction, the feature classification based on the machine learning algorithm, as well as the result of production, the BotGRABBER framework was employed. It is a multi-vector protection system that can perform network and host activity analyses. The BotGRABBER framework presents the tool, not only for botnet detection but also to produce the needed security scenario of the network reconfiguration according to the type of cyberattack performed by the detected botnet [11,13,43]. The mentioned tool includes several units aimed at traffic collection, packet processing, feature extraction, feature classification based on machine learning algorithms, and producing results. The feature classification unit of the framework is based on the scikit-learn library usage. It is a free software ML library for the Python programming language [58].

5.5. Results

Experimental results are presented in Tables 11–19.

As examples, comparisons of the different MLA efficiencies for Router/Mirai botnet detection (TCP attack, UDP attack, HTTP GET attack, and DNS tunneling) are presented in Figures 2–4.

As examples, comparisons of the different MLA efficiencies for Router/Mirai botnet detection (TCP attack, UDP attack, HTTP GET attack, and DNS tunneling) are presented in Figures 2–4.

Table 11. Classification results (router—Mirai).

Device/ Botnet	Attack	Algorithm	Accuracy	TP	FP	FN	TN	Precision	Recall	F1 Score	AUC
Router/ Mirai	TCP	RF	0.999479	3620	2	4	2024	0.9994748	0.999896	0.999572	0.999615
		DT	0.998584	3612	3	5	2030	0.99917	0.998618	0.998894	0.998994
		kNN	0.999469	3603	1	2	2044	0.999723	0.999445	0.999584	0.999692
		XGBoost	0.998938	3562	5	1	2082	0.998598	0.999719	0.999158	0.999573
		SVM	0.996991	3544	6	11	2089	0.99831	0.996906	0.997607	0.997881
	UDP	RF	0.999767	7531	5	2	2012	0.999937	0.999835	0.999935	0.999841
		DT	0.999267	7515	4	3	2028	0.999468	0.999601	0.999534	0.99975
		kNN	0.999476	7470	2	3	2075	0.999732	0.999599	0.999665	0.999821
		XGBoost	0.999686	7465	1	2	2082	0.999866	0.999732	0.999799	0.999827
		SVM	0.998534	7455	10	17	2068	0.998678	0.998678	0.998678	0.999174
	HTTP GET	RF	0.999694	6434	3	3	2060	0.999834	0.999734	0.999734	0.999839
		DT	0.999412	6419	1	4	2076	0.999844	0.999377	0.999611	0.999793
		kNN	0.999412	6387	1	4	2108	0.999843	0.999374	0.999609	0.999458
		XGBoost	0.999529	6340	2	2	2156	0.999685	0.999685	0.999685	0.999671
		SVM	0.997412	6381	5	14	2100	0.998636	0.99637	0.997502	0.999051
	DNS tunneling	RF	0.999624	5978	3	4	2005	0.999798	0.999731	0.999615	0.999944
		DT	0.999249	5935	2	4	2049	0.999663	0.999326	0.999495	0.999928
		kNN	0.999374	5920	3	2	2065	0.999493	0.999662	0.999578	0.999632
		XGBoost	0.998999	5903	5	3	2079	0.999154	0.999492	0.999323	0.999186
		SVM	0.997247	5899	5	14	2072	0.998649	0.99542	0.997032	0.997547

Table 12. Classification results (router—Gafgyt).

Device/ Botnet	Attack	Algorithm	Accuracy	TP	FP	FN	TN	Precision	Recall	F1 Score	AUC
Router/ Gafgyt	TCP	RF	0.999714	11984	2	2	2002	0.999833	0.999833	0.999833	0.999835
		DT	0.999571	11963	2	4	2021	0.999833	0.999666	0.999749	0.999757
		kNN	0.999357	11917	4	5	2064	0.999664	0.999581	0.999623	0.999792
		XGBoost	0.999643	11881	3	2	2104	0.999748	0.999832	0.99979	0.999734
		SVM	0.998713	11888	7	11	2084	0.999412	0.999076	0.999244	0.999523
	UDP	RF	0.999738	4498	2	1	1999	0.999656	0.999878	0.999667	0.999882
		DT	0.999077	4453	4	2	2041	0.999103	0.999551	0.999327	0.99947
		kNN	0.999385	4430	3	1	2066	0.999323	0.999774	0.999549	0.999648
		XGBOOST	0.999077	4391	5	1	2103	0.998863	0.999772	0.999317	0.999712
		SVM	0.998308	4433	6	9	4433	0.999056	0.998867	0.998961	0.998861
	HTTP GET	RF	0.999784	21082	2	3	2013	0.999905	0.999858	0.999881	0.999913
		DT	0.99987	21034	1	2	2063	0.999952	0.999905	0.999929	0.999912
		kNN	0.999697	20997	2	5	2096	0.999905	0.999762	0.999833	0.999971
		XGBoost	0.999827	20990	1	3	2106	0.999952	0.999857	0.999905	0.999845
		SVM	0.998961	20986	6	17	18684	0.998409	0.996144	0.997275	0.999221
	DNS tunneling	RF	0.999846	3191	2	4	2003	0.999674	0.999748	0.999561	0.999783
		DT	0.998269	3153	5	4	2038	0.998417	0.998733	0.998575	0.999548
		kNN	0.998654	3115	2	5	2078	0.999358	0.998397	0.998878	0.999539
		XGBoost	0.999615	3074	1	1	2124	0.999675	0.999675	0.999675	0.999882
		SVM	0.996154	3121	10	11	1485	0.998919	0.995688	0.997301	0.997861

Table 13. Classification results (router—Dark Nexus).

Device/ Botnet	Attack	Algorithm	Accuracy	TP	FP	FN	TN	Precision	Recall	F1 Score	AUC
Router/ Dark nexus	TCP	RF	0.999333	5490	4	1	2005	0.999272	0.999818	0.999545	0.999691
		DT	0.9992	5472	5	1	2022	0.999087	0.999817	0.999452	0.999982
		kNN	0.998933	5455	3	5	2037	0.99945	0.999084	0.999267	0.999836
		XGBOOST	0.9992	5417	2	4	2077	0.999631	0.999262	0.999446	0.999285
		SVM	0.9976	5394	9	9	2088	0.998334	0.998334	0.998334	0.999444
	UDP	RF	0.999344	10196	5	3	1996	0.99951	0.999706	0.999608	0.999488
		DT	0.999672	10171	1	3	2025	0.999902	0.999705	0.999803	0.999932
		kNN	0.999426	10146	3	4	2047	0.999704	0.999606	0.999655	0.999835
		XGBOOST	0.999426	10120	2	5	2073	0.999802	0.999506	0.999654	0.999844
		SVM	0.998279	10137	7	9	10137	0.998301	0.997736	0.998019	0.998421
	HTTP GET	RF	0.999771	19767	2	3	2018	0.999899	0.999848	0.999874	0.999853
		DT	0.999725	19746	3	3	2038	0.999848	0.999848	0.999848	0.999995
		kNN	0.999679	19716	2	5	2067	0.999899	0.999746	0.999823	0.999931
		XGBOOST	0.999771	19666	1	4	2119	0.999949	0.999797	0.999873	0.999794
		SVM	0.99899	19665	4	18	19665	0.99909	0.995918	0.997502	0.999452
	DNS tunneling	RF	0.999298	9351	3	5	2041	0.999679	0.999466	0.999572	0.999457
		DT	0.999474	9301	4	2	2093	0.99957	0.999785	0.999678	0.999974
		kNN	0.999737	9285	2	1	2112	0.999785	0.999892	0.999838	0.999859
		XGBOOST	0.999386	9243	5	2	2150	0.999459	0.999784	0.999621	0.999482
		SVM	0.997895	9302	7	10	9302	0.998109	0.99542	0.996763	0.998561

Table 14. Classification results (thermostat—Mirai).

Device/ Botnet	Attack	Algorithm	Accuracy	TP	FP	FN	TN	Precision	Recall	F1 Score	AUC
Thermostat/ Mirai	TCP	RF	0.999938	3623	1	5	2021	0.999724	0.999622	0.999773	0.999913
		DT	0.998938	3618	1	5	2026	0.999724	0.99862	0.999171	0.999446
		kNN	0.998938	3569	5	1	2075	0.998601	0.99972	0.99916	0.999017
		XGBOOST	0.999646	3528	1	1	2120	0.999717	0.999717	0.999717	0.999923
		SVM	0.996106	3535	5	17	2093	0.998588	0.995214	0.996898	0.999678
	UDP	RF	0.999986	7495	2	1	2052	0.999933	0.999897	0.9999	0.999865
		DT	0.999791	7451	1	1	2097	0.999866	0.999866	0.999866	0.999834
		kNN	0.999372	7446	1	5	2098	0.999866	0.999329	0.999597	0.999701
		XGBOOST	0.999476	7407	1	4	2138	0.999865	0.99946	0.999663	0.999991
		SVM	0.997906	7446	3	10	7446	0.999056	0.997172	0.998113	0.999816
	HTTP GET	RF	0.999859	6438	4	4	2054	0.999879	0.999779	0.999779	0.999861
		DT	0.999529	6391	2	2	2105	0.999687	0.999687	0.999687	0.999722
		kNN	0.999412	6369	4	1	2126	0.999372	0.999843	0.999608	0.999893
		XGBOOST	0.999294	6343	5	1	2151	0.999212	0.999842	0.999527	0.999791
		SVM	0.997176	6402	9	10	6402	0.998409	0.996144	0.997275	0.999465
	DNS tunneling	RF	0.999649	5976	1	5	2008	0.999833	0.999864	0.999498	0.999692
		DT	0.998874	5967	4	5	2014	0.99933	0.999163	0.999246	0.999617
		kNN	0.999249	5925	3	3	2059	0.999494	0.999494	0.999494	0.999828
		XGBOOST	0.999374	5876	4	1	2109	0.99932	0.99983	0.999575	0.999422
		SVM	0.996996	5890	10	15	5890	0.998379	0.995152	0.996763	0.998059

Table 15. Classification results (thermostat—Gafgyt).

Device/ Botnet	Attack	Algorithm	Accuracy	TP	FP	FN	TN	Precision	Recall	F1 Score	AUC
Thermostat/ Gafgyt	TCP	RF	0.999943	11973	3	2	2012	0.999849	0.999833	0.999891	0.999954
		DT	0.999714	11942	2	2	2044	0.999833	0.999833	0.999833	0.999876
		kNN	0.999571	11919	2	4	2065	0.999832	0.999665	0.999748	0.999744
		XGBOOST	0.999786	11903	1	2	2084	0.999916	0.999832	0.999874	0.999962
		SVM	0.99857	11885	7	13	2085	0.999411	0.998907	0.999159	0.998787
	UDP	RF	0.999815	4494	4	5	1997	0.999811	0.999889	0.999993	0.999972
		DT	0.998923	4459	2	5	2034	0.999552	0.99888	0.999216	0.998947
		kNN	0.999538	4435	2	1	2062	0.999549	0.999775	0.999662	0.999642
		XGBOOST	0.998923	4400	3	4	2093	0.999319	0.999092	0.999205	0.999741
		SVM	0.996769	4420	6	12	4420	0.99849	0.997548	0.998019	0.998866
	HTTP GET	RF	0.999784	21087	3	2	2008	0.999858	0.999905	0.999881	0.999862
		DT	0.99961	21042	4	5	2049	0.99981	0.999762	0.999786	0.999649
		kNN	0.99974	21025	1	5	2069	0.999952	0.999762	0.999857	0.999824
		XGBOOST	0.99974	20983	1	5	2111	0.999952	0.999762	0.999857	0.999743
		SVM	0.999351	20992	8	10	20992	0.998409	0.998182	0.998295	0.999371
	DNS tunneling	RF	0.999931	3187	3	1	2009	0.99976	0.999886	0.999773	0.999842
		DT	0.999231	3170	2	2	2026	0.999369	0.999369	0.999369	0.999636
		kNN	0.999231	3125	1	3	2071	0.99968	0.999041	0.99936	0.999325
		XGBOOST	0.998654	3084	2	5	2109	0.999352	0.998381	0.998866	0.998948
		SVM	0.995962	3137	7	9	3137	0.998649	0.995688	0.997166	0.998563

Table 16. Classification results (thermostat—Dark Nexus).

Device/ Botnet	Attack	Algorithm	Accuracy	TP	FP	FN	TN	Precision	Recall	F1 Score	AUC
Thermostat/ Dark nexus	TCP	RF	0.999067	5484	5	2	2009	0.999889	0.999735	0.999862	0.999807
		DT	0.999333	5457	4	1	2038	0.999268	0.999817	0.999542	0.999866
		kNN	0.999067	5440	2	5	2053	0.999632	0.999082	0.999357	0.999787
		XGBOOST	0.9988	5409	4	5	2082	0.999261	0.999076	0.999169	0.999392
		SVM	0.9972	5386	9	12	2093	0.998332	0.997777	0.998054	0.998168
	UDP	RF	0.99988	10188	5	5	2002	0.99988	0.99988	0.99988	0.99988
		DT	0.999344	10146	3	5	2046	0.999704	0.999507	0.999606	0.999544
		kNN	0.999262	10140	5	4	2051	0.999507	0.999606	0.999556	0.999830
		XGBOOST	0.999344	10130	5	3	2062	0.999507	0.999704	0.999605	0.999510
		SVM	0.998033	10073	3	16	10073	0.99849	0.996984	0.997736	0.998590
	HTTP GET	RF	0.999633	19765	3	5	2017	0.999848	0.999747	0.999798	0.999937
		DT	0.999541	19730	5	5	2050	0.999747	0.999747	0.999747	0.999730
		kNN	0.999725	19716	2	4	2068	0.999899	0.999797	0.999848	0.999950
		XGBOOST	0.999679	19680	3	4	2103	0.999848	0.999797	0.999822	0.999980
		SVM	0.999082	19644	4	13	19644	0.998182	0.997275	0.997728	0.999110
	DNS tunneling	RF	0.999649	9381	3	1	2015	0.99968	0.999893	0.999787	0.999683
		DT	0.999737	9345	1	2	2052	0.999893	0.999786	0.99984	0.999960
		kNN	0.999211	9335	4	5	2056	0.999572	0.999465	0.999518	0.999830
		XGBOOST	0.999561	9305	3	2	2090	0.999678	0.999785	0.999731	0.999730
		SVM	0.998421	9339	4	13	2044	0.998379	0.996763	0.99757	0.999860

Table 17. Classification results (camcorder—Mirai).

Device/ Botnet	Attack	Algorithm	Accuracy	TP	FP	FN	TN	Precision	Recall	F1 Score	AUC
Camcorder/ Mirai	TCP	RF	0.999292	3639	2	2	2007	0.999451	0.999451	0.999451	0.99907
		DT	0.998584	3623	4	4	2019	0.998897	0.998897	0.998897	0.999863
		kNN	0.999646	3598	1	1	2050	0.999722	0.999722	0.999722	0.999781
		XGBOOST	0.998938	3592	3	3	2052	0.999166	0.999166	0.999166	0.999396
		SVM	0.996106	3560	8	14	2068	0.997758	0.996083	0.99692	0.998166
	UDP	RF	0.999872	7545	2	4	1999	0.999835	0.99987	0.999603	0.99983
		DT	0.999267	7497	5	2	2046	0.999334	0.999733	0.999533	0.999701
		kNN	0.998953	7451	5	5	2089	0.999329	0.999329	0.999329	0.999833
		XGBOOST	0.999581	7444	1	3	2102	0.999866	0.999597	0.999731	0.999515
		SVM	0.997592	7412	6	9	7412	0.998867	0.996796	0.99783	0.998597
	HTTP GET	RF	0.999529	6479	2	2	2017	0.999691	0.999691	0.999691	0.99993
		DT	0.999412	6461	2	3	2034	0.999691	0.999536	0.999613	0.999737
		kNN	0.999412	6437	3	2	2058	0.999534	0.999689	0.999612	0.999959
		XGBOOST	0.999412	6421	2	3	2074	0.999689	0.999533	0.999611	0.999982
		SVM	0.998353	6405	7	10	6405	0.998409	0.998409	0.998409	0.999113
	DNS tunneling	RF	0.999249	5978	4	2	2006	0.999331	0.999666	0.999498	0.99968
		DT	0.998999	5959	4	4	2023	0.999329	0.999329	0.999329	0.999963
		kNN	0.999124	5942	4	3	2041	0.999327	0.999495	0.999411	0.999832
		XGBOOST	0.998874	5914	5	4	2067	0.999155	0.999324	0.99924	0.999737
		SVM	0.997121	5914	5	8	5914	0.99757	0.996225	0.996897	0.999861

Table 18. Classification results (camcorder—Gafgyt).

Device/ Botnet	Attack	Algorithm	Accuracy	TP	FP	FN	TN	Precision	Recall	F1 Score	AUC
Camcorder/ Gafgyt	TCP	RF	0.999971	11981	5	1	2003	0.999983	0.999917	0.99985	0.999889
		DT	0.999643	11941	1	4	2044	0.999916	0.999665	0.999791	0.999486
		kNN	0.999571	11924	5	1	2060	0.999581	0.999916	0.999748	0.999991
		XGBOOST	0.999571	11916	3	3	2068	0.999748	0.999748	0.999748	0.999364
		SVM	0.998713	11920	10	8	2052	0.999162	0.999329	0.999246	0.999484
	UDP	RF	0.999923	4496	4	3	1997	0.999811	0.999833	0.999822	0.999913
		DT	0.998615	4465	5	4	2026	0.998881	0.999105	0.998993	0.999888
		kNN	0.999077	4430	1	5	2064	0.999774	0.998873	0.999323	0.999442
		XGBOOST	0.999231	4387	2	3	2108	0.999544	0.999317	0.99943	0.999591
		SVM	0.997231	4365	8	11	1193	0.998301	0.998301	0.998301	0.997732
	HTTP GET	RF	0.99974	21056	3	3	2038	0.999858	0.999858	0.999858	0.999965
		DT	0.999827	21049	1	3	2047	0.999952	0.999857	0.999905	0.999425
		kNN	0.999784	21006	4	1	2089	0.99981	0.999952	0.999881	0.999628
		XGBOOST	0.99974	20958	5	1	2136	0.999761	0.999952	0.999857	0.999901
		SVM	0.999091	21005	8	20	21005	0.998636	0.996596	0.997615	0.999821
	DNS tunneling	RF	0.999038	3182	4	1	2013	0.999745	0.999686	0.999215	0.999491
		DT	0.998077	3176	5	5	2014	0.998428	0.998428	0.998428	0.999020
		kNN	0.998846	3165	2	4	2029	0.999368	0.998738	0.999053	0.999290
		XGBOOST	0.998462	3160	4	4	2032	0.998736	0.998736	0.998736	0.999390
		SVM	0.996923	3123	5	9	3123	0.998379	0.997301	0.99784	0.997460

Table 19. Classification results (camcorder—Dark Nexus).

Device/ Botnet	Attack	Algorithm	Accuracy	TP	FP	FN	TN	Precision	Recall	F1 Score	AUC
Camcorder/ Dark Nexus	TCP	RF	0.999722	5423	5	1	2071	0.999779	0.999816	0.999747	0.99986
		DT	0.998933	5403	3	5	2089	0.999445	0.999075	0.99926	0.999869
		kNN	0.9988	5378	4	5	2113	0.999257	0.999071	0.999164	0.999911
		EGB	0.9992	5359	1	5	2135	0.999813	0.999068	0.999441	0.999932
		SVM	0.997867	5336	10	6	2148	0.998129	0.998877	0.998503	0.999542
	UDP	RF	0.999918	10187	5	5	2003	0.999809	0.999839	0.999851	0.999861
		DT	0.99959	10139	4	1	2056	0.999606	0.999901	0.999753	0.998747
		kNN	0.999426	10132	4	3	2061	0.999605	0.999704	0.999655	0.999406
		EGB	0.999262	10124	4	5	2067	0.999605	0.999506	0.999556	0.999904
		SVM	0.998115	10105	7	16	2072	0.999308	0.998419	0.998863	0.999489
	HTTP GET	RF	0.999633	19769	3	5	2013	0.999848	0.999747	0.999798	0.999851
		DT	0.999587	19733	5	4	2048	0.999747	0.999797	0.999772	0.999996
		kNN	0.999862	19726	1	2	2061	0.999949	0.999899	0.999924	0.999952
		EGB	0.999633	19709	4	4	2073	0.999797	0.999797	0.999797	0.999766
		SVM	0.999036	19704	8	13	2065	0.999594	0.999341	0.999467	0.999123
	DNS tunneling	RF	0.999474	9385	4	2	2009	0.999574	0.999787	0.99968	0.999921
		DT	0.999386	9344	5	2	2049	0.999465	0.999786	0.999626	0.998696
		kNN	0.999649	9318	3	1	2078	0.999678	0.999893	0.999785	0.999282
		EGB	0.999386	9305	2	5	2088	0.999785	0.999463	0.999624	0.999645
		SVM	0.998421	9317	8	10	2065	0.999142	0.998928	0.999035	0.998664

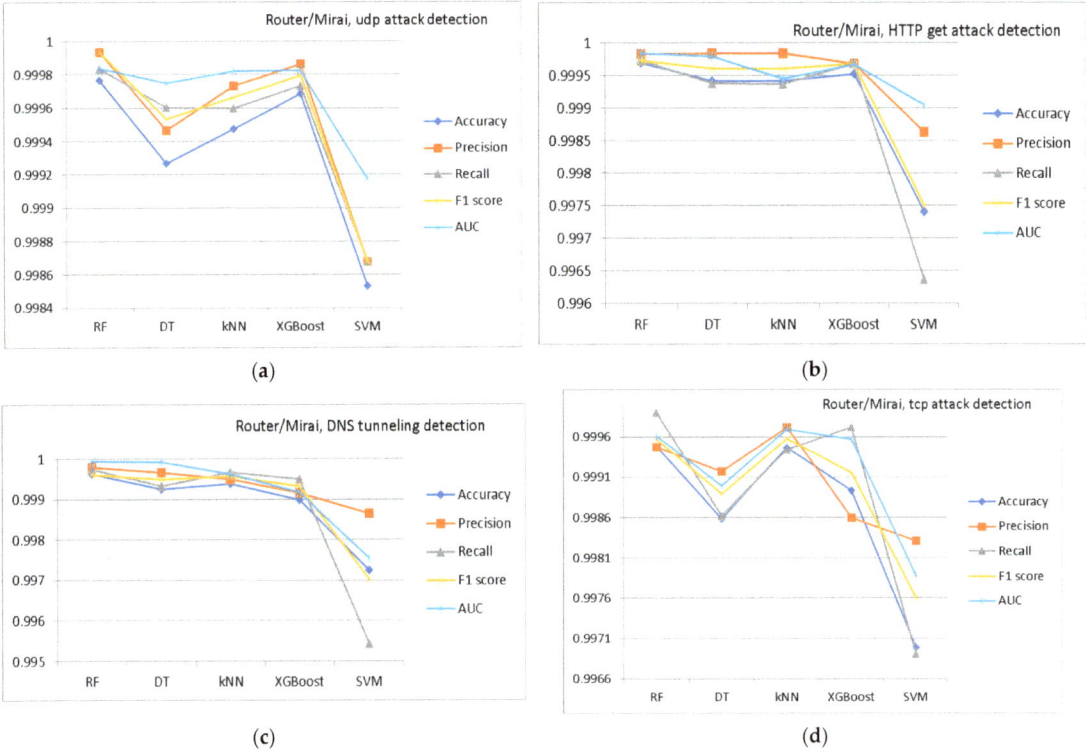

Figure 2. Comparison of different MLA efficiencies (decision tree—DT, random forest—RF, K-nearest neighbor—KNN, extreme gradient boosting—XGBoost, support vector machine—SVM) for Router/Mirai botnet detection: (**a**) TCP attack; (**b**) UDP attack; (**c**) HTTP GET attack; (**d**) DNS tunneling.

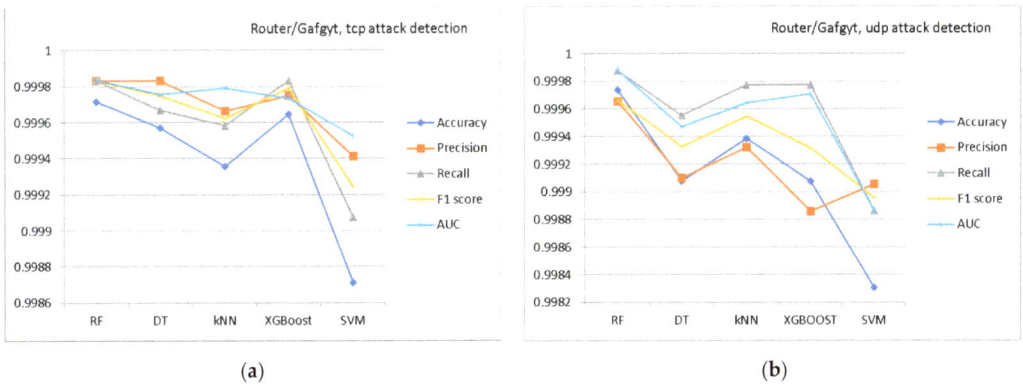

Figure 3. *Cont.*

(c)

(d)

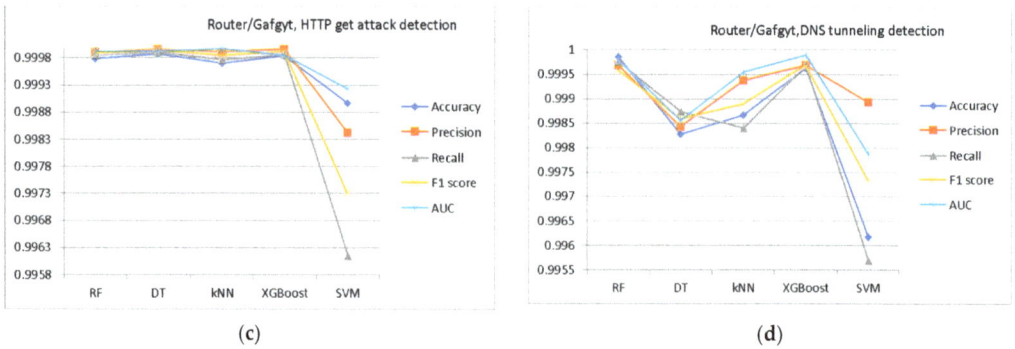

Figure 3. Comparison of different MLA efficiencies (decision tree—DT, random forest—RF, K-nearest neighbor—KNN, extreme gradient boosting—XGBoost, support vector machine—SVM) for Router/Gafgyt botnet detection: (**a**) TCP attack; (**b**) UDP attack; (**c**) HTTP GET attack; (**d**) DNS tunneling.

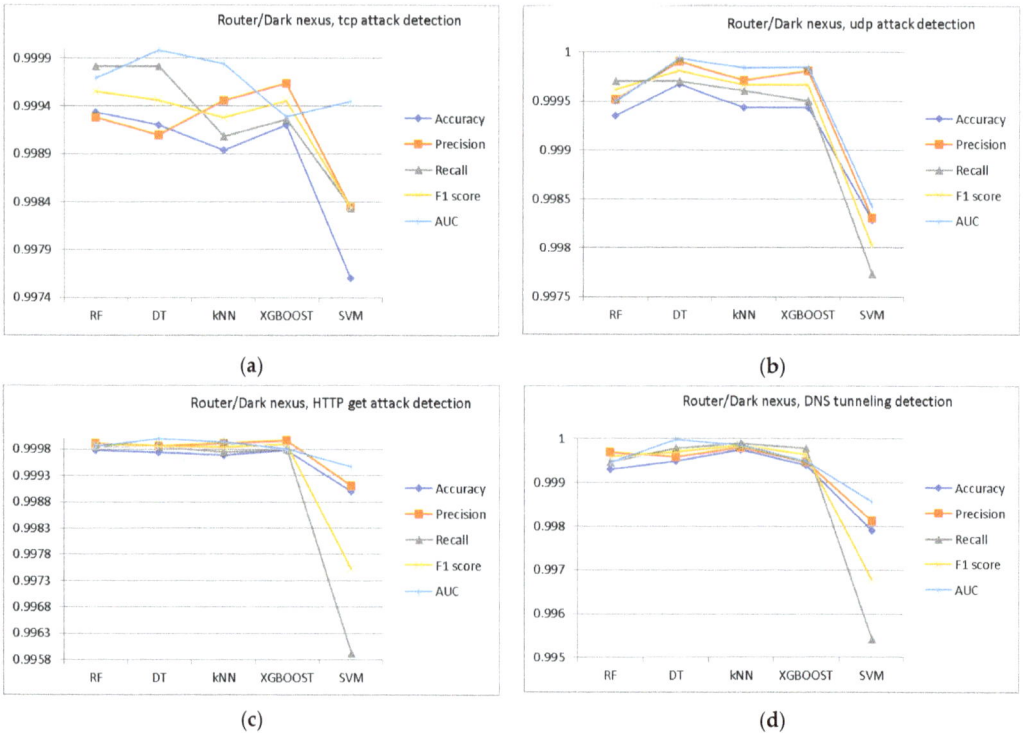

(a)

(b)

(c)

(d)

Figure 4. Comparison for different MLA efficiencies (decision tree—DT, random forest—RF, K-nearest neighbor—KNN, extreme gradient boosting—XGBoost, support vector machine—SVM) for Router/Dark Nexus botnet detection: (**a**) TCP attack; (**b**) UDP attack; (**c**) HTTP GET attack; (**d**) DNS tunneling.

6. Conclusions and Future Work

A flow-based traffic analysis allows detecting malicious behavior without the need for an in-depth packet analysis. Meanwhile, a packet content analysis provides an opportunity to decide whether the intercepted traffic belongs to the attack traffic or normal traffic in cases where the flow-based analysis does not give an unambiguous result. Attempting to cover features (as many as possible) that indicate the presence of attacks in the Internet of Things infrastructure has its weaknesses. Such an approach requires some time to analyze in-depth, and it is poorly scalable.

The main experiment results concerning MLA involvement showed that SVM demonstrated the worst results, while the RF algorithm demonstrated the best results.

In addition, the involvement of different IoT multi-vector cyberattack features based on flow analysis and features based on the most commonly used IoT protocols caused the detection of TCP, UDP, HTTP GET, and DNS tunneling attacks approximately at the same level.

In this paper, we reviewed the known approaches to detect attacks on the Internet of Things infrastructure based on machine learning and investigated their effectiveness. We investigated the possibility of detecting traffic attacks on the Internet of Things infrastructure based on flow analysis and the most commonly used IoT protocols, such as HTTP, MQTT, and DNS.

Traffic from well-known botnets, such as Mirai, Dark Nexus, and Gafgyt was taken from well-known databases that represent common attacks on the Internet of Things infrastructures, such as TCP, UDP, HTTP GET, and DNS tunneling, used as malicious traffic.

In addition, attack traffic was generated using known utilities, and benign IoT traffic was collected from devices such as a router, a thermostat, and a camcorder.

The features presented in the work were classified using various methods of machine learning and were removed from the received traffic.

The levels of detection of the multi-vector attacks on the Internet of Things infrastructure largely depend on the involved objects of training and test samplings/settings of machine learning algorithms. This important aspect is the subject of further research.

Therefore, future work will focus on the following issues:

1. Different Internet of Things protocols [64] to remove signs of traffic, which will improve the accuracy of attack detection in the lack of flow-based analysis cases;
2. Efficient ways to reduce the number of traffic features sufficient to detect attacks;
3. Development of ML-based methods for dependability assurance of IoT systems by combining attacks and intrusion detection, redundancy, and recovery procedures [65].

Author Contributions: Data curation K.B. and V.K.; formal analysis S.L.; investigation K.B. and O.S.; methodology K.B. and S.L.; project administration V.K.; Software K.B.; supervision V.K.; validation K.B. and O.S.; visualization K.B. and S.L.; writing—original draft K.B. and S.L. All authors have read and agreed to the published version of the manuscript.

Funding: This research received no external funding.

Institutional Review Board Statement: Not applicable.

Informed Consent Statement: Informed consent was obtained from all subjects involved in the study.

Data Availability Statement: The dataset used for this study is publicly available at [43–49].

Acknowledgments: This work was supported by the ECHO project, which has received funding from the European Union's Horizon 2020 research and innovation program under the grant agreement no 830943. The authors appreciate the scientific society of the consortium for creative analysis and discussion during the preparation of this paper.

Conflicts of Interest: The authors declare no conflict of interest.

References

1. Nozomi Networks Labs. New OT/IoT Security Report: Trends and Countermeasures for Critical Infrastructure Attacks. Available online: https://www.nozominetworks.com/blog/new-ot-iot-security-report-trends-and-countermeasures-for-critical-infrastructure-attacks/ (accessed on 3 February 2022).
2. Global Cyber Alliance. GCA Internet Integrity Papers: IoT Policy and Attack Report. Available online: https://www.globalcyberalliance.org/wp-content/uploads/IoT-Policy-and-Attack-Report_FINAL.pdf (accessed on 5 December 2021).

3.   Shaaban, A.M.; Chlup, S.; El-Araby, N.; Schmittner, C. Towards Optimized Security Attributes for IoT Devices in Smart Agriculture Based on the IEC 62443 Security Standard. *Appl. Sci.* **2022**, *12*, 5653. [CrossRef]
4.   Seo, S.; Kim, D. IoDM: A Study on a IoT-Based Organizational Deception Modeling with Adaptive General-Sum Game Competition. *Electronics* **2022**, *11*, 1623. [CrossRef]
5.   Makarichev, V.; Lukin, V.; Illiashenko, O.; Kharchenko, V. Digital Image Representation by Atomic Functions: The Compression and Protection of Data for Edge Computing in IoT Systems. *Sensors* **2022**, *22*, 3751. [CrossRef]
6.   Bliss, D.; Garbos, R.; Kane, P.; Kharchenko, V.; Kochanski, T.; Rucinski, A. Homo Digitus: Its Dependable and Resilient Smart Ecosystem. *Smart Cities* **2021**, *4*, 514–531. [CrossRef]
7.   Deorankar, A.V.; Thakare, S.S. Survey on Anomaly Detection of (IoT)- Internet of Things Cyberattacks Using Machine Learning. In Proceedings of the 2020 Fourth International Conference on Computing Methodologies and Communication (ICCMC), Erode, India, 11–13 March 2020; pp. 115–117. [CrossRef]
8.   Hristov, A.; Trifonov, R.A. Model for Identification of Compromised Devices as a Result of Cyberattack on IoT Devices. In Proceedings of the 2021 International Conference on Information Technologies (InfoTech), Varna, Bulgaria, 16–17 September 2021; pp. 1–4. [CrossRef]
9.   Lysenko, S.; Bobrovnikova, K.; Shchuka, R.; Savenko, O. A Cyberattacks Detection Technique Based on Evolutionary Algorithms. In Proceedings of the 2020 IEEE 11th International Conference on Dependable Systems, Services and Technologies (DESSERT), Kyiv, Ukraine, 14–18 May 2020; pp. 127–132.
10. Lysenko, S.; Pomorova, O.; Savenko, O.; Kryshchuk, A.; Bobrovnikova, K. DNS-based Anti-evasion Technique for Botnets Detection. In Proceedings of the 8th IEEE International Conference on Intelligent Data Acquisition and Advanced Computing Systems: Technology and Applications, Warsaw, Poland, 24–26 September 2015; pp. 453–458.
11. Savenko, B.; Lysenko, S.; Bobrovnikova, K.; Savenko, O.; Markowsky, G. Detection DNS Tunneling Botnets. In Proceedings of the 2021 IEEE 11th International Conference on Intelligent Data Acquisition and Advanced Computing Systems: Technology and Applications, Cracow, Poland, 22–25 September 2021; Volume 1, pp. 64–69.
12. Lysenko, S.; Savenko, O.; Bobrovnikova, K. DDoS Botnet Detection Technique Based on the Use of the Semi-Supervised Fuzzy c-Means Clustering. *CEUR-WS* **2018**, *2104*, 688–695.
13. Lysenko, S.; Bobrovnikova, K.; Matiukh, S.; Hurman, I.; Savenko, O. Detection of the botnets' low-rate DDoS attacks based on self-similarity. *Int. J. Electr. Comput. Eng.* **2020**, *10*, 3651–3659. [CrossRef]
14. Shire, R.; Shiaeles, S.; Bendiab, K.; Ghita, B.; Kolokotronis, N. Malware Squid: A Novel IoT Malware Traffic Analysis Framework Using Convolutional Neural Network and Binary Visualisation. In *Ininternet of Things, Smart Spaces, and Next Generation Networks and Systems*; Springer: Cham, Switzerland, 2019; pp. 65–76.
15. Elmrabit, N.; Zhou, F.; Li, F.; Zhou, H. Evaluation of machine learning algorithms for anomaly detection. In Proceedings of the 2020 International Conference on Cyber Security and Protection of Digital Services (Cyber Security), Dublin, Ireland, 15–19 June 2020; pp. 1–8.
16. Bagui, S.; Wang, X.; Bagui, S. Machine Learning Based Intrusion Detection for IoT Botnet. *Int. J. Mach. Learn. Comput.* **2021**, *11*, 399–406. [CrossRef]
17. Kumar, P.; Gupta, G.P.; Tripathi, R. Toward design of an intelligent cyberattack detection system using hybrid feature reduced approach for IoT networks. *Arab. J. Sci. Eng.* **2021**, *46*, 3749–3778. [CrossRef]
18. Ravi, N.; Shalinie, S.M. Learning-driven detection and mitigation of DDoS attack in IoT via SDN-cloud architecture. *IEEE Internet Things J.* **2020**, *7*, 3559–3570. [CrossRef]
19. Otoum, Y.; Liu, D.; Nayak, A. DL-IDS: A deep learning-based intrusion detection framework for securing IoT. *Trans. Emerg. Telecommun. Technol.* **2019**, *33*, e3803. [CrossRef]
20. Verma, A.; Ranga, V. Machine learning based intrusion detection systems for IoT applications. *Wirel. Pers. Commun.* **2020**, *111*, 2287–2310. [CrossRef]
21. Alrashdi, I.; Alqazzaz, A.; Aloufi, E.; Alharthi, R.; Zohdy, M.; Ming, H. Ad-IoT: Anomaly Detection of IoT Cyberattacks in smart City Using Machine Learning. In Proceedings of the 2019 IEEE 9th Annual Computing and Communication Workshop and Conference (CCWC), Las Vegas, NV, USA, 7–9 January 2019; pp. 305–310.
22. Krishna, E.S.; Thangavelu, A. Attack detection in IoT devices using hybrid metaheuristic lion optimization algorithm and firefly optimization algorithm. *Int. J. Syst. Assur. Eng. Manag.* **2021**, 1–14. [CrossRef]
23. Mihoub, A.; Fredj, O.B.; Cheikhrouhou, O.; Derhab, A.; Krichen, M. Denial of service attack detection and mitigation for internet of things using looking-back-enabled machine learning techniques. *Comput. Electr. Eng.* **2022**, *98*, 107716. [CrossRef]
24. Khan, M.A.; Khan Khattk, M.A.; Latif, S.; Shah, A.A.; Ur Rehman, M.; Boulila, W.; Ahmad, J. Voting classifier-based intrusion detection for IoT networks. In *Advances on Smart and Soft Computing*; Springer: Singapore, 2022; pp. 313–328.
25. Alharbi, A.; Alosaimi, W.; Alyami, H.; Rauf, H.T.; Damaševičius, R. Botnet attack detection using local global best bat algorithm for industrial internet of things. *Electronics* **2021**, *10*, 1341. [CrossRef]
26. Liu, H.; Lang, B. Machine learning and deep learning methods for intrusion detection systems: A survey. *Appl. Sci.* **2019**, *9*, 4396. [CrossRef]
27. Saia, R.; Carta, S.; Recupero, D.R. A Probabilistic-driven Ensemble Approach to Perform Event Classification in Intrusion Detection System. In Proceedings of the 10th International Joint Conference on Knowledge Discovery, Knowledge Engineering and Knowledge Management, Seville, Spain, 18–20 September 2018; pp. 141–148.

28. Abdulhammed, R.; Musafer, H.; Alessa, A.; Faezipour, M.; Abuzneid, A. Features dimensionality reduction approaches for machine learning based network intrusion detection. *Electronics* **2019**, *8*, 322. [CrossRef]
29. Abdulhammed, R.; Faezipour, M.; Abuzneid, A.; AbuMallouh, A. Deep and machine learning approaches for anomaly-based intrusion detection of imbalanced network traffic. *IEEE Sens. Lett.* **2018**, *3*, 1–4. [CrossRef]
30. Carta, S.; Podda, A.S.; Recupero, D.R.; Saia, R. A local feature engineering strategy to improve network anomaly detection. *Future Internet* **2020**, *12*, 177. [CrossRef]
31. Rokach, L.; Maimon, O. *Data Mining with Decision Trees: Theory and Applications*; World Scientific: Singapore, 2014; p. 81.
32. Flow of Decision Tree Algorithm. Available online: https://www.analyticsvidhya.com/blog/2022/04/complete-flow-of-decision-tree-algorithm/ (accessed on 10 December 2021).
33. Kotu, V.; Deshpande, B. *Data Science: Concepts and Practice*; Morgan Kaufmann: San Francisco, CA, USA, 2019; pp. 65–163.
34. Polamuri, S. How the Random Forest Algorithm Works in Machine Learning. Available online: https://dataaspirant.com/2017/05/22/random-forest-algorithm-machine-learing (accessed on 10 December 2021).
35. Biau, G.; Scornet, E.A. Random Forest Guided Tour. *Test* **2016**, *25*, 197–227. [CrossRef]
36. Scornet, E.; Biau, G.; Vert, J.-P. Consistency of random forests. *Ann. Statist.* **2015**, *43*, 1716–1741. [CrossRef]
37. Athey, S.; Tibshirani, J.; Wager, S. Generalized random forests. *Ann. Statist.* **2019**, *47*, 1148–1178. [CrossRef]
38. Ronaghan, S. The Mathematics of Decision Trees, Random Forest and Feature Importance in Scikit-Learn and Spark. Available online: https://towardsdatascience.com/the-mathematics-of-decision-trees-random-forest-and-feature-importance-in-scikit-learn-and-spark-f2861df67e3 (accessed on 10 December 2021).
39. Campos, G.O.; Zimek, A.; Sander, J.; Campello, R.J.; Micenková, B.; Schubert, E.; Assent, I.; Houle, M.E. On the evaluation of unsupervised outlier detection: Measures, datasets, and an empirical study. *Data Min. Knowl. Discov.* **2016**, *30*, 891–927. [CrossRef]
40. Chen, T.; He, T.; Benesty, M.; Khotilovich, V.; Tang, Y.; Cho, H.; Chen, K. Xgboost: Extreme gradient boosting. *R Package Version 0.4-2* **2015**, *1*, 1–4.
41. Weston, J.; Mukherjee, S.; Chapelle, O.; Pontil, M.; Poggio, T.; Vapnik, V. Feature selection for SVMs. *Advances in neural information processing systems* **2001**, *13*, 668–674.
42. Chapelle, O.; Vapnik, V.; Bousquet, O.; Mukherjee, S. Choosing multiple parameters for support vector machines. *Mach. Learn.* **2002**, *46*, 131–159. [CrossRef]
43. Lysenko, S.; Bobrovnikova, K.; Savenko, O.; Kryshchuk, A. BotGRABBER: SVM-Based Self-Adaptive System for the Network Resilience Against the Botnets' Cyberattacks. In *International Conference on Computer Networks*; Springer: Cham, Switzerland, 2019; pp. 127–143.
44. GoldenEye Is a HTTP DoS Test Tool. Available online: https://www.kali.org/tools/goldeneye/ (accessed on 11 December 2021).
45. hping3 Network Tool. Available online: https://github.com/antirez/hping (accessed on 11 December 2021).
46. DNS Tunneling Tool. Available online: https://github.com/yarrick/iodine (accessed on 11 December 2021).
47. Zeek. An Open Source Network Security Monitoring Tool. Available online: https://zeek.org/ (accessed on 11 May 2022).
48. UCI Machine Learning Repository. Available online: https://archive.ics.uci.edu/ml/index.php (accessed on 11 December 2021).
49. Kaggle. DS2OS Traffic Traces. Available online: https://www.kaggle.com/datasets/francoisxa/ds2ostraffictraces (accessed on 11 December 2021).
50. IEEEDataPort. The Bot-IoT Dataset. Available online: https://ieee-dataport.org/documents/bot-iot-dataset (accessed on 11 December 2021).
51. Kaggle. N-BaIoT Dataset to Detect IoT Botnet Attacks. Available online: https://www.kaggle.com/datasets/mkashifn/nbaiot-datasetURL (accessed on 11 December 2021).
52. Hochschule Coburg. CIDDS-Coburg Intrusion Detection Data Sets. Available online: https://www.hs-coburg.de/forschung/forschungsprojekte-oeffentlich/informationstechnologie/cidds-coburg-intrusion-detection-data-sets.html (accessed on 11 December 2021).
53. UNSW Sydney. The UNSW-NB15 Dataset. Available online: https://research.unsw.edu.au/projects/unsw-nb15-dataset (accessed on 11 December 2021).
54. UNB. University of New Brunswick. *NSL-KDD Dataset*. Available online: https://www.unb.ca/cic/datasets/nsl.html (accessed on 11 December 2021).
55. What Is the Mirai Botnet? Available online: https://www.cloudflare.com/learning/ddos/glossary/mirai-botnet/ (accessed on 11 May 2022).
56. Gafgyt Botnet Lifts DDoS Tricks from Mirai. Available online: https://threatpost.com/gafgyt-botnet-ddos-mirai/165424/ (accessed on 11 May 2022).
57. Dark Nexus, the Latest IoT Botnet Targets a Wide Range of Devices. Available online: https://crazygreek.co.uk/dark-nexus-iot-botnet-targets-devices/ (accessed on 11 May 2022).
58. Scikit-Learn. Machine Learning in Python. Available online: https://scikit-learn.org/stable/index.html (accessed on 11 May 2022).
59. Sklearn.Tree.DecisionTreeClassifier—Scikit-Learn 1.0.2 Documentation. Available online: https://scikit-learn.org/stable/modules/generated/sklearn.tree.DecisionTreeClassifier.html (accessed on 11 May 2022).

60. Sklearn.Ensemble.RandomForestClassifier—Scikit-Learn 1.0.2 Documentation. Available online: https://scikit-learn.org/stable/modules/generated/sklearn.ensemble.RandomForestClassifier.html (accessed on 15 May 2022).
61. Sklearn.Neighbors.KNeighborsClassifier—Scikit-Learn 1.0.2 Documentation. Available online: https://scikit-learn.org/stable/modules/generated/sklearn.neighbors.KNeighborsClassifier.html (accessed on 15 May 2022).
62. Sklearn.Neighbors.GradientBoostingClassifier—Scikit-Learn 1.0.2 Documentation. Available online: https://scikit-learn.org/stable/modules/generated/sklearn.ensemble.GradientBoostingClassifier.html (accessed on 11 May 2022).
63. Sklearn.Svm.SVC—Scikit-Learn 1.0.2 Documentation. Available online: https://scikit-learn.org/stable/modules/generated/sklearn.svm.SVC.html (accessed on 15 May 2022).
64. Kolisnyk, M. Vulnerability analysis and method of selection of communication protocols for information transfer in Internet of Things systems. *Radioelectron. Comput. Syst.* **2021**, *1*, 133–149. [CrossRef]
65. Illiashenko, O.; Kolisnyk, M.; Strielkina, A.; Kotsiuba, I.; Kharchenko, V. Conception and application of dependable Internet of Things based systems. *Radio Electron. Comput. Sci. Control* **2020**, *4*, 139–150. [CrossRef]

algorithms

MDPI

Article

An Adaptive Deep Learning Neural Network Model to Enhance Machine-Learning-Based Classifiers for Intrusion Detection in Smart Grids

Xue Jun Li [1,*,†,‡], Maode Ma [2,‡] and Yihan Sun [3,‡]

1   Department of Electrical and Electronic Engineering, Auckland University of Technology, Auckland 1010, New Zealand
2   College of Engineering, Qatar University, Doha P.O. Box 2713, Qatar; acadmmd@gmail.com
3   School of Electrical and Electronic Engineering, Nanyang Technological University, Singapore 639798, Singapore; suny0045@e.ntu.edu.sg
*   Correspondence: xuejun.li@aut.ac.nz; Tel.: +64-9-921-9999
†   Current address: Auckland University of Technology, Private Bag 92006, Auckland 1142, New Zealand.
‡   These authors contributed equally to this work.

Abstract: Modern smart grids are built based on top of advanced computing and networking technologies, where condition monitoring relies on secure cyberphysical connectivity. Over the network infrastructure, transported data containing confidential information, must be protected as smart grids are vulnerable and subject to various cyberattacks. Various machine learning based classifiers were proposed for intrusion detection in smart grids. However, each of them has respective advantage and disadvantages. Aiming to improve the performance of existing machine learning based classifiers, this paper proposes an adaptive deep learning algorithm with a data pre-processing module, a neural network pre-training module and a classifier module, which work together classify intrusion data types using their high-dimensional data features. The proposed Adaptive Deep Learning (ADL) algorithm obtains the number of layers and the number of neurons per layer by determining the characteristic dimension of the network traffic. With transfer learning, the proposed ADL algorithm can extract the original data dimensions and obtain new abstract features. By combining deep learning models with traditional machine learning-based classification models, the performance of classification of network traffic data is significantly improved. By using the Network Security Laboratory-Knowledge Discovery in Databases (NSL-KDD) dataset, experimental results show that the proposed ADL algorithm improves the effectiveness of existing intrusion detection methods and reduces the training time, indicating a promising candidate to enhance network security in smart grids.

Keywords: deep learning; machine learning; intrusion detection; smart grid; neural networks

Citation: Li, X.J.; Ma, M.; Sun, Y. An Adaptive Deep Learning Neural Network Model to Enhance Machine-Learning-Based Classifiers for Intrusion Detection in Smart Grids. *Algorithms* **2023**, *16*, 288. https://doi.org/10.3390/a16060288

Academic Editors: Frank Werner, Francesco Bergadano and Giorgio Giacinto

Received: 15 January 2023
Revised: 28 May 2023
Accepted: 30 May 2023
Published: 2 June 2023

1. Introduction

Aiming to provide secure and dependable electrical services, the smart grid integrates power generation, transmission and distribution through digital communication technologies to detect and react to local changes in usage. The smart grid has two core subsystems: Advanced Metering Infrastructure (AMI) and Supervisory Control And Data Acquisition (SCADA), where AMI realises bi-directional data exchange between the electricity supplier and the customer to improve the efficiency of electricity consumption, while SCADA enables real-time monitoring and controlling of the transmission network [1]. However, due to the dependence among components in smart grid, a cyber attack could still lead to catastrophic failure of the entire grid [2]. Obviously, it is important to ensure the security of smart grid. In the National Institute of Standards and Technology Interagency Report (NIS-TIR) 7628, Guidelines for Smart Grid Cyber Security [3], information security in the smart grid consists of three essential elements: confidentiality (*only authorised users can access the information*), integrity (*data must be accurate and consistent*) and availability (*information must be available with low-latency to authorized parties when needed*). Therefore, smart grid should have self-healing and recovery capabilities to ensure communication and data security.

Cyber attacks on smart grid networks include control signal attacks, measurement attacks, and control-signal-measurement attacks [4]. Typical threats that impede data availability include flooding, route destruction, selective forwarding, wormhole, Byzantine attacks and denial-of-service (DoS) attacks. In general, security solutions can be divided into two main techniques, called prevention techniques and detection techniques. Prevention techniques aim to protect network data from being intercepted and encryption is usually adopted. Detection techniques aim to detect intruders [5], which include signature-based detection and anomaly-based detection. The former compares the observed attack patterns with known ones. The latter compares network traffic parameters with normal ones, where a change from normal traffic simply declares the presence of an intruder.

This paper presents an adaptive deep learning (ADL) neural network model to improve the recognition efficiency of anomalous attacks in smart grids. The proposed algorithm determines the number of layers and neurons per layer of the model, depending on the size of the smart grid. The contribution of this paper is threefold: Firstly, we propose an adaptive deep learning algorithm with a data pre-processing module, a neural network pre-training module and a classifier module, which work together classify intrusion data types using their high-dimensional data features. Secondly, the proposed ADL algorithm complements existing classification methods, and it can deployed with any existing feature classification algorithms to improve the classification performance. Finally, through experiments using the NSL-KDD dataset, we show that the robustness and flexibility of the proposed ADL algorithm. Altogether, by adding the proposed ADL algorithm, existing classifier algorithms can effectively discriminate between large ranges of network traffic, improve the accuracy of intrusion detection, converge faster, and reduce detection time significantly. The rest of the paper is organised as follows. Section 2 discusses the related work, while Section 3 presents the proposed ADL neural network model. Section 4 discusses the results, and Section 5 concludes the paper. For the sake of readability, Table 1 lists the abbreviations used in this paper.

Table 1. Nomenclature.

Abbreviation	Term
ACE	asymmetric convolutional encoder
ADL	Adaptive deep learning
AMI	Advanced Metering Infrastructure
BPNN	Back Propagation Neural Network
CFS	correlation-based feature selection
DBN	deep belief network
DDoS	distributed denial of service
DoS	denial-of-service
DT	Decision Tree
FDIA	false data injection attacks
HAN	Home Area Network

Table 1. *Cont.*

Abbreviation	Term
HEMS	home energy management system
IDS	Intrusion Detection System
IoT	Internet of Things
KDD	Knowledge Discovery in Databases
KNN	K-nearest neighbour
LSTM	long-short-term-memory
NISTIR	National Institute of Standards and Technology Interagency Report
NSL-KDD	Network Security Laboratory - Knowledge Discovery in Databases
R2L	root-to-local
ReLU	rectified linear activation function
SCADA	Supervisory Control And Data Acquisition
SDF	symbolic dynamic filtering
SVM	Support Vector Machine
U2R	user-to-root
WAN	Wide Area Network

2. Related Work

A smart grid consists of Home Area Network (HAN), Neighborhood Area Network (NAN) and Wide Area Network (WAN). The HAN consists of smart sensors, actuators and a user interface like home energy management system (HEMS). The NAN collects data from multiple HANs and transmits the data to the corresponding High Level Control Centres [6]. The NAN is therefore a dedicated channel for information exchange between the HAN and the WAN. Finally, the WAN connects multiple NANs, controlling the power transmission.

The Intrusion Detection System (IDS) monitors and detects malicious behaviour by collecting data information from key host nodes, building assessment models and analysing the network for the presence of illegal behaviour [7,8]. The IDS detects the attack trajectory of an attacked host and reports warnings to ensure the integrity of the network's central host system. This will make the smart grid resistant to external network attacks [9].

Intrusion detection involves data acquisition, intrusion analysis and intrusion response. It reviews information from host logs, network segment protocol packets and gateways, and checks the network data using anomaly detection algorithms and discriminatory models [10]. The intrusion response module is used when anomalous attacks are reported by the intrusion analysis module. The module takes pre-defined measures, such as network disruption and alarm response, to prevent further deterioration of the situation.

Intrusion detection models are divided into host-based models, network-based models and feature-based models. Host-based models analyse the operating system's audit trail and log messages of a single host [11]. They can detect viruses, malicious programs and destructive intrusion attacks on hosts [12]. However, the model monitors a host's memory, which adversely affect the host's performance. Additionally, its high memory space requirement does not support the handling of multiple attacks. For network-based models, they protect hosts by monitoring the number of network packets in a gateway to determine the network communication traffic of multiple hosts. This model can monitor large network sections with less memory [13]. However, the model cannot analyse the information flow of an encrypted network. It results in low detection accuracy in large-scale high-speed networks, thus it cannot handle fragmentation attacks. For feature-based detection model, it matches network intrusions to defined attack features through the misuse detection analysis system, which usually defines a separate feature for each anomalous event and uses a database to store the features for maintenance and matching [14]. This model enables efficient detection of correlated intrusion without generating excessive warning reports. However, this model requires constant updating of the feature database to maintain the system security, and it is unable to prevent malformed network attacks.

Intrusion detection methods include anomaly detection and misuse detection. The former is behavioural detection, which assumes that all network attacks are anomalous behaviour, then builds a model to differentiate normal behaviours from anomalous ones by comparison. Anomaly detection requires a simplified and accurate amount of features and reasonable threshold settings to ensure the optimal performance [15]. Anomaly detection can quickly detect network intrusions, but it requires heavy computation, leading to relatively high resource requirements. The latter monitors data at the gateway, compares the data signature with those in the database to determine if an intrusion is present [16]. However, it is impossible to locate the intrusion. Additionally, digital signatures are system dependent, making it difficult to standardise the detection procedure.

The KDD99 dataset was the data set used for The Third International Knowledge Discovery and Data Mining Tools Competition, which was held in conjunction with The Fifth International Conference on Knowledge Discovery and Data Mining (KDD-99). The KDD99 dataset is the most widely used dataset for intrusion detection. It consists of network data collected by Lincoln Laboratory over 69 days simulating the US Air Force LAN system with various types of network hosts and attacks [11]. The Network Security Laboratory - Knowledge Discovery in Databases (NSL-KDD) dataset is an improved version of the KDD99 dataset. It removes most of the duplicate data from the original KDD99 dataset. Each data entry in NSL-KDD contains 41-dimensional features and 1-dimensional label feature. Four types of feature data are available in the NSL-KDD, whose data label can indicate whether the data is normal data or not [12], with the data tag indicating the attack type.

The NSL-KDD data set includes four main parts, KDDTrain+, KDDTest+, KDDTrain+_ 20Percent and KDDTest-21 [17], where KDDTrain+ and KDDTest+ contain 125,973 and 22,543 data sets, espectively. The redundant part of the KDD99 data set is eliminated, KDDTrain+_20Percent provides an additional subset for training. In this dataset, network data is divided into five types: Normal, DoS attacks, user-to-root (U2R) attacks, root-to-local (R2L) attacks and Probe attacks. The normal type represents normal data; DoS attacks prevent the destination host from responding to external requests and cause a waste of resources; U2R attacks are user-unauthorised attacks, which attempt to gain root access; R2L attacks are login and access attacks by unauthorised hosts on the system; Probe attacks are port monitoring or port scanning. These five types include a total of 39 subtypes of attack types [18]. The specific classifications are shown in the Table 2.

In this paper, NSL-KDD dataset is used for the experiments. The dataset is first normalised to generate a standard dataset. With classical machine learning methods, a classifier is built for the standard dataset as a control group, and then the data features are extracted by the proposed ADL algorithm, and the generated data features are used to build a classifier to evaluate the effectiveness and usefulness of the proposed ADL algorithm.

Table 2. Summary of attacks types labeled in the NSL-KDD dataset.

Type	Attack	Description
Normal	Normal data traffic	Normal data type
DoS	Back, Land, Neptune, Pod, Smurf, Teardrop, Mailbomb, Processtable, UDPstorm, Apache2, Worm	Denial of service attacks, which make computers and networks unable to provide normal services
Probe	Satan, Nmap, Mscan, Saint, IP sweep, Portsweep	Port attack, scan port vulnerabilities to attack
U2R	Buffer overflow, Sql attack, XtermLoadmodule, Rootkit, Perl, Ps	Unauthorized users obtain root vulnerabilities through network vulnerabilities and perform illegal operations
R2L	Guess password, Imap, Multihop, Ftp write, Phf, Warezmaster, Xclock, Xsnoop, Snmpguess, Snmpgetattack, Sendmail, Httptunnel, Named	Remote attack, users remotely log in operate illegally through accounts and passwords

2.1. Classification Algorithms

Intrusion detection scheme for smart grid based on machine learning refers to: converting the network intrusion problems into a packet type classification problems based on different intrusion types of packets, and using machine learning methods to train classification models to identify and classify intrusion packet types. However, due to the large number of network data features, if various features are used for training, it will increase the training time and model complexity, and the hardware requirements will also increase. To solve the problems of too many dimensions of network data features, there are various methods to extract data feature dimensions for reducing data feature dimensions. Consequently, intrusion detection can be treated as a packet type classification problem using machine learning. Feature extraction is usually adopted to reduce computation, whose common methods include correlation-based feature selection and encoding of data packets. The former uses a correlation function to select subsets of data, thereby reducing data size. The latter uses encoding to extract data features. Typical feature classification algorithms include K-nearest neighbour (KNN), Naïve Bayes (NB) classifier, Back Propagation Neural Network (BPNN) and Decision Tree (DT).

KNN Algorithm–The KNN algorithm first selects the value of K, which denotes the number of nearest neighbours. Between a given data point x and its neighbour y, their distance in the n-dimensional Euclidean space is

$$d_{xy} = \sqrt{\sum_{i=1}^{n} (x_i - y_i)^2} \tag{1}$$

Then it takes the K nearest neighbours as per the calculated Euclidean distance. Among these K neighbours, the algorithm counts the number of points in each class. Finally, it assigns x to that class for which the number of neighbours is maximum. The KNN algorithm is relatively accurate with simple implementation. Nevertheless, its efficiency will significantly decrease as the number of data points increases.

Naïve Bayes Algorithm—The Naïve Bayes classifier calculates conditional probability to perform classification.

$$y = \arg\max\left\{ p(y = C_k) \prod p(x|y = C_k) \right\} \tag{2}$$

With $x = (x_1, x_2, \ldots, x_n)$, assuming that all features x are mutually independent, from Bayesian theorem we have

$$p(C_k)p(x|C_k) = p(C_k) \prod_{i=1}^{n} p(x_i|C_k) \tag{3}$$

Therefore, $p(C_k|x) \propto p(C_k) \prod_{i=1}^{n} p(x_i|C_k)$. With Laplace Smoothing, the prior probability is given by (where λ is the smoothing parameter)

$$p_\lambda(C_k) = \frac{\sum_{i=1}^{N} I(y_i = C_k) + \lambda}{N + K\lambda} \tag{4}$$

The conditional probability is calculated using

$$p_\lambda(x_1 = a_j|y = C_k) = \frac{\sum_{i=1}^{N} I(x_1 = a_j, y_i = C_k) + \lambda}{\sum_{i=1}^{N} I(y_i = C_k) + A\lambda} \tag{5}$$

where K denotes the number of different values in y and A denotes the number of different values in a_j. Usually $\lambda = 1$.

BPNN–The BPNN consists of an input layer, a hidden layer and an output layer. Given training set $D = (x_1, y_1), (x_2, y_2), \ldots, (x_n, y_n)$, $x \in \mathbb{R}^d$, $y \in \mathbb{R}^l$, as shown in Figure 1, for the jth node (neuron), $x_1, x_2, \ldots, x_i$ are the inputs of the neuron, which are connected by the weights of $w_{j1}, w_{j2}, \ldots, w_{ji}$ to adjust the proportion of the input. Take the linear weighted sum as input and θ_j as decision variable, hidden layer y_j output is

$$y_j = f\left(\sum_{i=1}^{n} w_{ji}x_i - \theta_j\right) \tag{6}$$

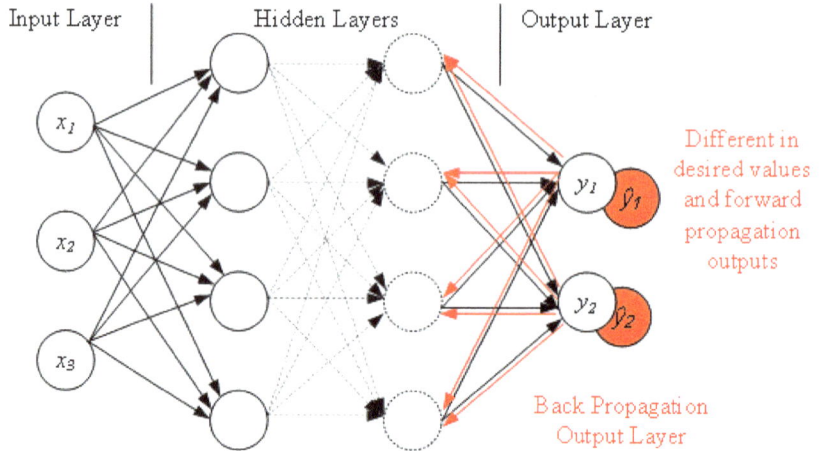

Figure 1. Illustration diagram shows how a Back Propagation Neural Network works.

The parameters are set through the training data to obtain a parametric model of the prediction error, and the parameters are updated using the Gradient Descent method.

DT Algorithm–A DT consists of a root node, internal nodes and leaf nodes. The root node contains the entire data set. The internal nodes use different features to make category judgements and each leaf node represents the final judgement category. The complexity of the DT model is related to the number of layers of the tree. Under DT, information gain is the expected reduction in entropy of target variable Y for data sample S, due to sorting on variable A

$$G(S, A) = H(A) - \sum_{v \in Values(A)} \frac{|S_v|}{|S|} H(S_v) \tag{7}$$

Next, the impurity (e.g., data partition) of S is given by

$$Gini(S) = 1 - \sum_{i=1}^{K} \left(\frac{|C_{i,S}|}{|S|} \right)^2 \tag{8}$$

2.2. Feature Extraction Methods

Correlation-Based Feature Extraction–Correlation-based feature selection (CFS) uses the evaluation function to select a feature subset. For two continuous random variables X and Y, their linear correlation coefficient is given by

$$r_{XY} = \frac{\sum_i (x_i - \overline{x_i})(y_i - \overline{y_i})}{\sqrt{\sum_i (x_i - \overline{x_i})^2}\sqrt{\sum_i (y_i - \overline{y_i})^2}} \tag{9}$$

Automatic Encoder–Asymmetric Convolutional Encoder (ACE) can be used with a convolutional neural network [19] for unsupervised feature learning to extract the local features of the original data. The output of a hidden layer can be used as the input of the next layer. In each round of training operation of the convolutional layer, the algorithm first initialises k convolution sum, each convolution with weight w and bias b,

$$h^k = f\left(x * w^k + b^k\right) \tag{10}$$

The convolutional layer output from the upper layer is reconstructed (with bias c) to obtain the output data characteristics, which are adjusted by comparing the input and output data.

$$y = f\left(\sum_{k=1}^{K} h^k * w^k + c^k\right) \tag{11}$$

This method uses multiple iterations of convolution, which increases the computational complexity.

Recently, intrusion detection was also studied for Internet of Things (IoT) and Jan et al. presented a lightweight intrusion detection method using supervised machine learning-based support vector machine (SVM) to detect malicious data injection [20]. However, it is difficult to apply it directly in smart grids due to different types of attacks. In [21], Karimipour et al. presented an unsupervised anomaly detection based on statistical correlation between measurements and time series partitioning to discover causal interactions between the subsystems. It adopted feature extraction utilising symbolic dynamic filtering (SDF) to reduce computational burden. In [22], Takiddin et al. presented an anomaly detector using stacked autoencoders with a long-short-term-memory (LSTM)-based sequence-to-sequence structure to detect electricity theft cyberattacks in smart grids. Inayat et al. presented an extensive survey on various cybersecurity enhancements of smart grids to detect false data injection attacks (FDIA), DoS attacks, distributed denial of service (DDoS) attacks, and spoofing attacks [23]. Interestingly, Zhou et al. presented a comprehensive survey for deep-learning-based abnormality detection in smart grids using multimodal image data [24], which include visible light, infrared, and optical satellite images. In [25], Berghout et al. reviewed different machine learning tools to detect cyberattacks in smart grids. In addition, it also highlighted various challenges, drawbacks and possible solutions of machine learning based cybersecurity applications in smart grids. A latest anomaly detection approach based on federated learning was proposed in [26], where machine learning models were trained locally in smart meters without sharing data with a central server, thus ensuring user Privacy. Table 3 compares our work with those machine learning based works found in the literature.

Table 3. Comparison of machine learning based intrusion detection techniques.

Works	Learning Type	Key Techniques	Datasets
[5]	Supervised	Particle swarm optimisation based neural network	KDD99 and NSL-KDD
[13]	Supervised	Work embedding-based deep learning	Intrusion Detection Evaluation Dataset (ISCX2012)
[14]	Semi-supervised	Long short-term memory and extreme gradient boosting with genetic algorithm	NSL-KDD
[18]	Unsupervised	Nonsymmetric deep autoencoder	KDD99 and NSL-KDD
[20]	Supervised	Support vector machine	Intrusion Detection Evaluation Dataset (CIC-IDS2017)
[21]	Unsupervised	Feature extraction using symbolic dynamic filtering	Data from testbed from Matpower
[22]	Supervised	Long short-term memory with stacked autoencoders	State Grid Corporation of China Dataset
[26]	Supervised	Federated Learning	KDD99, NSL-KDD and CIDDS-001 datasets
This work	Supervised	Adaptive deep learning using deep belief network	NSL-KDD

3. Proposed Adaptive Deep Learning

The proposed ADL algorithm consists of a data pre-processing module, a neural network pre-training module and a classifier module. In the data pre-processing module, the original dataset is normalised to generate a standard dataset. In the neural network pre-processing module, the algorithm is used to train the model and adjust the parameters to obtain a highly adaptive network model. The classifier module used high-dimensional data features to train a classifier to determine the intrusion data type on the test dataset.

With the proposed ADL algorithm, we extract data features through hidden layer neurons and change the distribution and structure of the data. The data features after each hidden layer are more accurate and essential. Transfer learning is embedded in the model so that it can be used for new tasks and improve the generalisation of the model. Next, deep belief networks (DBNs) enable compressed coding of raw data to accurately represent data features. A DBN consists of a multilayer Boltzmann machine network and a supervised back propagation network. The proposed algorithm combines DBNs to infer the appropriate number of hidden layers and the number of neurons per hidden layer for the neural network based on the original input data, allowing the pre-trained model to better match the size of the dataset and reduce the number of hidden layers. Too few hidden layers lead to under-reporting, while too many hidden layers lead to over-fitting. In the proposed ADL algorithm, parameters are used to control the training speed of the model and the accuracy of classification prediction. As shown in Figure 2, by adjusting the hidden layer, different data characteristics will be generated and transfer learning is adopted as shown in Figure 3.

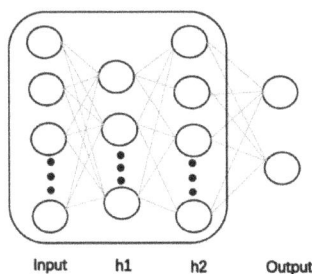

Figure 2. Illustration diagram of the adaptive deep learning framework.

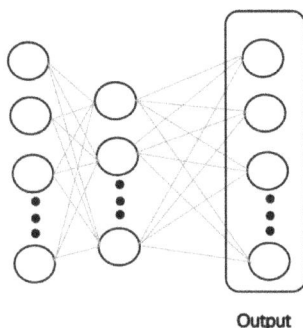

Figure 3. Illustration diagram of the model of transferring learning.

As explained in Algorithm 1, by the proposed ADL algorithm, the number of hidden layers and the number of neurons are determined by the dimensionality of the original training data and the parameter θ, which balances training time, output accuracy and convergence speed. The range of θ is set from 0 to 1, with a step size of 0.1.

The ADL algorithm defines the number of neurons in each hidden layer of the deep neural network. The neurons before the output layer are the dimensions of the neurons in the highest layer. When θ is set close to 1, the feature dimension of the last hidden layer of the pre-trained model is close to that of the original data; when θ is set close to 0, the feature dimension of the last hidden layer of the pre-trained model is lower. The number of neurons in the first layer of the neural network is pre-set to be the same as the original data, and the data features are transformed through the hidden layers. The number of hidden layers and the number of neurons are determined by the original training data and θ. After the pre-training of the model, the back-propagation algorithm is used to adjust the parameters of the preset network model. The error gradient between the input training data v_i and the model output data v'_i is adjusted, δ_h is the weight of the node from the hidden layer to the next layer, and δ_j is the error gradient of node j. The training results were obtained using the rectified linear activation function (ReLU), but experiments showed that the normal ReLU may result in the weights not being updated. Therefore, when $x \leq 0$, αx is used instead of 0 and the value of α is set to a smaller value to ensure that the weights can be updated correctly and speeds up the convergence of the network.

$$A(x) = \begin{cases} \alpha x, x \leq 0 \\ x, x > 0 \end{cases} \tag{12}$$

The output layer uses a sigmoid function to fit the output, ranging from 0.1 to 1, which determines the behaviour and legitimacy of the data.

$$s(x) = \frac{1}{1 + e^{-x}} \tag{13}$$

Algorithm 1 Proposed ADL algorithm: where v is number of training data samples, θ is the key parameter to balance the training speed and classification accuracy, η is the learning rate. N represents the set of neurons in the neural network; l is the number of neuron layers; D represents the dimension of the training data; and n_i is the ith neuron. δ_k, δ_h, W_{ij}, and $O_i \Delta W_{ij}$ are the intermediate variables, W_{ij} denotes the weights, and b_j is the bias.

---

  procedure ADL(v, θ, η) ▷ adaptive deep learning

      $N_{training\_data} \leftarrow v$

      if θ is empty **then**

          $\theta \leftarrow 0.3$

      end if

      if η is empty **then**

          $\eta \leftarrow 0.1$

      end if

      $N \leftarrow \varnothing$

      $D \leftarrow \text{sizeof}(N_{training\_data})$

      $l \leftarrow D/5$

      $n_i \leftarrow \theta * D$

      $i \leftarrow 2$

      while $i \leq l - 1$ **do**

          $n_i \leftarrow (D/i^2) + \theta * D$

          $N.append(n_i)$

          $i \leftarrow i + 1$

      end while

      $i \leftarrow 1$

      while $i \leq l$ **do**

          use N to build the current layer and the n_i neuron node

      end while

      output layer function $s(x) \leftarrow 1/(1 + e^{-x})$

      for each training sample v_i **do**

          calculate the actual output of the model v'_i

      end for

      $\delta_k \leftarrow v'_i * (1 - v'_i) * (v_i - v'_i)$

      $\delta_h \leftarrow v'_h * (1 - v'_h) * W_{hk} * \delta_k$

      $W_{ij} \leftarrow W_{ij} + \Delta W_{ij}$

      $O_i \Delta W_{ij} \leftarrow W_{ij} + \eta * O_i * \delta_j$

      $b_j \leftarrow \delta * b_j$

    return $x * W + b$

  end procedure

After the model has been trained, the remaining part of the network other than the output layer is removed and the resultant model for the network is used for pre-processing. The number of neurons in the last layer of the hidden layers is the feature dimension of the output data. The algorithm determines the structure of the neural network through θ and the data dimension. The features of the data output from the hidden layers are considered as a downscaling of the original data. The smaller the dimension of the features generated by the model, the faster the detection, at the expense of reduced accuracy.

Four performance metrics are evaluated, namely accuracy, precision, recall, and F1-score. TP denotes the number of intrusion network data is correctly identified as intrusion network data. TN denotes the number of normal network data is correctly identified as normal network data. FP denotes the number of normal network data is incorrectly identified as intrusion network data. FN denotes the number of intrusion network data is wrongly identified as normal network data.

Accuracy is given by

$$s_A = \frac{TP + TN}{TP + TN + FP + FN} \tag{14}$$

Precision is given by

$$s_P = \frac{TP}{TP + FP} \tag{15}$$

Recall rate is given by

$$s_R = \frac{TP}{TP + FN} \tag{16}$$

F1-score is given by

$$s_{F1} = \frac{2 s_P s_R}{s_P + s_R} \tag{17}$$

4. Results and Discussion

4.1. Preprocessing of Data

(a) IP Addresses and Port Numbers Removal—IP addresses and port numbers are removed from source and destination hosts because IP address and port numbers in the original dataset may lead to overtraining of neural networks and classifiers.

(b) Spaces Removal—Some tags in the dataset contain spaces that have no meaning in the actual data representation, but return different results in the data classification process, resulting in different classification of the packets. Thus, these spaces are removed.

(c) Label Encoding—The label of each piece of data is encoded. The label of each piece of data in the dataset contains the type of attack corresponding to that data, with different attack types corresponding to different specific strings. Encoding the strings into a specific value simplifies the learning process for the classifier. In the machine learning module, the classifier can learn the category values for each array.

(d) Data Normalisation—As the range of values taken from the data in the dataset does not meet the requirements of the classifier, the data range and format needs to be normalised to specify a minimum value for each data attribute. The normalisation and standardisation of the data provides a consistent value for the classifier, improving the correlation between the data and reducing the variability between the data features and improves the efficiency of the classifier.

4.2. Performance Evaluation

There are two types of classifiers considered in this paper–those based on four traditional machine learning models and those based on adaptive deep neural networks. As the training and test samples of the classifiers are the same, the interference of the data samples on the model results is effectively eliminated and the confidence of the comparison is improved.

80% of the data was used to train the proposed ADL model, while 80% of the remaining 20% data is used to train the classifier and 20% for testing. To verify the effectiveness of the ADL algorithm, we compare its performance with that of a traditional machine learning model for feature extraction. The experimental procedure uses the KNN, the DT, the NB algorithm and the BPNN to train the classifier. The raw data was passed through an adaptive deep neural network to obtain data features, which improved the classification accuracy of the classifier overall and also reduced the detection time of the network for abnormal data.

4.2.1. Two-Class Machine Learning Model

For the KNN algorithm, the value of K was set in the range of [3, 15] with a step size of 2. The average accuracy of the classifier with different parameters was tested. The experimental results show that the KNN algorithm produces better overall experimental results for the classifier, with the highest accuracy rates at $K = 3$ and $K = 5$. Hereafter, we use $K = 3$.

For the DT algorithm, the maximum depth of the tree and the minimum number of samples are needed for the leaf nodes. The range of values for depth is set to [10, 30] and the range of values for the minimum number of samples required by a leaf node is

[2, 20]. First, the accuracy of classification of the DT algorithm at different depths was tested. Experimental results show that by the DT algorithm has the highest accuracy when the depth is 26 and the minimum number of leaf nodes is 2.

The confusion matrix is also known as the error matrix. It uses a matrix to visualize the performance of a machine learning algorithm. The column data of the confusion matrix represents the predicted values, while the row data represents the actual values. The confusion matrix is introduced to indicate whether there is confusion between different categories, i.e. whether there is a misclassification. The results of the confusion matrix produced by different algorithms are shown in Figure 4. The X-axis represents the predicted values and the Y-axis represents the true values. The values in the first quadrant represent data where the predicted value is an attack and the true value is normal. The values in the second quadrant represent data where the predicted value is normal and the true value is normal. Values in the third quadrant represent data where the predicted value is normal and the true value is an attack. A value in the fourth quadrant represents data for which the predicted value is an attack and the true value is an attack. From Figure 4, we can see that KNN outperforms other three classification techinques, followed by DT and BPNN. The performance of Naïve Bayes is the worst in terms of two-class classification.

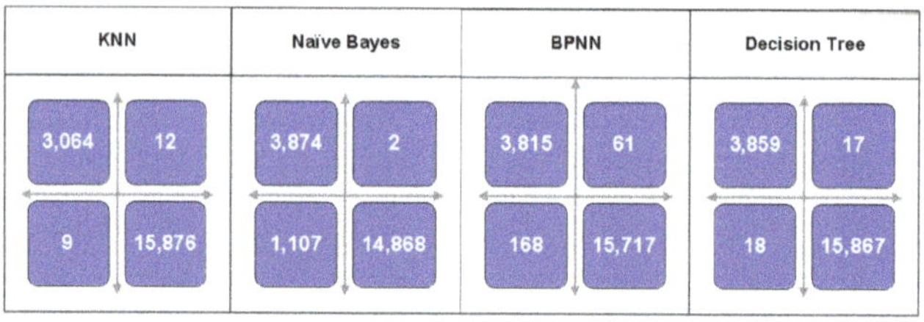

Figure 4. The confusion matrix results of two-class classification using the existing four classification methods: KNN, Naïve Bayes, BPNN and DT.

The experimental results are shown in Figure 5. Results show that with two classifications, all packets were divided into two categories–abnormal data and normal data. The abnormal data was not further classified and the sample imbalance was relatively less of a problem. However, in practical applications, the categories of abnormal data need to be further divided, so the effect of sample imbalance on the training results should be considered in the subsequent multi-classification cases.

Figure 5. Performance comparison of two-class classifications using the existing four classification methods: KNN, Naïve Bayes, BPNN and DT.

4.2.2. Multi-Class Machine Learning Model

In the same process as for the two classifications, dataset machine learning models were used for training and testing. In order to more accurately simulate network attacks in real life situations and to classify different attacks. This paper extends the two classifications into multiple classifications to identify different types of attacks.

(1) KNN algorithm parameter setting: In order to save computational effort and processing time, the range and step size of K values were set to be the same as in the case of two classification. The experimental results show that the highest classification accuracy is achieved with $K = 3$. The value of K is set in line with the case of two classification, which reduces the complexity of the comparison and the difficulty of the calculation to some extent.

(2) Decision Tree algorithm parameter setting: As in the case of two classification, the algorithm needs to determine the maximum depth and minimum number of samples required for the leaf nodes in the multi-level classification case. The range of values and step size of each parameter are set to the same as in the two classification case.

The experimental results showed that the classification accuracy was stable at 99.84% when the maximum depth of the tree was greater than or equal to 27. To reduce the computational effort and the complexity of machine learning as much as possible, the maximum depth of the tree was set to 27. Subsequently, the number of leaves and he minimum number of samples of nodes was tested. The experimental results showed that when the depth of the Decision Tree is 27, the classification accuracy of the classifier decreases as the minimum number of samples of the leaf nodes increases. So the maximum number of samples of the leaf nodes was set to 2.

For the KNN algorithm, $K = 3$ provides the best results. For the DT algorithms, result show that optimal results are achieved when the maximum depth of the tree is 27 and the maximum number of samples of the leaf nodes is 2. Figure 6 shows the confusion matrix for multi-class classification by KNN, Naive Bayes, BPNN and DT. Figure 7 shows the results for multiple classifications.

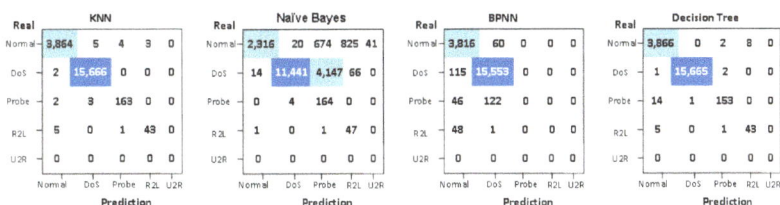

Figure 6. The confusion matrix results of multi-classification using the existing four classification methods: KNN, Naïve Bayes, BPNN and DT.

After 10 experiments comparing the s_{F1} of the ADL neural network model and the traditional machine learning classification model, it was demonstrated that the classification performance of the ADL neural network model was better. Since the s_{F1} refers to the weighted average of precision and recall, it can be concluded experimentally that the s_{F1} of the ADL classifier reaches its highest value when $\theta = 0.8$. By applying the ADL algorithm with the traditional machine learning algorithm, the performance metrics of the classifier were significantly improved. In particular, the s_A of the NB algorithm improved from 94.84% to 98.77% and the s_R improved from 95.06% to 99.75%. The improvements in accuracy and recall were more pronounced than those model without the ADL algorithm to extract features. The performance metrics are summarised in Table 4.

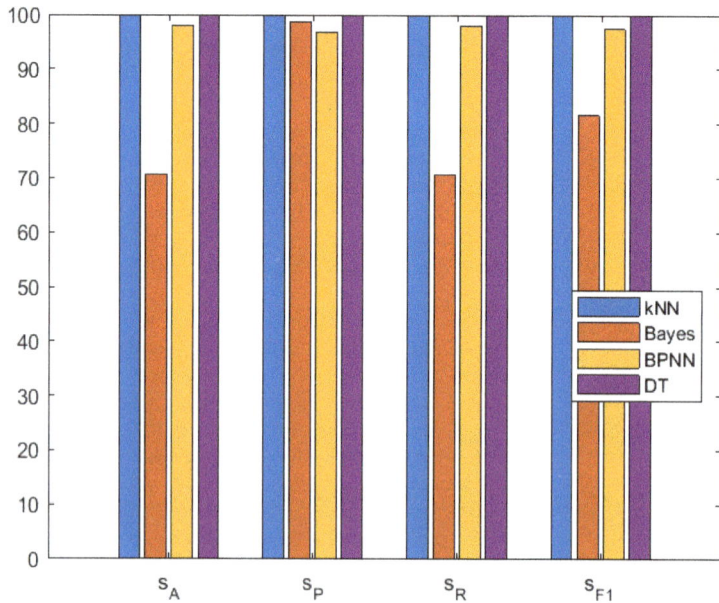

Figure 7. Performance comparison of multi-classification using the existing four classification methods: KNN, Naïve Bayes, BPNN and DT.

In traditional machine learning, the NB algorithm has the worst performance, whose performance is summarised in Table 5. Next, after adding the ADL algorithm to the NB algorithm, Table 6 shows the effect of using the ADL algorithm on the performance improvement of the NB algorithm. The reason for the negligible performance improvement for the DoS data is that the main attacks in the network come from DoS attacks, so the number of DoS attacks is large and better classification can be achieved without feature extraction. R2L and U2R are relatively small data in the dataset and the performance is greatly improved after feature extraction. The S_A of R2L is improved from 84.13% to 91.32%, the S_P increased from 91.23% to 96.44%, the S_A of U2R increased from 28.49% to 43.83%, and the S_P increased from 66.88% to 75.02%.

Table 4. Performance comparison of existing classifiers with and without the proposed adaptive deep learning algorithm.

Model	s_A (%)	s_P (%)	s_R (%)	S_{F1} (%)
kNN	99.89	99.89	99.89	99.89
kNN with ADL	99.23	99.49	98.74	99.15
Decision Tree	99.82	99.82	99.82	99.82
DT with ADL	99.58	99.72	99.58	99.65
Bayes	94.84	95.91	94.84	95.06
Bayes with ADL	98.77	95.25	99.75	97.70
BPNN	98.84	98.86	98.84	98.05
BPNN with ADL	99.15	99.36	99.82	99.13

Table 5. Results using Naïve Bayes classifier.

Data Type	s_A (%)	s_P (%)	s_R (%)	S_{F1} (%)
Normal	99.24	98.86	99.43	99.81
DoS	99.85	99.86	98.85	99.85
Probe	95.64	97.42	95.43	96.41
R2L	84.13	91.23	83.23	87.07
U2R	28.49	66.88	28.57	40.02

Table 6. Results using Naïve Bayes classifier together with the proposed ADL Algorithm.

Data Type	s_A (%)	s_P (%)	s_R (%)	S_{F1} (%)
Normal	99.91	98.65	99.90	99.78
DoS	99.85	99.88	98.85	99.87
Probe	98.83	99.75	99.87	99.34
R2L	91.32	96.44	90.39	93.31
U2R	42.83	75.02	42.85	54.45

To investigate the effect of ADL algorithm on the classifier performance at different θ values, parametric analysis of θ was carried out on selected datasets. The output efficiency of the model was determined by comparing the accuracy of the classifier at different values of θ. From the experimental results, θ can affect the accuracy of the classifier by adjusting the dimensionality of feature extraction. When θ was set to [0.1, 0.5], the S_A of the ADL algorithm improved gracefully. When $\theta > 0.6$, the S_A of the ADL algorithm is saturated and the stability of the ADL algorithm is high.

Through experiments using the NSL-KDD dataset, the performance of the naïve Bayesian algorithm classifier is greatly improved after processing by the ADL algorithm (see Table 6). The reason for the smaller performance improvement for the DoS data is that the main attacks in the network environment come from Dos attacks, so the number of Dos attacks is larger and better classification can be achieved without feature extraction. R2L and U2L are relatively small data in the dataset and the performance is greatly improved after feature extraction. The accuracy of R2L is improved from 84.13% to 91.32%, the accuracy rate increased from 91.23% to 96.44%, the accuracy rate of U2L increased from 28.49% to 43.83%, and the accuracy rate increased from 66.88% to 75.02%.

Based on the CFS method, the data features of the subset are extracted using specific measurement indicators, the correlation matrices of different feature subsets are established, and the function values of the subset matrices are solved to select the best correlation feature matrix A subset. The results of CFS are shown in Table 7. Next, coding-based feature extraction methods often use ACE for feature extraction. Thus, for the sake of comparison, we also present results of ACE in Table 8.

Table 7. Results using the Correlation Feature Selection Algorithm.

Data Type	s_A (%)	s_P (%)	s_R (%)	S_{F1} (%)
Normal	99.58	100	97.2	83.9
DoS	99.76	97.5	99.3	89.7
Probe	99.81	84.7	99.7	91.6
R2L	24.36	55.6	99.7	90.3
U2R	60.17	82.3	99.7	72.08

Table 8. Results using the Asymmetric Convolutional Encoder algorithm.

Data Type	s_A (%)	s_P (%)	s_R (%)	S_{F1} (%)
Normal	99.58	100	99.64	99.82
DoS	99.76	100	99.81	99.90
Probe	99.81	100	99.32	96.61
R2L	24.36	100	88.36	93.83
U2R	10.17	41.32	47.23	44.08

The experimental results show that the proposed ADL algorithm together with the NB algorithm (see Table 6) has a greater improvement in the classification accuracy of R2L and U2R compared to the ACE algorithm, and the overall performance of the algorithm is also better. This indicates that the algorithm is more efficient in intrusion detection for small sample data. Also, the use of encoders is avoided, as is the possibility of increased computational complexity. However, the s_{F1} of the ADL algorithm is relatively low and could be improved in future work.

Results showed that the classification accuracy of the ADL algorithm increased as θ increased in [0.1, 0.5]. The accuracy tends to saturate when θ is greater than 0.6, demonstrating the stability and feasibility of the proposed ADL algorithm. Meanwhile, comparison results proved that the ADL algorithm is better than the CFS feature extraction algorithm when the amount of sample data is large. Compared with the ACE feature extraction algorithm, the ADL algorithm improves the classification accuracy of R2L and U2R.

5. Conclusions

This paper proposes an adaptive deep learning algorithm that determines the number of hidden layers and neurons of a neural network by extracting the dimension of the original data. By setting parameters to balance the detection time and output accuracy, the proposed ADL algorithm can adapt to different network environments and network sizes. Moreover, by combining the ADL algorithm with traditional machine learning algorithms, they can effectively discriminate between large ranges of network traffic, improve the accuracy of intrusion detection, converge faster, and reduce detection time significantly. In particular, Naïve Bayes classification produces the worst performance as compared to KNN, DT and BPNN. After adding the proposed ADL to Naïve Bayes classification, its performance can be improved significantly. For example, the accuracy of R2L is improved from 84.13% to 91.32%, the accuracy rate increased from 91.23% to 96.44%, the accuracy rate of U2L increased from 28.49% to 43.83%, and the accuracy rate increased from 66.88% to 75.02%. The proposed ADL algorithm can also improve the performance of the other three traditional classifiers to different extent. For future work, a real-time packet capture platform can be adopted for analysis and further optimisation of the proposed ADL algorithm.

For future work, a real-time packet capture platform can be set up for further analysis and optimisation of the proposed ADL algorithm. In particular, the experiment utilised a combination of deep neural networks and traditional machine learning algorithms, with specific parameters set for different network sizes to reduce the training time of the network. However, the current process of parameter learning is resource-intensive in terms of computational power and computational time, and future research could optimise the learning time of parameters and the requirement for hardware computational power. In addition, to study the complex data patterns and low footprint stealth attacks of the contemporary network traffic, we plan to verify the performance of our proposed ADL algorithm using the UNSW-NB15 dataset [27].

Author Contributions: Conceptualization, X.J.L. and M.M.; methodology, M.M.; software, Y.S. and M.M.; validation, Y.S., M.M. and X.J.L.; formal analysis, Y.S., M.M. and X.J.L.; writing—original draft preparation, X.J.L.; writing—review and editing, X.J.L. and M.M.; supervision, M.M. All authors have read and agreed to the published version of the manuscript.

Funding: This research received no external funding.

Data Availability Statement: Not applicable.

Conflicts of Interest: The authors declare no conflict of interest.

References

1. Zhang, Y.; Wang, L.; Xiang, Y. Power System Reliability Analysis With Intrusion Tolerance in SCADA Systems. *IEEE Trans. Smart Grid* **2016**, *7*, 669–683. [CrossRef]
2. Nguyen, T.N.; Liu, B.-H.; Nguyen, N.P.; Chou, J.-T. Cyber Security of Smart Grid: Attacks and Defenses. In Proceedings of the ICC 2020—2020 IEEE International Conference on Communications (ICC), Dublin, Ireland, 7–11 June 2020.
3. Harvey, M.; Long, D.; Reinhard, K. Visualizing NISTIR 7628, Guidelines for Smart Grid Cyber Security. In Proceedings of the 2014 Power and Energy Conference at Illinois (PECI), Champaign, IL, USA, 28 February–1 March 2014; pp. 1–8.
4. Zhang, H.; Liu, B.; Wu, H. Smart Grid Cyber-Physical Attack and Defense: A Review. *IEEE Access* **2021**, *9*, 29641–29659. [CrossRef]
5. Khan, S.; Kifayat, K.; Bashir, A.K.; Gurtov, A.; Hassan, M. Intelligent intrusion detection system in smart grid using computational intelligence and machine learning. *Trans. Emerg. Telecommun. Technol.* **2021**, *32*, e4062. [CrossRef]
6. Meng, W.; Ma, R.; Chen, H.-H. Smart grid neighborhood area networks: A survey. *IEEE Netw.* **2014**, *28*, 24–32. [CrossRef]
7. Khoei, T.T.; Slimane, H.O.; Kaabouch, N. A Comprehensive Survey on the Cyber-Security of Smart Grids: Cyber-Attacks, Detection, Countermeasure Techniques, and Future Directions. *arXiv* **2022**, arXiv:2207.07738.
8. Ding, J.; Qammar, A.; Zhang, Z.; Karim, A.; Ning, H. Cyber Threats to Smart Grids: Review, Taxonomy, Potential Solutions, and Future Directions. *Energies* **2022**, *15*, 6799. [CrossRef]
9. Vaidya, B.; Makrakis, D.; Mouftah, H.T. Device authentication mechanism for Smart Energy Home Area Networks. In Proceedings of the 2011 IEEE International Conference on Consumer Electronics (ICCE), Las Vegas, NV, USA, 9–12 January 2011; pp. 787–788.
10. Bae, H.-S.; Lee, H.-J.; Lee, S.-G. Voice recognition based on adaptive MFCC and deep learning. In Proceedings of the 2016 IEEE 11th Conference on Industrial Electronics and Applications (ICIEA), Hefei, China, 5–7 June 2016; pp. 1542–1546.
11. Nicanfar, H.; Jokar, P.; Beznosov, K.; Leung, V.C.M. Efficient Authentication and Key Management Mechanisms for Smart Grid Communications. *IEEE Syst. J.* **2014**, *8*, 629–640. [CrossRef]
12. Hao, J.; Kang, E.; Sun, J.; Wang, Z.; Meng, Z.; Li, X.; Ming, Z. An Adaptive Markov Strategy for Defending Smart Grid False Data Injection From Malicious Attackers. *IEEE Trans. Smart Grid* **2018**, *9*, 2398–2408. [CrossRef]
13. Cui, J.; Long, J.; Min, E.; Mao, Y. WEDL-NIDS: Improving network intrusion detection using word embedding-based deep learning method. In *MDAI 2018: Modeling Decisions for Artificial Intelligence*; Springer: Berlin/Heidelberg, Germany, 2018; pp. 283–295.
14. Song, C.; Sun, Y.; Han, G.; Rodrigues, J.J.P.C. Intrusion detection based on hybrid classifiers for smart grid. *Comput. Electr. Eng.* **2021**, *93*, 107212. [CrossRef]
15. Hu, H.; Doufexi, A.; Armour, S.; Kaleshi, D. A Reliable Hybrid Wireless Network Architecture for Smart Grid Neighbourhood Area Networks. In Proceedings of the 2017 IEEE Wireless Communications and Networking Conference (WCNC), San Francisco, CA, USA, 19–22 March 2017; pp. 1–6.
16. Gobena, Y.; Durai, A.; Birkner, M.; Pothamsetty, V.; Varakantam, V. Practical architecture considerations for Smart Grid WAN network. In Proceedings of the 2011 IEEE/PES Power Systems Conference and Exposition, Phoenix, AZ, USA, 20–23 March 2011; pp. 1–6.
17. Mohi-ud-din, G. "NSL-KDD", IEEE Dataport, Published by IEEE, USA. Available online: https://dx.doi.org/10.21227/425a-3e55 (accessed on 29 December 2018).
18. Shone, N.; Ngoc, T.N.; Phai, V.D.; Shi, Q. A Deep Learning Approach to Network Intrusion Detection. *IEEE Trans. Emerg. Top. Comput. Intell.* **2018**, *2*, 41–50. [CrossRef]
19. Dong, S.; Wang, P.; Abbas, K. A survey on deep learning and its applications. *Comput. Sci. Rev.* **2022**, *40*, 100379. [CrossRef]
20. Jan, S.U.; Ahmed, S.; Shakhov, V.; Koo, I. Toward a Lightweight Intrusion Detection System for the Internet of Things. *IEEE Access* **2019**, *7*, 42450–42471. [CrossRef]
21. Karimipour, H.; Dehghantanha, A.; Parizi, R.M.; Choo, K.-K.R.; Leung, H. A Deep and Scalable Unsupervised Machine Learning System for Cyber-Attack Detection in Large-Scale Smart Grids. *IEEE Access* **2019**, *7*, 80778–80788. [CrossRef]
22. Takiddin, A.; Ismail, M.; Zafar, U.; Serpedin, E. Deep Autoencoder-Based Anomaly Detection of Electricity Theft Cyberattacks in Smart Grids. *IEEE Syst. J.* **2022**, *16*, 4106–4117. [CrossRef]
23. Inayat, U.; Zia, M.F.; Mahmood, S.; Berghout, T.; Benbouzid, M. Cybersecurity Enhancement of Smart Grid: Attacks, Methods, and Prospects. *Electronics* **2022**, *11*, 3854. [CrossRef]
24. Zhou, F.; Wen, G.; Ma, Y.; Geng, H.; Huang, R.; Pei, L.; Yu, W.; Chu, L.; Qiu, R. A Comprehensive Survey for Deep-Learning-Based Abnormality Detection in Smart Grids with Multimodal Image Data. *Appl. Sci.* **2022**, *12*, 5336. [CrossRef]

25. Berghout, T.; Benbouzid, M.; Muyeen, S.M. Machine learning for cybersecurity in smart grids: A comprehensive review-based study on methods, solutions, and prospects. *Int. J. Crit. Infrastruct. Prot.* **2022**, *38*, 100547. [CrossRef]
26. Jithish, J.; Alangot, B.; Mahalingam, N.; Yeo, K.S. Distributed Anomaly Detection in Smart Grids: A Federated Learning-Based Approach. *IEEE Access* **2023**, *11*, 7157–7179. [CrossRef]
27. Moustafa, N.; Slay, J. UNSW-NB15: A comprehensive data set for network intrusion detection systems (UNSW-NB15 network data set). In Proceedings of the 2015 Military Communications and Information Systems Conference (MilCIS), Canberra, Australia, 10–12 November 2015; pp. 1–6.

algorithms

MDPI

Article

Adaptive IDS for Cooperative Intelligent Transportation Systems Using Deep Belief Networks

Sultan Ahmed Almalki [1,*], Ahmed Abdel-Rahim [2] and Frederick T. Sheldon [1,*]

1   Department of Computer Science, University of Idaho, Moscow, ID 83844, USA
2   Department of Civil and Environmental Engineering, University of Idaho, Moscow, ID 83844, USA;
     ahmed@uidaho.edu
*   Correspondence: alma6989@vandals.uidaho.edu (S.A.A.); sheldon@ieee.org (F.T.S.)

Citation: Almalki, S.A.;
Abdel-Rahim, A.; Sheldon, F.T.
Adaptive IDS for Cooperative
Intelligent Transportation Systems
Using Deep Belief Networks.
Algorithms **2022**, *15*, 251. https://
doi.org/10.3390/a15070251

Academic Editors: Francesco
Bergadano and Giorgio Giacinto

Received: 11 May 2022
Accepted: 18 July 2022
Published: 20 July 2022

Abstract: The adoption of cooperative intelligent transportation systems (cITSs) improves road safety and traffic efficiency. Vehicles connected to cITS form vehicular ad hoc networks (VANET) to exchange messages. Like other networks and systems, cITSs are targeted by attackers intent on compromising and disrupting system integrity and availability. They can repeatedly spoof false information causing bottlenecks, traffic jams and even road accidents. The existing security infrastructure assumes that the network topology and/or attack behavior is static. However, the cITS is inherently dynamic in nature. Moreover, attackers may have the ability and resources to change their behavior continuously. Assuming a static IDS security model for VANETs is not suitable and can lead to low detection accuracy and high false alarms. Therefore, this paper proposes an adaptive security solution based on deep learning and contextual references that can cope with the dynamic nature of the cITS topologies and increasingly common attack behaviors. In this study, deep belief networks (DBN) modeling was used to train the detection model. Binary cross entropy was used as a loss function to measure the prediction error. Two activation functions were used, Relu and Softmax, for input–output mapping. The Relu was used in the hidden layers, while the Sigmoid was used in the last layer to map the real vector to output between 0 and 1. The adaptation mechanism was incorporated into the detection model using a moving average that monitors predicted values within a time window. In this way, the model can readjust the classification thresholds on-the-fly as appropriate. The proposed model was evaluated using the Next Generation Simulation (NGSIM) dataset, which is commonly used in such related works. The result is improved accuracy, demonstrating that the adaptation mechanism used in this study was effective.

Keywords: cooperative intelligent transportation systems (cITSs); IDS; vehicular ad-hoc networks (VANET); adaptive model; deep belief network (DBN)

1. Introduction

Cooperative intelligent transportation systems (cITSs) collect data from the end nodes (i.e., endpoints). These data are stored locally and shared with the other nodes [1–3]. The cITS adopts one of the two information-sharing standards, the European standard [4] and the American standard [4]. On the one hand, the European standard defines two types of messages, the Cooperative Awareness Message (CAM) and the Decentralized Environmental Notification Message (DENM) [5]. The CAMs are sent periodically and carry information about the vehicles such as their position, size, speed, and angle of steering wheel. The DENM messages carry information about events which occur on sections of road section such as lane changes and (sudden) braking. On the other hand, the American standard defines context information messages called basic safety messages (BSMs), which carry different information such as position, heading, speed, acceleration, steering wheel angle, vehicle role, vehicle size and status of vehicle lights [6]. If an event happens, then the BSM also carries those event-related information.

Notwithstanding, cITSs enable information sharing among neighboring nodes (i.e., vehicles). Unfortunately, this comes at the cost of needing to address several threats that target data and system integrity [7,8]. These threats could be imposed by either human-crafted attacks or malware [7,9–11]. Threats which target cITS systems can disable or disrupt the function of one or more components in the vehicle's navigation system [12]. For example, threats can spoof the exchanged data to inject false mobility information which is then exchanged among neighboring vehicles causing erroneous actions and calamitous outcomes.

Threat actors use sophisticated strategies and employ malware to carry out various attacks against cITSs [13,14]. These attacks could come from nodes inside or outside the network. Outside attacks by threat actors that are not part of the network are easy to detect, whereas inside attacks are usually carried out via legitimate but compromised vehicles. Such inside attacks are more challenging to detect. Typical cITS targeted attacks include jamming, replay, Sybil, and data falsification.

Jamming is carried out by overwhelming individual cITS nodes by an enormous amount of messages, which disrupt the connectivity with the cITS, a denial-of-service attack type [15]. The consequences include message loss within the cITS, causing a data insufficiency situation that adversely affects the accuracy of the intrusion detection systems (IDS) trained on such data. Replay attacks occur if the attacker can impersonate an original node enabling the interception of messages exchanged between the vehicles and thereby injecting false data by re-sending them to a victim node [16]. Likewise, a Sybil attack creates several identities and uses them to poison (fake) BSM messages that deceive victim nodes; as such, a Sybil attack compromises network services when an attacker subverts the service's reputation system by creating a large number of pseudonymous identities and then using them to gain a disproportionately large influence. Thus, false data injection can be used to share and promote false information about the current traffic situation on the road for the purpose of disrupting traffic flow and triggering congestion.

Data falsification is another type of attack that can be conducted to compromise BSM messages exchanged between cITS nodes. The first step is to compromise a legitimate node and employ it to share false data with neighboring vehicles. Since the compromised node has been previously authenticated, a trust relationship was established with other nodes in the cITS network. Attackers can utilize this fact to spread the false data using the compromised node [5]. Attackers thus manipulate the BSM and inject false data which is then share with neighboring nodes [17]. The false data may cause a vehicle to take unexpected actions such as sudden braking, lane changing, and/or sudden acceleration. Therefore, taking security measures to protect BSM messages is crucial [6].

2. Related Works

The current solutions proposed for protecting the cITSs can be categorized into node-centric and data-centric IDSs. Some of these solutions tried to protect the system against threats coming from the outside caused by Sybil, malware, and DoS attacks. By comparing the patterns from incoming traffic with the patterns of normal applications, those solutions can detect suspicious threats and raise alarms. Moreover, other solutions focus on detecting misbehaving nodes in cITSs. These solutions aim to protect the system against threats carried out by legitimate yet compromised nodes, which is more challenging as those nodes are trusted and thus less suspicious [18]. Nonetheless, most of these solutions assume that the cITS is stationary. Such an assumption is not realistic as the ephemeral nature of cITSs make it a very dynamic constantly changing topology. Developing data-driven detection solutions on presumed stationary data prohibits handling the numerous and rapid changes typical inside the cITS. These solutions quickly become outdated and consequently, their accuracy decreases. Some studies have tried to rectify the issue by adopting solutions with the dynamic nature of the operating environment in mind [8]. These solutions, again, are typically categorized into node-centric and data-centric.

The existing IDS proposal for cITS relies on the BSM messages exchanged between the communicating vehicles as well as the contextual metadata that describes the operating environment. Such data in many studies are static, which might not be suitable for such a dynamic cITSs where the node's operational environment changes continuously. Therefore, static security thresholds become outdated more often. This represents a major issue for existing IDS solutions. To address this issue, some studies have proposed solutions, such as the context-aware data-centric misbehavior detection scheme (CA-DC-MDS) developed by [13]. This solution overcomes the aforementioned drawbacks. Static thresholds are replaced by a dynamic threshold statistically determined using a contextual model, which is constructed and updated online. The sequential analysis of temporal and spatial correlation is conducted using Kalman and Hampel filters to assess the consistency of mobility data exchanged between neighboring vehicles. The Kalman filter tracks mobility data from the neighboring vehicles, while the Hampel filter assesses the consistency of these data. Based on the proximity from the threshold, the message containing the data is classified as either normal or suspicious. However, the scheme assumes that data collected at the early phases after the model has updated its profile are sufficient for consistency assessment. This is not realistic in most cases, as the contextual data that describe the new situation are not yet ready for a variety of reasons as described below.

Node-centric IDSs determine whether a vehicle is malicious based on how it behaves on the road section [19]. The trustworthiness of legitimate vehicles is also assessed based on such behavior, which can be perceived by observing the number and validity of BSM messages shared by the vehicle [20,21]. Reputation-based evaluation is usually adopted for the trustworthiness estimation of each node in the cITS. The estimation is performed by a voting strategy whose outcome relies on the majority concept. However, relying on node behavior is sub-optimal because the cITS is non-stationary and since nodes change their behavior as the topology changes [22,23]. Moreover, relying on a voting approach for the trustworthiness estimation is always biased towards the majority, which in some cases, can be compromised when the attacker gains a majority foothold. A case in point occurs when attackers use advanced and sophisticated attack strategies such as malware and botnets to create a majority of rogue nodes enabling them to control the trustworthiness estimation. Consequently, such reputation-based mechanisms used by node-centric solutions cannot be trusted for the early identification of misbehaving or faulty vehicles [6].

Another set of IDSs for cITS adopt the data-centric detection approach by inspecting the BSM messages exchanged between the neighboring vehicles. These solutions perform several checks to determine whether the messages are falsified. BSM messages are checked against several criteria such as consistency and plausibility to determine whether they are trustworthy [6]. The consistency checks that BSM messages undergo in data-centric solutions determine whether the data shared by the node are consistent with the general context from the particular cITS. By vetting these BSMs, data-centric solutions can also identify the plausibility of the shared data to help in determining validity (i.e., whether they are in-line with those coming from other nodes in the cITS system).

The node-centric and data-centric approaches adopted in existing IDS solutions for cITS rely on estimating the reputation of the nodes and trustworthiness of the data they share with each other. However, both approaches have inherent weaknesses and may not be suitable for tumultuous environments such as cITSs. In such dynamic systems, the nodes join and leave the network frequently, which creates an unstable topology. This makes it difficult to capture sufficient and consistent patterns that represent all behavioral aspects of the nodes. Therefore, existing security solutions with rigid thresholds are not suitable as they do not have the sufficient data needed for accurate decisions. Therefore, these solutions suffer from a high rate of false alarms. Thus, data insufficiency makes it difficult for adaptive mechanisms used by some solutions to accurately calculate the new thresholds, which also have a negative effect on IDS accuracy.

The contribution of this study is two-fold:

- A bi-variate moving average (BiMAV) technique was proposed. Unlike existing methods that only rely on the values estimated at the output layer, BiMAV correlates the changes of the output layer with the averaged input variables. Such an approach provides precise change detection by avoiding the instantaneous changes that could compromise the stability of the detection model.
- The proposed method was incorporated into the detection model, which helps to prevent the unnecessary re-adjustment of security thresholds at the output layer of the DBN classifier thanks to the bivariate-based moving average used to monitor and detect the change in the classification accuracy estimation.

The rest of the paper is organized as follows. Section 3 presents the methodology in which we describe the proposed solution. The results are analyzed and discussed in Section 4 along with a comparison with existing related work. Section 5 concludes the paper with a summary of the contribution and findings.

3. Methodology

Given the literature reviewed above, we have concluded that the ephemeral nature of cITSs is a major challenge that makes many existing solutions ineffective. To overcome such a challenge, herein we propose an adaptive IDS for cITS. Our adaptive approach has the ability to cope with the dynamical nature of the cITS operating environment. A bi-variate moving average (BiMAV) method was developed to detect the (potential) diversion, in practice, from the existing threshold used by the detection model. Unlike existing methods that rely only on the values estimated at the output layer, BiMAV correlates the change of output layer with the averaged input variables. Such an approach provides precise change detection by avoiding the instantaneous changes that will eventually compromise the stability of the detection model. The proposed method prevents the unnecessary re-adjustment of security thresholds at the output layer of the DBN classifier thanks to the bivariate-based moving average used to monitor and detect the change in the classification accuracy estimation. This is important for dynamic environments such as cITSs where sufficient data might not be available. Based on the amount of change, adaptation can be triggered. in other words, if the difference exceeds a certain limit (i.e., according to the standard deviation), retraining the model is triggered. Model retraining will be performed based on the new data. If the difference does not exceed the threshold, there is no need for retraining.

The proposed solution here relies on the supervised learning approach. The deep belief network (DBN), one of the famous deep learning algorithms, is used to train the IDS based on data collected from the BSM messages. Before training, the data are pre-processed to make them suitable for ingestion by the DBN. As part of the preparation, noise data are removed, and data normalization is carried out. During data normalization, the values of all attributes are converted to a range of 0–1. This ensures that all attributes are in the same scale and prevents those with higher ranges from having undue influence over the model's output decision.

The data are now ready for the mutual information feature selection (MIFS) process that selects out discriminative features to reduce data dimensionality. This avoids the over-fitting problem that negatively affects the accuracy of the IDS [24,25]. By selecting the most relevant features, the model also generates less false alarms, which contributes to higher precision. Furthermore, reducing data dimensionality helps decrease the model complexity, which is more favorable for ephemeral environments such as cITSs. The MIFS ranks the features based on the entropy, such that those with higher entropy value correspond to a lower rank. Then, the MIFS selects the n-top ranking features (n experimentally chosen to give higher accuracy). The selected features are then used as input for the DBN algorithm.

During the model's training phase, the DBN is trained using the data and features selected by the MIFS. The DBN model is composed of several layers, namely input, output and hidden. The number of input layer nodes is determined by the number of features selected by the MIFS. These nodes receive data and process them into the hidden layers, after being scaled (i.e., multiply) by an input weight. In our methodology, the hidden part of the DBN is constructed from three layers. The number of hidden layers is determined based on an overfitting factor during the training phase. The number of nodes in the hidden layer is thus determined based on the bias factor during the training phase as well. The value of the bias factor was set to 0.25, multiplied by the standard deviation $\sigma(W)$ of the previous window. Therefore, the number of nodes in hidden layers were taken as a percentage of the original number. As we start with 18 nodes (because the number of nodes in a hidden layer should be lower than then nodes in input layer), in the hidden layers, the data are processed based on the activation function used by the hidden nodes. The Relu function is used as the activation function in all nodes in the hidden layers of the DBN, except the layer that precedes the output, where the sigmoid function was used. These activation functions are used to map the output of nodes into values between 0 and 1, which are needed for prediction. The output layer receives the data from the sigmoid functions in the last hidden layer and determines whether the instance is malicious or normal based on a threshold σ, where values greater than σ are considered as attacks.

Training and Testing

The DBN model was trained using the 10-fold cross-validation method, wherein data are divided into two sets. During the training/testing process, the data were divided into two sets, i.e., training and testing. The training builds the model while testing evaluates its accuracy. The size of training set was 90% of the data and, naturally, the testing set was 10% of the data. This process was repeated 10 times and the accuracy of the model was recorded. At the end of the training/testing process, the averaged accuracy was calculated, which determines the overall model accuracy.

4. Model Adaptation Using Bi-Variate Moving Average

Our proposed model, as described above, is aimed at improving detection within the dynamic cITS environment. Therefore, here we describe an adaptation capability needed to ensure that the model can better handle the constantly changing network topology. We propose a bi-variate moving average (BiMAV) model adaptation method that observes the model performance and adapts to the change in the operating environment. The proposed method follows the progressive modeling used by works that rely on time series data [26]. The method uses a two-dimensional window for change detection. That is, the window defines two variables, the aggregated input values and the estimated output. Within this window, the accuracy trend is monitored against a threshold calculated based on the standard deviation from previous windows. Equation (1) implements the BiVAM method:

$$\text{BiMAV} = \frac{\sum_{i=0}^{i=n-1} X_i}{n} \times \frac{\sum_{j=0}^{j=l-1} Y_j}{l} \tag{1}$$

where X_i and Y_j are the input features and estimated output values, respectively. The variable n represents number of features while l represents number of instances in the window. The retraining is triggered if the value of BiMAV is higher than the standard deviation of the previous windows, as expressed by Equation (2):

$$\text{BiMAV} = \begin{cases} if < \sigma(W) & \text{then No retraining} \\ if > \sigma(W) & \text{then Retraining} \end{cases} \tag{2}$$

where $\sigma(W)$ represents the standard deviation of the previous windows. The decision that Equation (2) makes is binary as it determines whether the re-training is needed or not based on the threshold $\sigma(W)$.

5. The Dataset

The dataset used for this study was the Next Generation Simulation (NGSIM) Vehicle Trajectories Dataset [7]. NGSIM is an open source publicly available dataset with a collection of real-world vehicles' trajectories collected by smart vehicles. It contains a detailed vehicle trajectory data on southbound US 101 and Lankershim Boulevard in Los Angeles, CA, eastbound I-80 in Emeryville, CA and Peachtree Street in Atlanta, Georgia. Data in NGSIM were collected through a network of synchronized digital video cameras. NGVIDEO, a customized software application developed for the NGSIM program, transcribed the vehicle trajectory data from the video. This vehicle trajectory data provides the precise location of each vehicle within the study area every one-tenth of a second, resulting in detailed lane positions and locations relative to other vehicles. Moreover, NGSIM consists of many patterns representing different drive situations and driver behavior [7]. In addition, NGSIM provides high-quality contextual data that describe realistic real-world scenarios on different road sections [19]. Particularly, NGSIM was built by collecting data from vehicles moving on a road section with 500 m-long and seven-lane highway. For each vehicle, the data are collected (recorded) for 45 min using 16 sensors. Each record in the dataset contains s set of basic elements regarding the vehicle like position, speed, time, direction, and acceleration. Although there are similar datasets such as the Connected Vehicles Pilot (CVP), the NGSIM dataset was chosen in this study to be consistent when comparing with the related works as they used the NGSIM as well.

The dataset represents the ground truth information and each vehicle represents a cITS node. In real-world deployment, the dataset needs to be fed each cITS node. That is, each node should have a copy of the dataset to run its own applications and adjust its communication or driving behavior. As such, the collection of accurate and reliable context information is crucial. The context information in the dataset combines two types of messages, cooperative awareness message (CAM) and decentralized environmental notification message (DENM) into a basic safety message (BSM). While CAMs are sent periodically, DENMs are event-driven that only sent when an event has occurred. The CAM consists of information about the vehicles such as the position, size, speed, and steering wheel angle.

In contrast, DENM contains information about a certain event such as lane changing and sudden braking. BSM combines CAM and DENM messages. The first part of BSM, as well as CAM in the European standard, carries information about position, heading, speed, acceleration, steering wheel angle, vehicle role, vehicle size, and the status of vehicle lights [4,27,28]. Unlike the first part of BSM that is included in all BSM messages, the second part of BSM (which corresponds to DENM in the European standard) is only included when an event happens, to carry information about such an event.

6. Experimental Environment Setup

To implement the different components of the proposed mode and evaluate its performance, the development and experimental evaluation will be conducted using several tools and software including Python, TensorFlow, Scikit Learn, SKFeature, and Numpy. These tools and libraries are all included in the Anaconda development platform. Meanwhile, the preparation of data samples, implementation of algorithms, and the analysis of the results will be carried out on a machine with Intel(R) Core (TM) i7-4790 CPU @ 3.60 GHZ and 16 GB RAM.

Evaluation Metrics

To evaluate the performance of the proposed IDS for cITS, this paper uses the accuracy, detection rate, and the false alarms rate as they are common metrics widely used by the extant research. Equations (3)–(6) are used to calculate the detection accuracy, detection rate, precision, false positive rate, and the F measure, respectively.

$$\text{ACC} = \frac{TP + TN}{TP + TN + FP + FN} \tag{3}$$

$$\text{DR} = \frac{TP}{TP + FN} \tag{4}$$

$$\text{FPR} = \frac{FP}{FP + TN} \tag{5}$$

$$\text{F1} = \frac{2 \times \text{Precision} \times \text{Recall}}{\text{Precision} + \text{Recall}} \tag{6}$$

where $TP, TN, FP,$ and FN denote the true positive, true negative, false positive, and false negative, respectively.

7. Experimental Results

Table 1 shows the accuracy (ACC), detection rate (DR), false positive rate (FPR), and F1 measure of the proposed adaptive deep belief network-based IDS (ADBN-IDS). In addition, Tables 2 and 3 show the results of the IDS built using conventional machine learning classifiers, namely the support vector machines (SVMs), and the logistic regression (LR). As pointed out previously, the ACC, DR, FPR, and F1 were calculated based on Equations (3)–(6). In the tables, the first column in each table lists the accuracy of the proposed; while the second lists the detection rat; the third column lists the false positive rate; and the fourth column lists the F1 measure of the proposed and related models. The tables' rows are used to list feature sets with different sizes. The feature sizes range between 5 and 25 incremented by 3. The results show that the proposed ADBN-IDS achieved higher accuracy over the other two classifiers (i.e., SVM and LR) [28,29]. This is attributed to the ability of the BiMAV method (incorporated into ADBN-IDS) to detect the degradation in the model's performance and trigger the training on the right time. This contributes to keeping the model up to date and prevent the concept drift from affecting the accuracy of the model.

The results also show that the accuracy increased when more features were added, until the number of features reached 20. After that, the model experienced a decrease in the accuracy. This also can be observed from the other evaluation metric, namely DR, FPR, and F1. The same trend was observed not only for the ADBN-IDS, but also for SVM and LR. The reason is that the model needs sufficient features to make correct decisions. However, when the number of features exceed a certain limit, the model would suffer from high variance that makes it prone to overfitting. The situation exacerbates when the coming observations lack the sufficient attack patterns necessary for clear and accurate decision. This would result to a model that can only recognize the patterns that it has seen, and if new patterns that have less similarity with the known ones are encountered, the likelihood that the model could miss the true classification becomes high.

Figures 1–4 show the comparison between the proposed ADBN-IDS and the models built using the SVM and LR, in terms of accuracy, detection rate, false positive rate, and F measure, respectively. The x axis represents the number of features used for training, and the y axis represents the value of performance measure achieved. The comparison was conducted between the ADBN-IDS that employed the BiMAV for adaptation and the conventional approach used in the existing studies [28,29]. As depicted in the figures, the proposed ADBN-IDS outperformed the related techniques in terms of accuracy, detection rate, false positive rate, and the F measure. It can also be observed that the ADBN-IDS maintain a stable increment in the performance for the four measures when the number of features increase until it reaches 20 features where the performance shows declining trend. This is attributed to the efficacy of the BiMAV incorporated for the model adaptation and the reliance on the combination of output and averaged inputs for proximity calculation from the threshold. Such an approach makes the change detection mechanism robust, which avoids unnecessary re-training and only triggers it if the change in the cITS topology or attack behavior is significant. It is also worth noting that the frequency of adaptation varies based on the threshold. When the threshold is set to a higher value, the rate of adaptation becomes less frequent. When the threshold value is set to low, the adaptation frequency increases. Moreover, Figure 5 shows the area under the curve of the proposed model under several thresholds. The x axis represents the false positive rate while the y axis represents the true positive rate. It can be observed that the false positive rate decreases when the detection rate increases.

Table 1. The experimental evaluation results for the proposed ADBN-IDS in terms of accuracy, detection rate, false positive rate, and F measure.

Metric and Number of Features	ACC	DR	FPR	F1
5	0.92	0.924	0.132	0.927
8	0.929	0.926	0.128	0.931
11	0.946	0.937	0.113	0.947
14	0.968	0.965	0.084	0.969
17	0.97	0.968	0.076	0.973
20	0.974	0.972	0.071	0.978
23	0.973	0.97	0.072	0.975
25	0.969	0.971	0.077	0.972

Table 2. The experimental evaluation results for the proposed SVM-IDS in terms of accuracy, detection rate, false positive rate, and F measure.

Metric and Number of Features	ACC	DR	FPR	F1
5	0.892	0.89	0.176	0.894
8	0.9	0.894	0.179	0.892
11	0.91	0.913	0.15	0.915
14	0.951	0.95	0.132	0.954
17	0.956	0.953	0.129	0.958
20	0.957	0.953	0.122	0.958
23	0.951	0.948	0.13	0.953
25	0.947	0.942	0.154	0.951

Table 3. The experimental evaluation results for the proposed LR-IDS in terms of accuracy, detection rate, false positive rate, and F measure.

Metric and Number of Features	ACC	DR	FPR	F1
5	0.898	0.894	0.162	0.9
8	0.904	0.902	0.157	0.907
11	0.919	0.917	0.144	0.918
14	0.943	0.94	0.14	0.946
17	0.958	0.952	0.131	0.96
20	0.954	0.951	0.137	0.956
23	0.95	0.948	0.139	0.952
25	0.945	0.943	0.142	0.948

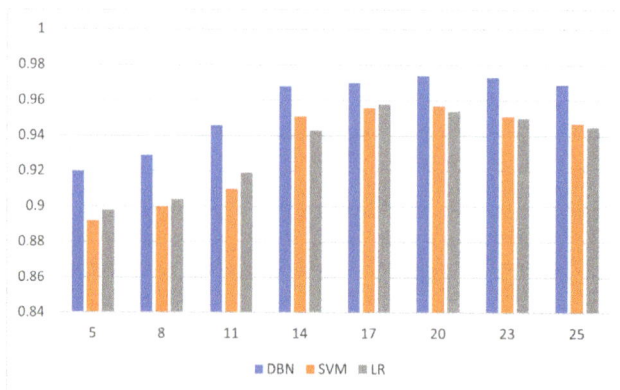

Figure 1. Comparison of the proposed ADBN-IDS with SVM and LR in terms of detection accuracy.

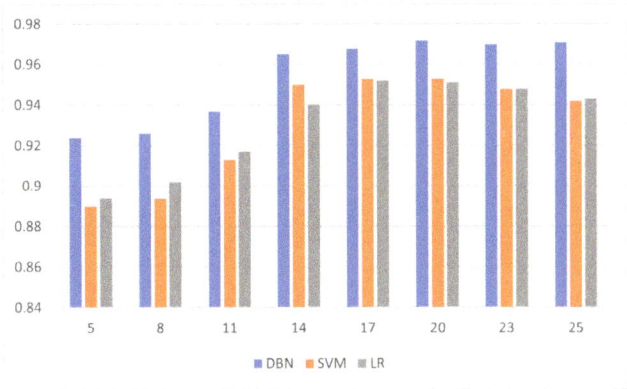

Figure 2. Comparison of the proposed ADBN-IDS with SVM and LR in terms of detection rate.

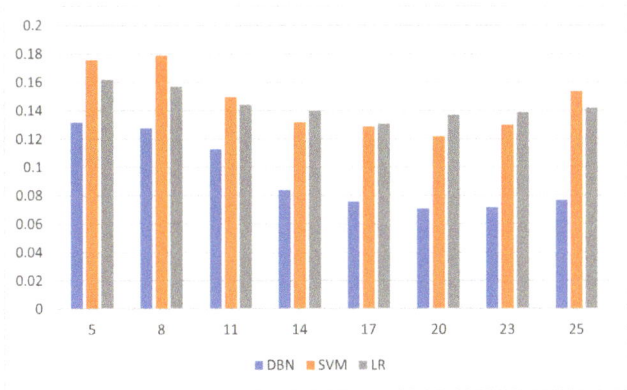

Figure 3. Comparison of the proposed ADBN-IDS with SVM and LR in terms of false positive rate.

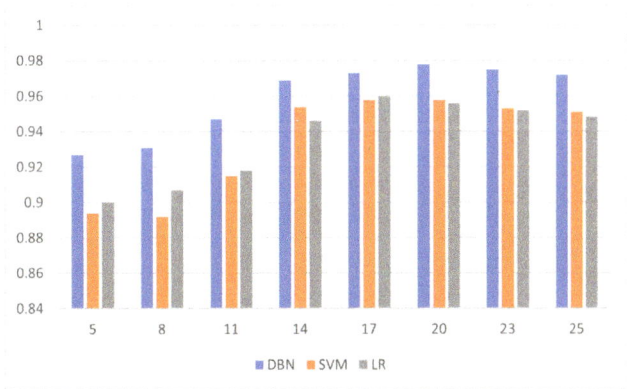

Figure 4. Comparison of the proposed ADBN-IDS with SVM and LR in terms of F measure.

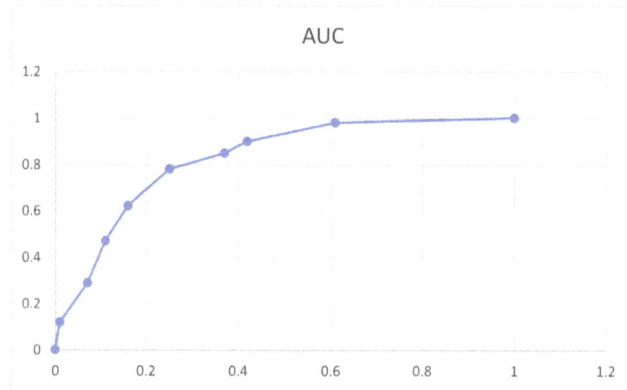

Figure 5. Area under the curve comparison for several detection thresholds.

8. Conclusions and Summary

In this paper, our adaptive deep belief network-based intrusion detection system (ADBN-IDS) for cITS is described. ADBN-IDS is composed of three components: pre-processing, feature selection, and training/testing. Thus, the model is created from the deep belief network (DBN) classifier, and includes the bi-variate moving average (BiMAV) method as our adaptation technique. This inclusion allows the model to cope with the dynamic nature of the cITS environment and has never been tested using the NGSIM dataset.

The classifier was trained using the NGSIM dataset and tested using 10-fold cross validation. The performance of the model was evaluated using several metrics including accuracy, detection rate, false positive rate, and the F1 measure. The evaluation of our results demonstrate that the proposed ADBN-IDS achieved higher performance in terms of accuracy, detection rate, false positive rate, and F1, which indicates the importance of the BiMAV adaptation mechanism in achieving and maintaining a safer more resilient cITS.

In summary, our proposed ABDN-IDS model, for the NGSIM dataset, showed on average, an improvement of 2.35%, 2.47%, and 42% in terms of accuracy, detection and false positive rate, respectively.

Author Contributions: Conceptualization, S.A.A.; methodology, S.A.A.; software, S.A.A.; validation, S.A.A., F.T.S. and A.A.-R.; formal analysis, S.A.A.; investigation, S.A.A. and F.T.S.; resources, S.A.A.; data curation, S.A.A.; writing—original draft preparation, S.A.A.; writing—review and editing, S.A.A., F.T.S. and A.A.-R. visualization, S.A.A.; supervision, F.T.S. project administration, S.A.A.; funding acquisition, F.T.S. All authors have read and agreed to the published version of the manuscript.

Funding: This research received no external funding.

Institutional Review Board Statement: Not applicable.

Informed Consent Statement: Not applicable.

Data Availability Statement: The Next Generation Simulation (NGSIM) dataset that was used in this study is publicly available online at the following link: https://ops.fhwa.dot.gov/trafficanalysistools/ngsim.htm (accessed on 10 May 2022), and can be downloaded directly from the following link: https://data.transportation.gov/Automobiles/Next-Generation-Simulation-NGSIM-Vehicle-Trajector/8ect-6jqj (accessed on 10 May 2022).

Conflicts of Interest: The authors declare no conflict of interest.

References

1. Ghaleb, F.A.; Al-Rimy, B.A.S.; Almalawi, A.; Ali, A.M.; Zainal, A.; Rassam, M.A.; Shaid, S.Z.M.; Maarof, M.A. Deep Kalman Neuro Fuzzy-Based Adaptive Broadcasting Scheme for Vehicular Ad Hoc Network: A Context-Aware Approach. *IEEE Access* **2020**, *8*, 217744–217761. [CrossRef]

2. Almalki, S.A.; Song, J. A review on data falsification-based attacks in cooperative intelligent transportation systems. *Int. J. Comput. Sci. Secur. (IJCSS)* **2020**, *14*, 22.

3. Talal, M.; Ramli, K.N.; Zaidan, A.; Zaidan, B.; Jumaa, F. Review on car-following sensor based and data-generation mapping for safety and traffic management and road map toward ITS. *Veh. Commun.* **2020**, *25*, 100280. [CrossRef]

4. *ETSI TS 102 636-4-2*; Intelligent Transport Systems (ITS); Vehicular Communications; GeoNetworking; Part 4: Geographical Addressing and Forwarding for Point-to-Point and Point-to-Multipoint Communications; Sub-Part 2: Media-Dependent Functionalities for ITS-G5. ETSI: Valbonne, France, 2013; Volume 102, p. 636-4.

5. Ghaleb, F.A.; Maarof, M.A.; Zainal, A.; Al-rimy, B.A.S.; Alsaeedi, A.; Boulila, W. Ensemble-based hybrid context-aware misbehavior detection model for vehicular ad hoc network. *Remote Sens.* **2019**, *11*, 2852. [CrossRef]

6. van der Heijden, R.W.; Dietzel, S.; Leinmüller, T.; Kargl, F. Survey on misbehavior detection in cooperative intelligent transportation systems. *IEEE Commun. Surv. Tutor.* **2018**, *21*, 779–811. [CrossRef]

7. Maseer, Z.K.; Yusof, R.; Mostafa, S.A.; Bahaman, N.; Musa, O.; Al-rimy, B.A.S. DeepIoT. IDS: Hybrid deep learning for enhancing IoT network intrusion detection. *CMC-Comput. Mater. Contin.* **2021**, *69*, 3945–3966. [CrossRef]

8. Al-rimy, B.A.S.; Kamat, M.; Ghaleb, F.A.; Rohani, F.; Razak, S.A.; Shah, M.A. A user mobility-aware fair channel assignment scheme for wireless mesh network. In *Computational Science and Technology*; Springer: Berlin/Heidelberg, Germany, 2020; pp. 531–541.

9. Ahmed, Y.A.; Huda, S.; Al-rimy, B.A.S.; Alharbi, N.; Saeed, F.; Ghaleb, F.A.; Ali, I.M. A Weighted Minimum Redundancy Maximum Relevance Technique for Ransomware Early Detection in Industrial IoT. *Sustainability* **2022**, *14*, 1231. [CrossRef]

10. Urooj, U.; Al-rimy, B.A.S.; Zainal, A.; Ghaleb, F.A.; Rassam, M.A. Ransomware detection using the dynamic analysis and machine learning: A survey and research directions. *Appl. Sci.* **2021**, *12*, 172. [CrossRef]

11. Olaimat, M.N.; Maarof, M.A.; Al-rimy, B.A.S. Ransomware anti-analysis and evasion techniques: A survey and research directions. In Proceedings of the 2021 3rd international cyber resilience conference (CRC), Langkawi Island, Malaysia, 29–31 January 2021; pp. 1–6.

12. Ercan, S.; Ayaida, M.; Messai, N. Misbehavior detection for position falsification attacks in VANETs using machine learning. *IEEE Access* **2021**, *10*, 1893–1904. [CrossRef]

13. Ghaleb, F.A.; Maarof, M.A.; Zainal, A.; Rassam, M.A.; Saeed, F.; Alsaedi, M. Context-aware data-centric misbehaviour detection scheme for vehicular ad hoc networks using sequential analysis of the temporal and spatial correlation of the consistency between the cooperative awareness messages. *Veh. Commun.* **2019**, *20*, 100186. [CrossRef]

14. Ghaleb, F.A.; Maarof, M.A.; Zainal, A.; Al-Rimy, B.A.S.; Saeed, F.; Al-Hadhrami, T. Hybrid and multifaceted context-aware misbehavior detection model for vehicular ad hoc network. *IEEE Access* **2019**, *7*, 159119–159140. [CrossRef]

15. Azam, F.; Kumar, S.; Priyadarshi, N. Privacy and Authentication Schemes in VANETS Using Blockchain: A Review and a Framework to Mitigate Security and Privacy Issues. In *AI Enabled IoT for Electrification and Connected Transportation*; Springer: Singapore, 2022; pp. 127–145.

16. Alharthi, A.; Ni, Q.; Jiang, R. A privacy-preservation framework based on biometrics blockchain (BBC) to prevent attacks in VANET. *IEEE Access* **2021**, *9*, 87299–87309. [CrossRef]

17. Ghaleb, F.A.; Zainal, A.; Rassam, M.A.; Mohammed, F. An effective misbehavior detection model using artificial neural network for vehicular ad hoc network applications. In Proceedings of the 2017 IEEE conference on application, information and network security (AINS), Miri, Malaysia, 13–14 November 2017; pp. 13–18.

18. Pandiangan, T.; Bali, I.; Silalahi, A. Early lung cancer detection using artificial neural network. *Atom Indones.* **2019**, *45*, 9–15. [CrossRef]

19. Ghaleb, F.A.; Saeed, F.; Alkhammash, E.H.; Alghamdi, N.S.; Al-Rimy, B.A.S. A Fuzzy-Based Context-Aware Misbehavior Detecting Scheme for Detecting Rogue Nodes in Vehicular Ad Hoc Network. *Sensors* **2022**, *22*, 2810. [CrossRef]

20. Alsoufi, M.A.; Razak, S.; Siraj, M.M.; Nafea, I.; Ghaleb, F.A.; Saeed, F.; Nasser, M. Anomaly-based intrusion detection systems in iot using deep learning: A systematic literature review. *Appl. Sci.* **2021**, *11*, 8383. [CrossRef]

21. Qafzezi, E.; Bylykbashi, K.; Ampririt, P.; Ikeda, M.; Matsuo, K.; Barolli, L. A fuzzy-based approach for resource management in SDN-VANETs: Effect of trustworthiness on assessment of available edge computing resources. *J. High Speed Netw.* **2021**, *27*, 33–44. [CrossRef]

22. Sultan, S.; Javaid, Q.; Malik, A.J.; Al-Turjman, F.; Attique, M. Collaborative-trust approach toward malicious node detection in vehicular ad hoc networks. *Environ. Dev. Sustain.* **2022**, *24*, 7532–7550. [CrossRef]

23. Alghofaili, Y.; Albattah, A.; Alrajeh, N.; Rassam, M.A.; Al-rimy, B.A.S. Secure Cloud Infrastructure: A Survey on Issues, Current Solutions, and Open Challenges. *Appl. Sci.* **2021**, *11*, 9005. [CrossRef]

24. Al-Rimy, B.A.S.; Maarof, M.A.; Alazab, M.; Shaid, S.Z.M.; Ghaleb, F.A.; Almalawi, A.; Ali, A.M.; Al-Hadhrami, T. Redundancy coefficient gradual up-weighting-based mutual information feature selection technique for crypto-ransomware early detection. *Future Gener. Comput. Syst.* **2021**, *115*, 641–658. [CrossRef]

25. Khalaf, B.A.; Mostafa, S.A.; Mustapha, A.; Mohammed, M.A.; Mahmoud, M.A.; Al-Rimy, B.A.S.; Abd Razak, S.; Elhoseny, M.; Marks, A. An adaptive protection of flooding attacks model for complex network environments. *Secur. Commun. Netw.* **2021**, *2021*, 5542919. [CrossRef]

26. Cook, A.A.; Mısırlı, G.; Fan, Z. Anomaly detection for IoT time-series data: A survey. *IEEE Internet Things J.* **2019**, *7*, 6481–6494. [CrossRef]

27. Ghaleb, F.A.; Saeed, F.; Al-Sarem, M.; Ali Saleh Al-rimy, B.; Boulila, W.; Eljialy, A.; Aloufi, K.; Alazab, M. Misbehavior-aware on-demand collaborative intrusion detection system using distributed ensemble learning for VANET. *Electronics* **2020**, *9*, 1411. [CrossRef]

28. Akshaya, K.; Sarath, T. Detecting Sybil Node in Intelligent Transport System. In *Innovative Data Communication Technologies and Application*; Springer: Berlin/Heidelberg, Germany, 2022; pp. 595–607.

29. Alsarhan, A.; Alauthman, M.; Alshdaifat, E.; Al-Ghuwairi, A.R.; Al-Dubai, A. Machine Learning-driven optimization for SVM-based intrusion detection system in vehicular ad hoc networks. *J. Ambient. Intell. Humaniz. Comput.* **2021**, 1–10. [CrossRef]

algorithms

MDPI

Article

Reducing the False Negative Rate in Deep Learning Based Network Intrusion Detection Systems

Jovana Mijalkovic [†] and Angelo Spognardi [*,†]

Department of Computer Science, Sapienza University, 00198 Rome, Italy;
mijalkovic.1908929@studenti.uniroma1.it
* Correspondence: spognardi@di.uniroma1.it
† These authors contributed equally to this work.

Abstract: Network Intrusion Detection Systems (NIDS) represent a crucial component in the security of a system, and their role is to continuously monitor the network and alert the user of any suspicious activity or event. In recent years, the complexity of networks has been rapidly increasing and network intrusions have become more frequent and less detectable. The increase in complexity pushed researchers to boost NIDS effectiveness by introducing machine learning (ML) and deep learning (DL) techniques. However, even with the addition of ML and DL, some issues still need to be addressed: high false negative rates and low attack predictability for minority classes. Aim of the study was to address these problems that have not been adequately addressed in the literature. Firstly, we have built a deep learning model for network intrusion detection that would be able to perform both binary and multiclass classification of network traffic. The goal of this base model was to achieve at least the same, if not better, performance than the models observed in the state-of-the-art research. Then, we proposed an effective refinement strategy and generated several models for lowering the FNR and increasing the predictability for the minority classes. The obtained results proved that using the proper parameters is possible to achieve a satisfying trade-off between FNR, accuracy, and detection of the minority classes.

Keywords: NIDS; deep learning; false negative rate; machine learning; artificial neural network

Citation: Mijalkovic, J.; Spognardi, A. Reducing the False Negative Rate in Deep Learning Based Network Intrusion Detection Systems. *Algorithms* **2022**, *15*, 258. https://doi.org/10.3390/a15080258

Academic Editors: Francesco Bergadano and Giorgio Giacinto

Received: 30 June 2022
Accepted: 22 July 2022
Published: 26 July 2022

Publisher's Note: MDPI stays neutral with regard to jurisdictional claims in published maps and institutional affiliations.

1. Introduction

Since the introduction of the first Intrusion Detection Systems, one of the biggest challenges they faced was a high False Positive Rate (FPR) which means that they generate many alerts for non-threatening situations. Security analysts have a massive amount of threats to analyze, which can result in some severe attacks being ignored or overlooked [1]. Another challenge was the False Negative Rate (FNR), which was still not low enough. A high FNR presents an even bigger problem than a high FPR because it is more dangerous to falsely classify an attack as regular network traffic than vice versa. Because of the constant technological improvements and network changes, new and more sophisticated types of attacks emerge, creating the need for continuous improvement of Intrusion Detection Systems.

One way of improving IDSs, on which the researchers have been working in the last years, is using machine learning techniques to reduce the FPR and FNR and improve general detection capabilities [2]. A good example can be found in [3], where the authors developed a prototype IDS which aimed to detect data anomalies by using the k-means algorithm implemented in Sparks MLib. The reason behind using ML algorithms is that they can analyze massive amounts of data and gather any information which can then be used to enhance the capabilities of IDSs [1]. Another reason for using ML algorithms is that they are not domain-dependent and are very flexible- functional for multiple problems [4].

Researchers identified two primary issues in the literature regarding the already-existing deep learning models used for IDS [5]. The first issue is that some have low detection accuracy, especially when dealing with unbalanced data [6]. Most of the research that focuses on the problem of machine learning and deep learning intrusion detection systems uses the same publicly available datasets. After analyzing the extensive work available on this topic, it emerges that some classes had meager detection rates when it comes to multiclass classification, as will be presented in Section 3. The second issue is that some models have somewhat high accuracy but also high False Positive and False Negative Rates, which can lead to lower detection efficiency and weaken the network security [7,8]. Aside from these problems, the datasets used in some research are very aged and might not reflect the modern-day network traffic, so the question arises: can these Intrusion Detection Systems detect modern-day attacks? Moreover, how much can NIDS based on Deep Neural Networks reduce the quite dangerous false negatives?

The objective of this research is tackle the above mentioned problems, and propose a robust solution to improve the detection quality of Network Intrusion Detection Systems using deep learning techniques, namely artificial neural networks. More specifically, the idea is to lower the FNR and FPR and increase the attack predictability of the less represented attack types. We state that, from a security perspective, we could tolerate a slight increase in the FPR if this is a price for nullifying the FNR because it is more dangerous to wrongly classify an attack as benign traffic than the other way around.

We start building a deep neural network for network intrusion detection purposes. The deep neural network will be fed using two different datasets for binary and multiclass network traffic classification. The models will be able to differentiate between regular network traffic and attacks, as well as between different categories of attacks. We then propose a strategy to lower the False Negative Rate of the models by doing various experiments with different methods to reduce the FNR while keeping the False Positive Rate low and the other metrics such as accuracy, precision, and recall high. The strategy we use can be summarized in three steps: modifying the distribution of the training and testing datasets, reducing the number of dataset features, and using class weights. For our purposes, we used two different datasets, NSL-KDD [9] and UNSW-NB15 [10]. The idea was to train the neural network models using an older and a more recent dataset and, in that way, include a more extensive range of network attacks that the network will be able to detect.

The rest of the paper is structured as follows: the next section provides the theoretical background, with the introduction of the building blocks of our research. Section 3 describes the used dataset and surveys the literature's main results related to the deep learning approach for NIDS. This section will show the reduced concern of the related work about lowering false positives and false negatives. Section 4 provides the details of our approach, while Sections 5 and 6 report the descriptions and the results of the experimental campaign. Finally, Section 7 concludes the paper with some overall observations and some future investigation directions.

2. Theoretical Background

In this section we present the fundamental elements to have a reference background, namely Intrusion Detection Systems and artificial neural networks for deep learning.

2.1. Intrusion Detection Systems

An Intrusion Detection System is a software application or a device that monitors network traffic and computer usage intending to detect any suspicious actions that go against regular or expected use of the system, for example, a harmful activity or a policy breach, in order to allow for system security to be maintained. Once the system detects such actions, it alerts the user and collects information on the suspicious activity [11].

Network Intrusion Detection Systems are designed to protect the whole network, not just a single host. NIDS are placed in strategic positions, for example, at the edge of the network, where a network is most vulnerable to attacks [12]. NIDS analyze inbound and outbound traffic to see if it fits the expected average behavior or matches known attack patterns. One positive aspect of this type of IDS is that it can be tough to detect its presence in a system, which means that, usually, an attacker will not even realize that NIDS is scrutinizing his actions. On the other hand, one negative aspect is that this type of IDS analyzes enormous amounts of traffic, which leaves space for making mistakes and generating an excess of false positives, or even some false negatives [13]. To avoid this, they need more fine-tuning done by the administrator to ensure that they are working correctly and not missing anything that might be crucial to the network's security.

IDS need to know how to differentiate between suspicious and regular behavior. For this purpose, there are different methods that they can use. The two main detection approaches are called *signature-based* detection and *anomaly-based* detection [11]. The signature-based approach, also known as *knowledge-based* or *definition-based*, uses a database of known vulnerabilities, signatures (byte combinations), or attack patterns. It identifies attacks by comparing them to this database [12]. The underlying idea is to have a database of anomalies recognized as attacks so that IDS can detect, promptly alert, and possibly avoid the same (or similar) events in that database.

The anomaly-based approach, also known as the behavior-based, focuses on identifying instances of malicious behavior, or in other words, system or network activity that does not fit the expected behavior. These instances are called outliers, and once the IDS detects an outlier, it is supposed to warn the administrator about it. Unlike signature-based IDS, anomaly-based IDS can detect and alert the administrator when they discover a suspicious behavior unknown to them. Instead of searching through a database of known attacks, anomaly-based IDS use machine learning to train their detection system to recognize a normalized baseline, which typically represents how the system behaves. Once the baseline is determined, all the activity is compared to this baseline to see what stands out from the typical network behavior [14].

2.2. Artificial Neural Networks for Deep Learning

Machine Learning (ML) is a specific branch of computer science and artificial intelligence (AI) that focuses on using existing data and algorithms to mimic how people think, learn and make decisions while gradually improving the accuracy of the decision-making process and its results [15]. ML algorithms build a mathematical model using sample data, also known as training data, aiming to make decisions that they are not explicitly programmed to make [16].

Artificial neural networks (ANN), usually only called neural networks (NN), are computing systems that contain a group of artificial neurons used to process data and information. The architecture of the ANNs, and the idea behind building them, is based on the biological neural networks found in human brains. The artificial neurons (nodes) are a collection of connected units loosely modeled on the human brain's neurons. The idea is that these neurons should replicate how the human brain works. At its core, the neuron is a mathematical function, which takes an input, does a calculation and transformation on it, and gives an output.

Deep learning is essentially a subfield of machine learning, and it represents a particular case of an artificial neural network having more than one hidden layer. As previously mentioned, these types of neural networks aim to simulate the human brain and learn from large amounts of data [17]. The idea behind adding additional hidden layers is to increase accuracy and optimize the model. The difference between deep learning and machine learning is in the type of data they use and the methods they use to learn. Machine learning usually uses structured, labeled data to make predictions. Even if the data are not structured, they usually go through the data preparation phase to be organized in a way that the learning model can use. On the other hand, deep learning can use data that are not structured, such as images and text, which means that these algorithms can shorten the processing phase or even remove it altogether [18].

In recent years, machine learning methods have been extensively used to build efficient network intrusion detection systems [1]. The use of machine learning methods has significantly impacted and improved the detection accuracy of these intrusion detection systems. However, there are still some downsides and limitations to using shallow machine learning methods. In particular, they still require a high level of human interaction and a significant amount of expert knowledge to process data and identify patterns [19], making them expensive, time-consuming, and unsuitable for high-dimensional learning with a large amount of data. Another negative side of using shallow machine learning techniques is that their learning efficiency decreases as the network complexity increases. When there are many multi-type variables, logistic regression can underfit with low accuracy, decision trees can overfit, and Support Vector Machines are inefficient, mainly when dealing with large amounts of data [20].

To address these limitations, researchers have identified Deep Learning as a valid alternative to shallow learning techniques in the above mentioned situations. Advantages of DL over ML are, for example, automatic feature learning and flexible adaptation to novel problems, making it easier to work with big data [20].

3. Related Work

Deep Learning for NIDS is an emerging topic that has generated a new research branch. There have been many novel approaches proposed by authors, such as in [21], where the authors have proposed a modified bio-inspired algorithm, which is the Grey Wolf Optimization algorithm (GWO), that enhances the efficacy of the IDS in detecting both normal and anomalous traffic in the network. Another example is [22], where the researchers analyzed the evolutionary sparse convolution network (ESCNN) intrusion and threat activities in the Internet of things (IoT) with the goal to improve the overall attack detection accuracy with a minimum false alarm rate. In this section, we report our analysis of the main proposals found in the literature. The discussion will include an analysis of the deep learning methods and the datasets used, the models produced, and the results obtained for each research paper. We separated the different research proposals according to the dataset they adopted to build their deep neural network models, namely the NSL-KDD [9] and NSW-NB15 [10] datasets. The selection process of the related literature was based on the following criteria:

1. Usage of the NSL-KDD and UNSW-NB15 datasets
2. Being relevant to Network Intrusion Detection Systems
3. Usage of deep learning algorithms

3.1. Datasets for Training Deep Learning Based NIDS

Training machine learning algorithms requires huge amounts of data, and the quality of these data is crucial. Since most problems are very dependent on the type and the quality of data, high quality datasets need to be used. Both NSL-KDD and UNSW-NB15 datasets have been used in many previous IDS researches, as described in the following Sections 3.2 and 3.3.

The original researchers produced the NSL-KDD dataset to try to solve the shortcomings and problems of the KDD Cup 99 dataset, once the most widely used dataset for the evaluation of anomaly detection methods, prepared by Stolfo et al. [23]. The KDD Cup 99 dataset's biggest problem was biased results due to redundant and duplicate records. The NSL-KDD dataset consists of selected records from the complete KDD Cup 99 dataset. This new dataset removes the identical records, resulting in around 78% of the training dataset records and around 75% of the test dataset records [24]. Moreover, the number of selected records from each difficulty level group is inversely proportional to the percentage of the records in the original KDD Cup 99 dataset [25]. The NSL-KDD dataset contains both regular traffic and traffic representing network attacks, so all the data in the dataset are labeled as either normal or attack.

The NSL-KDD dataset is divided into four datasets: KDDTest+, KDDTrain+, KDDTest-21, and KDDTrain+20%, where the latter are subsets of the former two, respectively. The KDDTest-21 is a subset of the KDDTest+ and excludes the most challenging records. Similarly, the KDDTrain+20% is a subset of the KDDTrain+ and contains 20% of the records in the entire training dataset [26].

The training dataset consists of 21 different attack types, while the testing dataset has 39 different types of attacks. The attack types in the training dataset are considered known attacks, while the testing dataset consists of the known attacks, plus the additional, novel attacks. The attacks are grouped into DoS, Probe attacks, U2R, and R2L. More than half of the records are regular traffic, while the distribution of the R2L and U2R attacks is low. On the other hand, a lower distribution corresponds to real-life internet traffic attacks, where these types of traffic are very rarely seen [26]. The dataset includes a total of 43 features. The first 41 are related to the traffic input and are categorized into three types: basic features, content-based features, and traffic-based features.

The distribution of the above mentioned attack types is skewed and the breakdown of the data distribution can be seen in Table 1. More than half of the records are normal traffic, while the distribution of the R2L and U2R attacks is low.

Table 1. NSL-KDD record distribution.

	Total	Normal	DoS	Probe	U2R	R2L
KDDTrain+	125,973	67,343 (53%)	45,927 (37%)	11,656 (9.11%)	52 (0.04%)	995 (0.85%)
KDDTest+	22,544	9711 (43%)	7458 (33%)	2421 (11%)	200 (0.9%)	2654 (12.1%)

The UNSW-NB15 dataset is a relatively new network dataset, released in 2015 and used in developing NIDS models [10]. The authors reported several main reasons for making this new dataset. They wrote, in fact, that available datasets were too old, did not reflect modern network traffic, and did not include some essential modern-day attacks. The original dataset consists of 2,540,044 records, which can be classified as regular traffic and network attacks. The authors have also made two smaller subsets, the training, and testing subsets, consisting of 175,341 and 82,332 records, respectively. The original dataset distinguishes a total of 49 features, and the authors arranged 35 in four categories: flow, basic, content, and time features. These 35 features hold the integrated gathered information about the network traffic. The following 12 features are additional and grouped into two groups based on their nature and purpose. The first group contains features 36–40, considered general-purpose features, while the remaining 41–47 are considered connection features [10]. Each of the general-purpose features has its purpose from the defense point of view, while the connection features give information in different connection scenarios. The remaining two features, 48 and 49, are the label features, and they represent the attack category and the traffic label, which shows whether the record is regular traffic or an attack, respectively.

Similarly to the NSL-KDD dataset, the UNSW-NB15 dataset is also very unbalanced. The breakdown of the data distribution can be seen in Table 2.

Table 2. UNSW-NB15 record distribution. Normal traffic accounts for 87% of the total, while some attacks are <0.00005% (i.e., Shellcode and Worms).

Normal	Fuzzers	Analysis	Backdoors	DoS
2,218,761	24,262	2677	2329	16,353
Exploits	Generic	Reconnaissance	Shellcode	Worms
44,525	215,481	13,987	1511	174
		Total		
		2,540,044		

3.2. Related Research Using the NSL-KDD Dataset

This section surveys the research papers which used the NSL-KDD dataset for training and testing of the model.

Jia et al. [27] considered the two datasets, KDD Cup 99 and NSL-KDD, and proposed a network intrusion detection system based on a deep neural network with four hidden layers. Each hidden layer has 100 neurons and uses the ReLU activation function. The output layer is fully connected and uses the softmax activation function. The authors have built a multiclass classifier with the final aim to increase the model's accuracy. In the end, they have obtained an accuracy of >98% on all the classes except the U2R and R2L attacks. The authors claimed that the main reason is the severely unbalanced nature of the datasets, since there are too few records for these classes. We can observe two main downsides of this research: it uses a very old dataset (KDD Cup 99), and the two used datasets are very similar. This last point could mean that, even though this model performs well on these datasets, it might not perform as well when detecting in a real network environment.

Vinayakumar et al. [28] proposed an intrusion detection system based on a hybrid scalable deep neural network. They tested their model using six different datasets: KDD Cup 99, NSL-KDD, Kyoto, UNSW-NB15, WSN-DS, and CICIDS 2017. The proposed model consists of an artificial neural network with five hidden layers using the ReLU activation function. Each hidden layer has a different number of neurons ranging from 1024 in the first hidden layer to 128 in the last. The authors evaluated both binary and multiclass classification, obtaining broadly varied results. Depending on the dataset used, the proposed models obtained the best accuracy for the KDD Cup 99 and the WSN-DS, and the worst for the NSL-KDD and the UNSW-NB15. The authors' main goal was to develop a flexible model that can detect and classify different kinds of attacks, which is why they used multiple datasets. The downsides of the proposed approach are that the obtained model is very complex and has a lower detection rate for some of the classes.

Another research on this topic was done by Yin et al. [29]. Their study proposed a network intrusion detection system based on a Recurrent Neural Network (RNN) model. The dataset used in this research is the NSL-KDD dataset, and the authors have trained an RNN model to do both binary and multiclass classification. The idea behind the study was to build a model that will achieve higher performance in attack classification than the models using the more traditional machine learning algorithms, such as Random Forest, Support Vector Machine, etc. After the data preparation phase, the dataset used to train the model consisted of 122 features, while the final model consisted of 80 hidden nodes. The accuracy results obtained when testing the model were: 83.28% for binary classification and 81.29% for multiclass classification. The authors state that these results are better than those of other machine learning algorithms. Some downsides of this approach are that the detection rates for the R2L and U2R classes are still low, and the model's performance is lower than other deep learning IDS models.

Potluri et al. [30] propose a DNN architecture of a feed-forward network where each hidden layer is an auto-encoder, trained with the NSL-KDD dataset. Using auto-encoders as hidden layers allows the training process to be done one layer at a time. The network has three hidden layers: the first two are auto-encoders, with 20 and 10 neurons, respectively; the third layer uses the softmax activation function and has five neurons. The first two layers are used in the pre-training process: they perform a feature extraction phase and reduce the number of features used by the DNN first to 20 and in the end to 10. The third hidden layer selects five features out of 10 as a fine-tuning phase. The experiments considered binary and multiclass classification: the detection accuracy for the binary classification is high (>96%). In contrast, the detection accuracy for multiclass classification varied considerably: it was satisfactory (>89%) for DoS, Probe, and regular traffic and low for U2R and R2L. Similar to other research papers mentioned, the low detection accuracy for some classes is a downside of this model.

Kasongo et al. [31] also proposed a network intrusion detection system based on a feed-forward deep neural network using the NSL-KDD dataset. The goal of the research was to build a model that would perform better, meaning it would have a higher detection accuracy than the existing machine learning models used for intrusion detection. The authors divided the original training dataset into two subsets: one for training and one for the evaluation after the training process. The initial test dataset was used to test the performance of the model. The experiment included binary and multiclass classification in two scenarios: the first used all 41 features of the dataset, and the second used a reduced number of features (21 features) extracted during the feature selection phase. The model with all the features showed a detection accuracy of 86.76% for binary and 86.62% for multiclass classification. On the other hand, when using the reduced number of features, the detection accuracy was 87.74% for binary and 86.19% for multiclass classification. Among the downsides of this model were lower detection rates for R2L and U2R classes and lower accuracy compared to other deep learning models used for intrusion detection.

The research paper by Shone et al. [19] also focuses on building a network intrusion detection system based on a deep learning model using the KDD Cup 99 and the NSL-KDD datasets. The proposed model is constructed by stacking non-symmetric deep auto-encoders and combining them with the Random Forest classification algorithm. One of the research purposes is to develop a technique for unsupervised feature learning, and the authors have done this by using another non-symmetric deep auto-encoder. The authors proposed two classifications: a 5-class classification for both datasets and a 13-class classification for the NSL-KDD dataset. The average detection accuracy for the 5-class classification was 97.85% with the KDD Cup 99 dataset, and 85.42% for the NSL-KDD dataset, while achieving 89.22% for the 13-class classification with the NSL-KDD dataset. The downside of this model is that it has low detection accuracy for classes with a lower number of records.

The research paper by Fu et al. [32] proposes a deep learning model for network intrusion detection with the goal to address the issue of low detection accuracy in imbalanced datasets. The authors have used the NSL-KDD dataset for the training and testing of the model. The model combines an attention mechanism and the bidirectional long short-term memory (Bi-LSTM) network, by first extracting sequence features of data traffic through a convolutional neural network (CNN) network, then reassigning the weights of each channel through the attention mechanism, and finally using Bi-LSTM to learn the network of sequence features. This paper employs the method of adaptive synthetic sampling (ADASYN) for sample expansion of minority class samples, in order to address data imbalance issues. The experiments included both binary and multiclass classification and the accuracy and F1 score of the proposed network model reached 90.73% and 89.65% on the KDDTest+ test set, respectively.

3.3. Related Research Using the UNSW-NB15 Dataset

This section discusses the research papers which used the UNSW-NB15 dataset for training and testing of the model.

In the research by Kanimozhi et al. [33], the authors proposed a network intrusion detection system based on an artificial neural network, trained and tested on the UNSW-NB15 dataset. The authors used deep learning in combination with other machine learning algorithms to extract the most relevant features of the dataset and use them for training the model. The goal was to increase the detection accuracy and decrease the False Alarm Rate. During the feature extraction phase, the authors used a combination of the Random Forest and the Decision Tree algorithms for feature extraction. In the end, they selected four features out of 45 in the original dataset. The authors have decided to do only binary classification, meaning that the model will only classify a record as an attack or regular traffic. The accuracy obtained in the testing phase was 89%, which is still lower than the accuracy of other proposals with deep learning approaches.

Mahalakshmi et al. [34] have implemented an intrusion detection system based on a convolutional neural network (CNN). The goal was to make a model that would overtake the existing machine learning models used for intrusion detection concerning detection accuracy. The proposed algorithm is a CNN used for binary classification, with an accuracy of 93.5%.

The research done by Al-Zewairi et al. [35] uses the whole dataset, with all 2,540,044 records, instead of the separate training and testing datasets prepared by the authors of the UNSW-NB15 dataset. The proposed model is a deep artificial neural network consisting of five hidden layers and a total of 50 neurons. The neural network is feed-forward and uses backpropagation and stochastic gradient descent. The research aimed to find the optimal network hyperparameters to achieve the best performance for binary classification. The authors conducted experiments to find the best activation function for their model and the optimal features to be used for training. The activation function that proved optimal for this research was the rectifier function without the dropout method. The second experiment regarding the optimal features showed that using the top 20% features, which were selected during feature extraction, gave the best results. After testing the proposed model, the evaluation showed high accuracy (98.99%) and a low false alarm rate (0.56%).

We can note that few researchers, from the ones mentioned in this section, included the FPR and FNR as an evaluation metric in their research. However, most of them focused on calculating the accuracy. The main problem with this approach is that the datasets used are significantly unbalanced. Therefore the accuracy is not a good metric because it does not distinguish between the records of different classes that were correctly classified. With this concern in mind, in this paper we propose to focus on lowering the FNR and increasing the predictability for the minority classes.

3.4. Summary and Comparison of the Related Research

A summary and comparison of all of the surveyed research papers are in Table 3. We can observe that only half of the authors included the FPR and FNR as an evaluation metric in their research since most of them focused on improving the accuracy. Moreover, only two of the authors that considered the False Rates also proposed a multiclass classification.

The main problem of focusing on the accuracy metric is that the datasets used are significantly unbalanced. Therefore the accuracy is not a good metric because it does not distinguish between the records of different classes that were correctly classified. Thus, in the next we focus on a strategy to improve the FNR and FPR, while improving the detection of the less represented attack classes. In order to provide a better overview and the possibility to compare the related work with the results which were achieved by the model proposed in this research, we have included a brief summary of the proposed model as the last row in Table 3.

Table 3. Summary and comparison of related works.

Researchers	Year	Dataset(s)	Algorithm(s)	Classification Type	Accuracy	FPR and FNR
Jia et al. [27]	2019	KDD Cup 99 and NSL-KDD	Deep neural network	Multiclass	>98% on all classes except U2R and R2L	FNR = 0.5%, FPR = 0.3%
Vinayakumar et al. [28]	2019	KDDp Cup 99, NSL-KDD, Kyoto, UNSW-NB15, WSN-DS and CICIDS 2017	Deep neural network	Binary and multiclass	Big variations between datasets	Big variations between datasets
Yin et al. [29]	2017	NSL-KDD	Recurrent neural network	Binary and multiclass	83.28% for binary and 81.29% for multiclass	N/A
Potluri et al. [30]	2016	NSL-KDD	Deep neural network with auto-encoders as hidden layers	Binary and multiclass	>96% for binary; >89% for multiclass	N/A
Kasongo et al. [31]	2019	NSL-KDD	Deep neural network	Binary and multiclass	All features: 86.76% (binary), 86.62% (multiclass); 21 features: 87.74% (binary) and 86.19% (multiclass)	N/A

Table 3. *Cont.*

Researchers	Year	Dataset(s)	Algorithm(s)	Classification Type	Accuracy	FPR and FNR
Kanimozhi et al. [33]	2019	UNSW-NB15	Deep neural network	Binary	89%	FNR = 15%
Mahalakshmi et al. [34]	2021	UNSW-NB15	Convolutional neural network	Binary	93.5%	N/A
Shone et al. [19]	2018	KDD Cup 99 and NSL-KDD	Stacked non-symmetric deep auto-encoder network with Random Forest classification algorithm	Multiclass (5 and 13 classes)	97.85% (5-class KDD Cup 99); 85.42% (5-class NSL-KDD) and 89.22% (13-class NSL-KDD)	Only FPR considered, big variations between experiments (from 2.15% to 14.58%)
Al-Zewairi et al. [35]	2017	UNSW-NB15	Deep neural network	Binary	98.99%	FPR = 0.56%
Fu et al. [32]	2022	NSL-KDD	Deep neural network	Binary and multiclass	90.73%	Lowest FPR for U2R class (1.73%), highest for Normal class (13.44%)
Mijalkovic J., Spognardi A. (proposed model)	2022	NSL-KDD and UNSW-NB15	Deep neural network	Binary and multiclass	>99% for NSL-KDD and >97% for UNSW-NB15	Lowest FNR = 0.049%; lowest FPR = 0.33%

4. Materials and Methods

In this section, we present the strategy we propose to achieve our research goals, while in the next Section 5, we report the experimental campaign that confirms our approach.

Our strategy to reduce FNR and FPR and increase the detection of low-represented attack categories consists of three points, as depicted in Figure 1. The first point, *distribution alteration*, refers to the idea of altering the distribution of the original datasets. The rationale is that the split proposed by the original dataset's authors is sub-optimal, limiting the final accuracy of the trained model. Our idea is that by reshuffling the datasets, it is possible to improve the detection rate of most of the attack categories.

The second point, *feature reduction*, is the canonical approach of reducing the number of features [36], selecting the more suitable for the primary goal.

The final point, *class weight*, refers to the idea of altering the importance of the different categories of the data samples used in the network. The rationale is that we can reduce the number of false negatives and improve the detection of the less common attacks at the price of a low increase in the number of false positives.

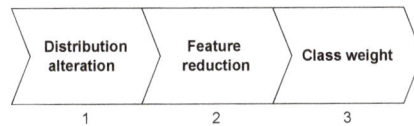

Figure 1. An overview of the proposed strategy.

In the Experiment section (Section 5), we reported and evaluated all the intermediate results to show the improvement introduced by each of the points of our strategy.

4.1. Strategy Implementation

Figure 2 shows the details of the phases we took to construct and evaluate the generated models. In the following, we give an overall description of each phase, and the details of the data preparation and model architecture in Sections 4.3 and 4.2, respectively.

The first step was to collect the data. As mentioned in Section 3.1, we selected NSL-KDD and UNSW-NB15 to have two different datasets considered among the most suitable for our research. Since both datasets are divided into smaller datasets, the following were chosen for our research: KDDTrain+, KDDTest+, and the full UNSW-NB15 dataset, which we split into 4 CSV files.

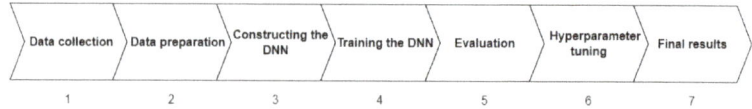

Figure 2. An overview of the steps taken to build and evaluate the Deep Learning model.

The next step is to prepare the datasets so they can be ready to be used for training our model. In this phase, we processed the datasets by removing missing and redundant values, normalizing the numerical data, and encoding the categorical data into numerical. Section 4.2 gives a detailed explanation of this step.

The third phase is constructing the deep neural network used in the research and setting all of its parameters. A detailed explanation of the architecture and the parameters chosen for the model is given in Section 4.3.

The fourth step is essential in deep learning and consists of training the neural network since the dataset is used to train the model and enhance its ability to make predictions.

After the training of the model, the fifth step is to evaluate the model to see how it performs. The testing datasets are used in this step, in order to see how well the model will perform on the data that it has never seen before.

After the evaluation process, the next step is to tune the hyperparameters to see if it would be possible to improve the learning process and achieve better results. The hyperparameters are the parameters used to control the learning process, as opposed to other parameters, such as node weights, whose values come from training. Some of the parameters modified in this phase to obtain better results are, for example, the number of epochs and the learning rate. We detail all the values and the obtained results of this step in Section 4.3.

The final step is the prediction step, in which we achieve the final results of the model. In this step, we conclude how well the model performed and if it reached the experiment's goal. The predictions of each experiment and the evaluation of their results are in Sections 5 and 6, respectively.

4.2. Data Preparation

Preparing the data is a crucial step and can significantly impact the model's learning process. If we do not give appropriate input to the model, it might not give us the result that we want to obtain. As mentioned earlier, we have two datasets used in this research, the NSL-KDD and the UNSW-NB15 datasets. Both of these datasets need to be processed, and since they have a similar structure, we used the same preparation process.

4.2.1. Preparation of the NSL-KDD Dataset

As a starting point in the data preparation of the NSL-KDD dataset, we have two subsets of data already divided by the authors, the KDDTrain+ and the KDDTest+. These subsets have 43 features, while the KDDTrain+ subset has 125,793 records and the KDDTest+ subset 22,544. We processed and verified that both subsets do not contain any missing values. Therefore, we could proceed with doing the rest of the data preparation on the subsets as they are.

The goal of multiclass classification is to correctly classify records that represent a network attack as the attack category they belong to. Therefore, it is necessary to change the label for every record from the attack type to the class to which that attack type belongs. This step is repeated for both subsets. For the model to learn from this data, we need to transform it into numerical values. For this transformation, we employed one-hot encoding. One-hot encoding is a technique used for categorical features where no ordinal relationship exists. Therefore it is not enough to just do integer encoding (assign each category an integer). One-hot encoding creates new binary columns for each possible unique categorical feature value. In other words, it converts the categorical data into a binary vector representation. We applied one-hot encoding to training and test subsets specifically for the following features: protocol_type, service, and flag. Ultimately, we removed the original categorical columns and obtained a dataset with 124 columns.

The next step was the encoding of the label. For binary classification, the 'normal' value was represented by a 0, while all the others 'abnormal' were given the value 1. For multiclass classification, again, the 'normal' value was given the value 0, and the rest of the values were integer encoded. The multiclass values range from 0 to 4. This was done for both subsets.

The next step was to strip the label and attack category columns from the train and test datasets, building the effective subsets used to generate the model. The combination of the original subset with the label column is used for the binary classification, while the combination with the attack category column is used for the multiclass classification. Thus, we divided the training and the testing subsets into 6 subsets: $train_f, train_\ell, train_c, test_f, test_\ell$, and $test_c$. The subsets $train_f$ and $test_f$ contain all the columns with the features of the original training and testing datasets except for label and attack category columns: they will be given to the model as the input. The label column for training and testing for binary classification went in $train_\ell$ and $test_\ell$, respectively, while the attack category column went in $train_c$ and $test_c$.

The last preparation step was to normalize the data in the $train_f$ and $test_f$ subsets using the min-max method. For every feature, the minimum value is changed to 0, the maximum value is changed to 1, and every other value is transformed into a decimal value between 0 and 1 using the following formula $\frac{value-min}{max-min}$. The final subsets used, $train_f$ and $test_f$, now contain 123 columns each, and all the data is encoded into numerical values and normalized.

4.2.2. Preparation of the UNSW-NB15 Dataset

Unlike the NSL-KDD dataset, we opted to use the original full UNSW-NB15 dataset, which contains 2,540,044 records, instead of using the two subsets pre-divided by the authors. The authors have provided four separate CSV files which contain the records of this dataset. The first step was to load all four CSV files and merge them into one dataset.

The next step was to check if there were any duplicate records and remove them. The removal of the duplicates is essential to avoid having the same records in the training and testing subsets because the testing subset should contain only the records that were not previously seen by the neural network. During this phase, we removed 480,625 duplicate records.

The next step was to check if the dataset contains any missing values. Three features contained missing values: 'ct_flw_http_mthd', 'is_ftp_login' and 'ct_ftp_cmd'. The missing values were then replaced with '0'. It has been noted that the dataset contains the value '-' for the feature 'service' in a significant number of records, so this value was renamed as 'undefined' to give more meaning to it. Then, we removed the columns 'srcip' and 'dstip'. We also fixed some white-space inconsistencies among records with the same values and other minor typos (i.e., 'Backdoors' instead of 'Backdoor' in the 'attack_cat' field).

We repeated the one-hot encoding for the whole dataset, changing the categorical features 'proto', 'service', and 'state'. At the end of this process, the dataset contained 202 columns.

While the column 'label' used for binary classification already contained 0 for regular traffic and 1 for abnormal, the 'attack category' required an encoding for the multiclass classification. Thus, in the next step, we encoded with a 0, the 'normal' (no-attack) value, and assigned values from 1 to 9 to the other attack categories.

The next step was to split the dataset into training and testing subsets. The training subset was a random sample with 80% of the original records, while the testing subset contained a random sample with 20%.

As for the NSL-KDD dataset, we separated the feature data columns ($train_f$ and $test_f$) from the label ($train_\ell$ and $test_\ell$) and attack category ($train_c$ and $test_c$) columns.

As for the NSL-KDD dataset, the final step was the normalization of the numerical variables of the $train_f$ and $testf$ subsets of the features with the min-max normalization method. In the end, these subsets contain 200 columns.

4.3. Model Architecture

After the data preparation phase, we started training the deep neural network. We adopted the same model architecture for both datasets to evaluate which would perform better. Different activation functions are used for different layers of the neural network. We differentiated the model for the binary classification and the one for multiclass classification, changing the number of nodes in the output layer and the activation function for the output layer. The hyperparameters related to the training algorithm are:

- Batch size. This is a training parameter that indicates the number of records passed and processed by the algorithm before updating the model.
- Number of epochs. This is also a training parameter which indicates the number of passes done through the complete training dataset.
- Optimizer. Optimizer is an algorithm, or a method, which is used to change the attributes of the network such as weights and learning rate in order to reduce the loss. The most used optimizers, among the others, are gradient descent, stochastic gradient descent, adagrad, and adaptive moment estimation (Adam) [37]. The optimizer used for the model is stochastic gradient descent (SGD) with Nesterov momentum.
- Momentum. This parameter is used to help predict the direction of the next step, based on the previous steps. It is used to prevent oscillations. The usual choice is a number between 0.5 and 0.9.
- Learning rate. The learning rate is a parameter which controls the speed at which the neural network learns. It is usually a small positive value in range between 0.0 and 1.0. This parameter controls how much we should change the model in order to respond to the estimated error each time the weights of the model are updated [38].
- Loss function. The loss function in a neural network is used to calculate the difference between the expected output and the output that was generated by the model. This function allows acquiring the gradients that the algorithm will use to update the neural network's weights. The loss function used for this model for binary classification is the binary cross-entropy loss function. On the other hand, we used a sparse categorical cross-entropy loss function for multiclass classification.

At the end of our experiments, the final values chosen for the training are provided in Table 4. These final values were reached after a process of manual hyperparameter tuning which included a series of trials with different values. The number of epochs shown in Table 4 indicates the maximum number of epochs, but Early Stopping is used in the experiments in order to prevent overfitting.

The neural network used for the experiment is a feed-forward neural network, which means that the connections between the nodes do not form any cycles and the data in the network moves only forward from the input nodes, going through the hidden nodes, and in the end reaching the output nodes. The algorithm used to train the network is the backpropagation algorithm. As mentioned earlier, backpropagation is short for "backward propagation of errors". Given an error function and an artificial neural network, the backpropagation algorithm calculates the gradient of the error function with respect to the weights of the neural network [39].

Table 4. Final values chosen for the training phase.

Hyperparameter	Value
Batch Size	64
Epochs	100
Optimizer	Stochastic Gradient Descent (SGD) with Nesterov momentum
Momentum	0.9
Learning rate	0.01
Regularization	1×10^{-6}

Moreover, the number of layers in the network is six: one input layer, one output layer and four hidden layers. The input layer takes the input dimension which is equal to the number of features used in the training dataset. The first hidden layer uses the Parametric Rectified Linear Unit (PReLU) activation function and it has 496 neurons. The PReLU activation function generalizes the traditional rectified unit with a slope for negative values and it is formally defined as [40]:

$$f(y_i) = \begin{cases} y_i & \text{if } y_i > 0 \\ a_i y_i & \text{if } y_i \leq 0 \end{cases} \tag{1}$$

The other hidden layers use the Rectified Linear Unit (ReLU) activation function. This function was designed to overcome the vanishing gradient problem and it works in the way that it returns 0 for any negative input, but for a positive input, it returns the value of the input back. It can be defined as:

$$f(x) = max(0, x) \tag{2}$$

The second, third and fourth hidden layers have 248, 124 and 62 nodes, respectively. The output layer has a different activation function and a different number of neurons based on the type of classification which is being done. For binary classification, the output layer uses the sigmoid activation function and has only one neuron. The sigmoid function takes a value as the input, and outputs another value between 0 and 1. It can be defined as:

$$f(x) = \frac{1}{1 + e^{-x}} \tag{3}$$

On the other hand, for the multiclass classification, the output layer has the number of neurons which is equal to the number of the attack categories in the dataset, and the activation function which is used is the softmax function. This function converts a vector of K real values into a vector of K real values that sum to 1 [41]. It can be defined as:

$$f_i(\vec{x}) = \frac{e^{x_i}}{\sum_{j=1}^{J} e^{x_j}} \text{ for } i = 1, ..., J \tag{4}$$

Additionally, to prevent overfitting during the training phase, we implemented the dropout on all the hidden layers. Dropout is a regularization method that causes some of the neurons of a layer to be randomly dropped out (ignored) during the training of the network. Dropping out the neurons means that they will not be considered during the specific forward or backward passing through the neural network. The dropout rate chosen for this network, for each hidden layer, was equal to 0.1. This means that 10% of the units will be dropped (set to 0) at each step. The units that are not dropped are scaled up by $\frac{1}{(1-rate)}$ so that the sum of all the units remains unchanged. A graphical representation of the architecture of the neural network can be seen in Figure 3.

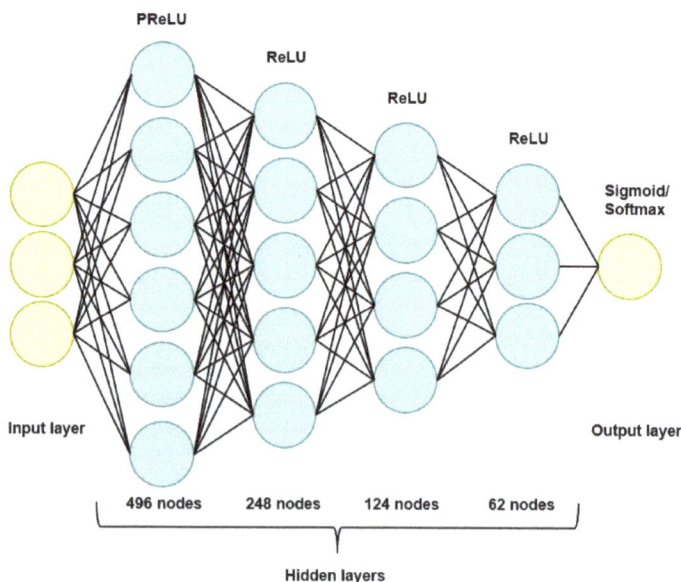

Figure 3. A graphical representation of the DNN architecture.

4.4. Development Tools

The data preparation, model implementation, training, testing, and evaluation were all done in *Python* using the following libraries:

- NumPy.This is a *Python* library which provides support for working with large multi-dimensional arrays. It allows the user to perform different mathematical operations on such arrays and it guarantees efficient calculations [42].
- Pandas. Pandas is a *Python* library used for data analysis and manipulation. It provides support for manipulating numerical tables and time series [43].
- Matplotlib. This is a *Python* library that provides support for data visualization. It is used to create static, animated and interactive graphs and other visualizations [44].
- Scikit-learn. This is a machine learning *Python* library used for predictive analysis. It is built on *NumPy, SciPy* and *Matplotlib* and it can provide features for classification, regression, model selection, clustering, preprocessing and so on. Another name for it is sklearn [45].
- Tensorflow. This is a *Python* library for machine learning. It provides features for building and training machine learning models and it allows users to create large scale neural networks with many layers [46].
- Keras. This is a *Python* library which provides an interface for artificial neural networks. It is built on top of *Tensorflow* and it acts as a frontend for it [47].
- Jupyter notebook. This is an interactive computational environment which allows the user to edit the code live, create equations, visualizations, and much more. It is practical for research because it allows the researcher to combine code, output, explanations, and multimedia resources in one document [48].
- PyCaret. This is an open-source *Python* library used for automation of the machine learning processes. It gives the user many options which include automatic data preparation, automatic model construction, training of the models, and evaluation and comparison of the models [49]. For this experiment, *PyCaret* was used to automate the data preparation and feature selection process.

All the experiments were conducted on a HP Pavilion Power laptop with the Intel(R) Core(TM) i7-7700HQ CPU @ 2.80 GHz processor. The rest of the hardware specifications of the laptop used for the experiment can be seen in Table 5.

Table 5. Hardware specifications of the computer used for training.

Hardware	Specification
GPU	NVIDIA GeForce GTX 1050
Memory	16 GB system memory
Storage	256 GB SSD
GPU Memory	4 GB GPU memory

5. Experiments

Our experimental campaign aimed to achieve the lowest False Negative Rate (FNR) while keeping the False Positive Rate (FPR) low. When it comes to multiclass classifications, an additional goal was to improve the accuracy of some of the classes which have a smaller number of records. The purpose of the experiments has been to find which architecture and hyperparameters give us the lowest FNR. Additionally, other performance metrics mentioned in Section 1 were compared for each experiment. A total of 14 experiments were conducted: 4 for binary classification on the NSL-KDD dataset, 4 for multiclass classification on the NSL-KDD dataset, three for binary classification on the UNSW-NB15 dataset, and three experiments for multiclass classification on the UNSW-NB15 dataset.

5.1. Experiments on the NSL-KDD Dataset

Since both binary and multiclass classification were done on this dataset, the first part of the experiments which will be explained were conducted for binary classification, and the second part for the multiclass classification.

5.1.1. NSL-KDD Binary—Full Features

The first binary classification experiment considered the training of the model with all the features extracted during the data preparation phase (Section 4.2). Since the NSL-KDD subsets used for training and testing ($train_f$ and $test_f$) had a total of 123 columns each, the neural network's input layer has 123 nodes.

The first step is to train the neural network on this version of the training subset and assess the results achieved. We used the *Keras* library to build the model and fine-tune the hyperparameters, as mentioned in Section 4.3. We used Early Stopping (ES) to prevent overfitting the network. One problem which can lead to overfitting is using too many epochs to train the network. Hence, ES allows the user to set many training epochs, but it stops the process once the model performance reaches the best possible result and before it drops. The confusion matrix for this model can be seen in Table 6. The confusion matrix makes it easier to see which classes are easily confused by the model, and from this matrix it can be seen that the number of False Negatives (FN) is 3296, which is very high. This means that the model wrongly classified 3296 attack records as normal traffic. The number of false positives is equal to 662, which means that the model wrongly classified 662 records as attacks.

Table 6. Confusion matrix for NSL-KDD "full features" binary classification experiment.

		Predicted	
		Normal	**Attack**
Actual	Normal	9049	662
	Attack	3296	9537

5.1.2. NSL-KDD Binary—Modified Distribution

The second binary classification experiment considered using the same 123 features of the data preparation stage but slightly changing the training and testing subsets distribution. The idea behind this was that, maybe, the neural network could not learn from the training subset prepared by the authors. This experiment aims to see if a different distribution of records in the training and the testing subsets will give better results.

To obtain the new subsets, we combined the two subsets together into one dataset, shuffling the data and then splitting them again so that 80% of the records are used for the training of the network, and 20% of the records are used for testing. The training subset contained 118,813 records and the testing subset 29,704 records. The architecture of the neural network was the same as for the first experiment, and again, Early Stopping was used. The confusion matrix for this experiment can be seen in Table 7. It can be seen that the number of false negatives in this experiment is equal to 32, which is significantly lower than in the "NSL-KDD binary—full features" experiment. The number of false positives is is 52, which is also lower when compared to the previous experiment.

Table 7. Confusion matrix for NSL-KDD "modified distribution" binary classification experiment.

		Predicted	
		Normal	**Attack**
Actual	Normal	15,371	52
	Attack	32	14,249

5.1.3. NSL-KDD Binary—Reduced Features

For the third binary classification experiment, we used feature selection to reduce the False Positive and the False Negative Rates. The testing and training subsets used for this experiment were the same ones which were used for the second experiment. The feature selection process was automatized by using the *Python* library *PyCaret*. This library makes feature selection on a dataset by combining several supervised feature selection methods to select a subset of features that contribute the most to the prediction of the target variable [50].

After the feature selection process, the total number of features selected as the most important was 41 out of 123. This means that the model for this experiment had 41 neurons in the input layer. The rest of the architecture remained unchanged, including the use of the Early Stopping method. The confusion matrix for this experiment can be seen in Table 8. The number of false negatives in this experiment is 147, which is slightly higher than in the previous experiment, but still significantly lower than in the first NSL-KDD binary experiment. The number of false positives is 215, which is higher than in the previous experiment, but again, lower than in the first experiment.

Table 8. Confusion matrix for NSL-KDD "reduced features" binary classification experiment.

		Predicted	
		Normal	**Attack**
Actual	Normal	15,185	215
	Attack	147	14,157

5.1.4. NSL-KDD Binary—Class Weights

The fourth binary classification experiment included the use of class weights. When dealing with an imbalanced dataset, assigning weights to different classes can help the model make more accurate predictions. For our research, we consider the false negatives more dangerous than false positives. Hence, we needed a way to make the model penalize the false negatives by assigning different class weights. We assigned a weight of 1 for the normal class (which has the label 0) and 2 for the attack class (which has the label 1). Aside from assigning weights to the classes, this experiment uses the same hyperparameters as the first and the second. The input dimension is equal to the second experiment since we considered the same 41 features of the feature selection phase. We used early Stopping in this experiment as well. The confusion matrix for this experiment can be seen in Table 9. From the confusion matrix it can be seen that the number of false negatives is 115, which is slightly lower than in the previous experiment, but still higher than in the "NSL-KDD binary—modified distribution" experiment. The number of false positives is higher than in the previous experiment.

Table 9. Confusion matrix for NSL-KDD "class weights" binary classification experiment.

		Predicted	
		Normal	Attack
Actual	Normal	15,024	376
	Attack	115	14,189

5.1.5. NSL-KDD Multiclass—Full Features

The first multiclass classification experiment included the usage as an input in the neural network of all 123 features produced in the data preparation phase. The initial training and testing subsets provided by the authors were used. After the division of the subsets into input and output subsets during the data preparation phase, $train_f$ and $test_f$ contain the 123 features and will be used as inputs in the training and testing of the network. As described in Section 4.2, $train_c$ and $test_c$ subsets, which contain the attack category, will be used as the output in the training and the testing phase. Hence, the neural network's input layer has 123 nodes, like the "NSL-KDD binary—full features" experiment. The output layer has five nodes, one for each of the four attack categories and an additional one for the records which represent regular traffic. As mentioned earlier, the loss function used for the multiclass classification is the sparse categorical cross-entropy function. We opted for this function since it is recommended when the output is made of integers. The other hyperparameters are the same as explained in Section 4.3. As for binary classification, we used Early Stopping to prevent the model from over-fitting. Table 10 shows the confusion matrix for this example. To calculate the False Negative Rate, the class 'Normal' will be considered as the negative class, and the others as the positive classes. By taking a look at the confusion matrix, it can be concluded that the last column of the matrix shows the classes which were predicted as the 'Normal' (negative) class, so in the intersection of the last column and the last row, we have the number of True Negatives (TN). The TN in this case are the records which actually belong to the 'Normal' class and were correctly classified as the 'Normal' class. The other elements of the last column are false negatives (FN), meaning that they are records which actually belong to other classes and were wrongly classified as the 'Normal' class. Furthermore, the other elements in the last row are false positives (FP) since they actually belong to the 'Normal' class but were wrongly classified as attacks. All the other elements can be considered as true positives (TP) in this case. Taking this into account, the False Negative Rate can be calculated using these values and it is equal to 31.17%, which is a very unsatisfactory value.

Table 10. Confusion matrix for NSL-KDD "full features" multiclass classification experiment.

		Predicted				
		DoS	**Probe**	**R2L**	**U2R**	**Normal**
Actual	DoS	5836	223	71	0	1330
	Probe	266	1835	1	0	319
	R2L	105	289	148	2	2341
	U2R	2	25	22	8	10
	Normal	387	243	3	0	9078

5.1.6. NSL-KDD Multiclass—Modified Distribution

The second multiclass classification experiment was conducted using the same logic as for the second binary classification experiment. Again, all 123 features were used as the input in the neural network, therefore the input layer has 123 nodes. The two original subsets provided by the authors are mixed into one, shuffled, and split, to obtain a different distribution of the testing and training subsets. After the split, the training subset contains 80% of the records, while the testing subset contains 20%. After the split, the training subset contains 118.813 records and the testing subset contains 29.704 records. Again, the Early Stopping method was used. The confusion matrix for this experiment can be seen in Table 11. Using the same logic as in the previous experiment, for calculating the False Negative Rate, the "Normal" class will be considered as the negative class. The FNR in this experiment is equal to 0.17%. By looking at the confusion matrices in Tables 10 and 11, it can be seen that the classes R2L and U2R have less records in the testing subset than in the previous experiment. By having more records in the training subset, and less in the testing one, the network learned to better classify records belonging to these classes. In fact, the testing subset used for this experiment contains 716 records belonging to the R2L class, and 20 belonging to the U2R class, and in case of the "full features" experiment, the testing subset contained 2885 R2L and 64 U2R records.

Table 11. Confusion matrix for NSL-KDD "modified distribution" multiclass classification experiment.

		Predicted				
		DoS	**Probe**	**R2L**	**U2R**	**Normal**
Actual	DoS	10,666	2	1	0	7
	Probe	4	2806	1	0	9
	R2L	0	2	706	5	6
	U2R	0	1	5	12	2
	Normal	3	4	25	0	15,437

5.1.7. NSL-KDD Multiclass—Reduced Features

This experiment considered the use of the same subsets generated in the previous experiment, preforming feature selection with *PyCaret* and then training the network by using only the selected features. Out of 123 features, only 35 features were selected for this model. The next step was to train the neural network by using these 35 selected features, which means that the input layer of the neural network in this case had 35 nodes, one for every feature used for training. The confusion matrix for this experiment is shown in Table 12. The FNR for this experiment was equal to 0.133%. By looking at Tables 11 and 12 it can be concluded that the "NSL-KDD multiclass—modified distribution" model achieves better performance for the minority classes. On the other hand, the "NSL-KDD multiclass—reduced features" model achieves a lower FNR.

Table 12. Confusion matrix for NSL-KDD "reduced features" multiclass classification experiment.

		Predicted				
		DoS	**Probe**	**R2L**	**U2R**	**Normal**
Actual	DoS	10,666	3	1	0	6
	Probe	1	2803	8	1	7
	R2L	1	10	702	2	4
	U2R	0	0	9	9	2
	Normal	8	26	62	0	15,373

5.1.8. NSL-KDD Multiclass—Class Weights

For this experiment, we used the training and testing subsets of the previous experiment and the 35 features selected using feature selection. In addition, we introduced the class weights. The goal of setting specific class weights, in this case, is to make the neural network learn to better differentiate between the classes with a smaller number of records (U2R and R2L), and that is done by giving those classes a higher weight. Moreover, by correctly classifying records that belong to those classes, the FNR should also be lowered. We resolved to use the *Scikit-learn* method compute_class_weight for computing the class weights. Because very few records belong to the U2R class, the weights returned by this function needed to be slightly altered to avoid overfitting and falsely classifying many records belonging to the U2R class. The final class weights used were: 0.55 for the DoS class, 2.11 for the Probe class, 7.51 for the R2L class, 24.03 for the U2R class, and 0.38 for the Normal class. Again, the network's input layer had 35 nodes for the 35 selected features. The confusion matrix is shown in Table 13. The FNR for this experiment was the lowest, and it was equal to 0.049%.

Table 13. Confusion matrix for NSL-KDD "class weights" multiclass classification experiment.

		Predicted				
		DoS	**Probe**	**R2L**	**U2R**	**Normal**
Actual	DoS	10,629	19	23	0	5
	Probe	1	2799	13	6	1
	R2L	0	5	706	7	1
	U2R	0	0	6	14	0
	Normal	27	76	88	0	15,278

5.2. Experiments on the UNSW-NB15 Dataset

As for the NSL-KDD dataset, for this dataset we built binary and multiclass classification models.

5.2.1. UNSW-NB15 Binary—Full Features

The first binary classification experiment included using all 200 features obtained during the data preparation phase by splitting the original dataset. The neural network architecture, and the training hyperparameters, are explained in Section 4.3. The neural network's input layer has 200 nodes, one for each feature used. As for the NSL-KDD experiments, we used Early Stopping for the UNSW-NB15 experiments, to prevent the neural network from overfitting. The confusion matrix is shown in Table 14. The number of false negatives is equal to 1662 and the number of false positives is 2965.

Table 14. Confusion matrix for UNSW-NB15 "full features" binary classification experiment.

		Predicted	
		Normal	**Attack**
Actual	Normal	389,003	2965
	Attack	1662	18,254

5.2.2. UNSW-NB15 Binary—Reduced Features

The second binary classification experiment considered using a minimized set of features obtained using a combination of several feature selection methods implemented by *PyCaret*. The feature selection picked 53 features out of 200, which were labeled as the most relevant for the classification process. Hence, the neural network's input layer consisted of 53 nodes, while the training hyperparameters were the same as in the other experiments. The rest of the network architecture was the same as the one presented in Section 4.3. The whole process included, again, the Early Stopping mechanism. The confusion matrix is shown in Table 15. The number of false negatives is equal to 1431 and the number of false positives is 3257. The number of FN is slightly lower than in the previous experiment, while the number of FP slightly increased.

Table 15. Confusion matrix for UNSW-NB15 "reduced features" binary classification experiment.

		Predicted	
		Normal	**Attack**
Actual	Normal	388,177	3257
	Attack	1431	18,485

5.2.3. UNSW-NB15 Binary—Class Weights

In this experiment, we incorporated the class weights into the network of the previous experiment. We used the same subsets for training and testing and the 53 features selected during the feature selection process. The training hyperparameters and the rest of the network architecture are the same as in the previous experiment. The class weights were assigned in the following manner: 1 for the normal class (labeled with 0) and 3 for the attack class (labeled with 1). The confusion matrix is shown in Table 16. The number of false negatives is equal to 56, which is the lowest value in all three binary experiments. The number of false positives increased, and is equal to 5536. The increase was expected, because there is a trade-off between the false positives and false negatives.

Table 16. Confusion matrix for UNSW-NB15 "class weights" binary classification experiment.

		Predicted	
		Normal	**Attack**
Actual	Normal	386,432	5536
	Attack	56	19,860

5.2.4. UNSW-NB15 Multiclass—Full Features

The first multiclass classification experiment included the usage of all 200 features obtained during the data preparation process, as well as the subsets generated by splitting the main dataset. The input layer has 200 nodes, one for each feature, and the output layer has ten nodes, one for each possible class representing normal traffic and the other nine for each attack category. The loss function used for this model is the sparse categorical cross-entropy function, and the activation function in the output layer is the softmax function.

Early Stopping was used. The confusion matrix can be seen in Table 17. The confusion matrix shows that the model was not able to correctly predict any of the attacks that belong to the class 'Worms'. The reason for this is the fact that the dataset is very unbalanced, and there were only 38 records belonging to this class in the testing dataset. Other classes, besides the class 'Worms' which have a low number of records are: Shellcode, Backdoor, and Analysis. Considering the 'Normal' class as the negative class, looking at the Table 17, the number of true negatives can be found in the intersection of the 7th row and the 7th column, and it is equal to 390,912. The false positives are all the records that belong to the 'Normal' class, but were wrongly classified as an attack, and they can be seen in the 7th row. The false negatives are the records that represent an attack, but were wrongly classified as belonging to the 'Normal' class, and they can be seen in the 7th column. Based on this, the FNR can be calculated, and it is equal to 16.48%.

Table 17. Confusion matrix for UNSW-NB15 "full features" multiclass classification experiment.

		Predicted								
	Analysis	**Backdoor**	**DoS**	**Exploits**	**Fuzzers**	**Generic**	**Normal**	**Recconnaissanse**	**Shellcode**	**Worms**
Analysis	7	20	4	25	310	0	89	0	0	0
Backdoor	8	11	10	49	261	3	11	41	0	0
DoS	9	8	28	496	359	53	62	62	15	0
Exploits	9	21	21	4232	459	70	345	427	12	0
Fuzzers	11	20	2	127	1560	20	2536	123	12	0
Generic	0	0	6	406	356	4175	52	38	5	0
Normal	0	0	0	190	540	11	390,912	180	13	0
Recconnaissanse	0	0	1	102	309	4	172	2122	0	0
Shellcode	0	0	0	22	20	4	34	140	84	0
Worms	0	0	0	27	3	5	2	1	0	0

(Row label on left: *Actual*)

5.2.5. UNSW-NB15 Multiclass—Reduced Features

For this experiment, we used the same data subsets as the previous one and included a feature selection process to use only the most relevant features for the classification of the records in each of the ten categories. By using the feature selection method from *PyCaret*, 44 features were selected as the most important. The input layer has 44 nodes, one for each of the selected features. Early Stopping was used. The confusion matrix is shown in Table 18. In comparison to the results obtained in the previous experiment, in this experiment, the model had worse performance when it comes to correctly classifying the minority classes. None of the records belonging to the 'Analysis', 'Backdoor' and 'Worms' classes were correctly classified. However, the FNR for this experiment was equal to 12.69%, which was a bit lower than in the "UNSW-NB15 multiclass—full features" experiment.

Table 18. Confusion matrix for UNSW-NB15 "reduced features" multiclass classification experiment.

	Predicted									
	Analysis	**Backdoor**	**DoS**	**Exploits**	**Fuzzers**	**Generic**	**Normal**	**Recconnaissanse**	**Shellcode**	**Worms**
Analysis	0	0	15	167	131	0	82	60	0	0
Backdoor	0	0	19	138	116	6	16	99	0	0
DoS	0	0	28	625	144	62	42	173	18	0
Exploits	0	0	28	4218	212	111	163	849	15	0
Fuzzers	0	0	14	413	1346	63	2101	456	18	0
Generic	0	0	8	518	164	4186	31	125	6	0
Normal	0	0	3	393	582	47	390,217	586	18	0
Recconnaissanse	0	0	5	149	129	8	91	2328	0	0
Shellcode	0	0	0	13	17	5	17	178	74	0
Worms	0	0	0	25	0	7	0	5	1	0

(Actual)

5.2.6. UNSW-NB15 Multiclass—Class Weights

This experiment included the usage of the same subsets as for the previous experiment (44 selected features) but with the addition of the class weights. The weights were calculated using the function `compute_class_weight` from the *Python* library *Scikit-learn*. We further refined the obtained weights to avoid over-fitting. The final weights used for training were the following: 9 for class 0 (no-attack), 10 for class 1, 5 for class 2, 3 for class 3, 3 for class 4, 3 for class 5, 1 for class 6, 4 for class 7, 15 for class 8 and 20 for class 9. The neural network architecture was the same as in the second experiment, and Early Stopping was used. The confusion matrix can be seen in Table 19. When compared to the first two experiments, the "UNSW-NB15 multiclass—class weights" experiment has seen an improvement in the performance metrics for these classes. The FNR is equal to 0.77%, which is the lowest of all three experiments.

Table 19. Confusion matrix for UNSW-NB15 "class weights" multiclass classification experiment.

	Predicted									
	Analysis	**Backdoor**	**DoS**	**Exploits**	**Fuzzers**	**Generic**	**Normal**	**Recconnaissanse**	**Shellcode**	**Worms**
Analysis	14	349	0	23	0	0	69	0	0	0
Backdoor	0	322	0	11	9	0	3	46	0	3
DoS	0	376	30	456	45	52	12	92	28	1
Exploits	8	442	13	3964	209	73	53	682	142	10
Fuzzers	0	387	2	202	3076	103	8	443	190	0
Generic	1	339	1	390	41	4154	6	59	32	15
Normal	38	11	6	382	3886	78	386,630	684	131	0
Recconnaissanse	0	323	2	12	101	6	4	2237	25	0
Shellcode	0	0	0	1	21	4	0	142	136	0
Worms	0	0	0	26	0	1	0	5	1	5

(Actual)

6. Results

This Section provides a detailed explanation of the results which were obtained from the experimental campaign, with a comparison of the results.

6.1. Results of the NSL-KDD Experiments

The results which were obtained in the 4 binary experiments done on the NSL-KDD dataset can be seen in Table 20.

Table 20. Comparison of the results achieved in the NSL-KDD binary classification experiments.

Experiment	Training Accuracy	Prediction Accuracy	Precision	Recall	F1 Score	ROC AUC Score	FPR	FNR
Full features	99.77%	82.44%	93.51%	74.32%	82.82%	83.75%	6.82%	25.68%
Modified distribution	99.76%	99.72%	99.64%	99.78%	99.71%	99.72%	0.33%	0.22%
Reduced features	98.94%	98.78%	98.5%	98.97%	98.74%	98.79%	1.4%	1.03%
Class weights	98.32%	98.35%	97.42%	99.2%	98.3%	98.38%	2.44%	0.8%

Observing the table, we can see that the "NSL-KDD binary—modified distribution" experiment achieved the best results, with the lowest FPR and FNR. The "NSL-KDD binary—full features" experiment achieved the lowest results, which could mean that the initial training and testing subsets distribution was not appropriate. The "NSL-KDD binary—reduced features" and the "NSL-KDD binary—class weights" experiments achieved more or less similar results, with the fourth one having a slightly lower FNR, which was the goal. On the other hand, the FPR in the "NSL-KDD binary—class weights" experiment was higher than in the "NSL-KDD binary—reduced features" one, which was expected because there is a trade-off between the false positives and false negatives.

Table 21 reports the results obtained in the 4 multiclass experiments with the NSL-KDD dataset.

Table 21. Comparison of the results achieved in the NSL-KDD multiclass classification experiments.

Experiment	Training Accuracy	Prediction Accuracy	Precision	Recall	F1 Score	ROC AUC Score	FNR
Full features	99.82%	74.99%	74.68%	74.99%	70.89%	93.55%	31.17%
Modified distribution	99.82%	99.74%	99.74%	99.74%	99.74%	99.9%	0.17%
Reduced features	99.59%	99.49%	99.51%	99.49%	99.49%	99.99%	0.133%
Class weights	99.09%	99.06%	99.14%	99.06%	99.09%	99.97%	0.049%

Observing the results in Table 21, we can see that the lowest FNR was reached in the "NSL-KDD multiclass—class weights" experiment and the highest in the "NSL-KDD multiclass—full features" experiment. In fact, all the evaluation metrics from the "NSL-KDD multiclass—full features" experiment show very poor performance, which again might mean that the datasets which were pre-made by the authors need a feature selection when facing the FNR minimization problem. The "NSL-KDD multiclass—modified distribution" experiment has slightly higher precision, recall, and F1 score than the "NSL-KDD multiclass—reduced features" and the "NSL-KDD multiclass—class weights" experiments. Overall, the "NSL-KDD multiclass—modified distribution", "NSL-KDD multiclass—reduced features", and "NSL-KDD multiclass—class weights" experiments all have performance metrics that are >99%, and that can be considered a satisfactory result. When it comes to the specific performance of the classes with a lower number of records, the U2R and R2L classes, the best performance for them was achieved in the "NSL-KDD multiclass—modified distribution" experiment.

As mentioned earlier, another one of the goals of this research is to increase the detection rates of some specific classes which were shown to have low detection rates in previous works by other authors, as shown in Section 3. For this dataset, the classes that had the lowest detection rates were R2L and U2R, so we report the following performance metrics specifically for these two classes: precision, recall, and F1 score. For the NSL-KDD multiclass experiments, the detailed results of those metrics are in Table 22. The column "No. of records" refers to the number of records belonging to those classes in the testing dataset.

Table 22. Performance metrics for U2R and R2L classes in NSL-KDD multiclass experiments.

Experiment	Class	Precision	Recall	F1 Score	No. of Records
Full features	R2L	60%	5%	9%	2885
	U2R	80%	12%	21%	67
Modified distribution	R2L	96%	98%	97%	719
	U2R	71%	60%	65%	20
Reduced features	R2L	90%	98%	94%	719
	U2R	75%	45%	56%	20
Class weights	R2L	84%	98%	91%	719
	U2R	52%	70%	60%	20

When it comes to the specific performance of the classes with a lower number of records, the U2R and R2L class, the best performance for them was achieved in the "NSL-KDD multiclass—modified distribution" experiment. Since Early Stopping was used in order to prevent the model from overfitting, the average number of epochs needed to reach the optimal result while training the model on the NSL-KDD dataset was 25. The average time needed to train the network for this dataset was approximately 3 min for each experiment.

6.2. Results of the UNSW-NB15 Experiments

The results achieved in the three binary classification experiments with the UNSW-NB15 dataset are in Table 23. We can observe that the prediction accuracy is very similar in all three experiments. However, there is a considerable variation between the precision and recall, especially between the "UNSW-NB15 binary—full features" and the "UNSW-NB15 binary—class weights" experiments. The observed reduction is likely because the "UNSW-NB15 binary—class weights" experiment produced more false positives and fewer false negatives. After all, there is a trade-off between those two when using class weights. The goal was to lower the FNR as much as possible, and the model used in the "UNSW-NB15 binary—class weights" experiment was the most successful.

Table 23. Comparison of the results achieved in the UNSW-NB15 binary classification experiments.

Experiment	Training Accuracy	Prediction Accuracy	Precision	Recall	F1 Score	ROC AUC Score	FPR	FNR
Full features	98.87%	98.88%	86.03%	91.65%	88.75%	95.45%	0.76%	8.34%
Reduced features	98.86%	98.86%	85.02%	92.81%	88.75%	96%	0.83%	7.18%
Class weights	98.63%	98.64%	78.2%	99.72%	87.66%	99.15%	1.41%	0.28%

Three multiclass classification experiment results for the UNSW-NB15 dataset are in Table 24. We can observe that the lowest FNR was achieved in the "UNSW-NB15 multiclass—class weights" experiment, jointly with the best performance for minority classes. The same experiment also reached the highest precision. However, it is noticeable as there was not a significant variation between the other metrics among all the performed experiments. Since the dataset has a minimal number of records representing attacks compared to records representing regular traffic, even introducing the class weights, it is hard for the network to learn how to distinguish between the different classes since there are too few samples.

Table 24. Comparison of the results achieved in the UNSW-NB15 multiclass classification experiments.

Exp. Name	Training Accuracy	Prediction Accuracy	Precision	Recall	F1 Score	ROC AUC Score	FNR
Full features	97.91%	97.87%	97.55%	97.87%	97.65%	99.87%	16.48%
Reduced features	97.91%	97.7%	97.44%	97.7%	97.48%	99.85%	12.69%
Class weights	97.25%	97.25%	98.2%	97.25%	97.55%	99.83%	0.77%

The dataset's lowest number of records classes are Worms, Shellcode, Backdoor, and Analysis. The goal is to try to increase the prediction capability for these classes, so we report in Table 25 the class-specific precision, recall, and F1 score for these four classes. As for the other dataset, the column "No. of records" refers to the number of records belonging to those classes in the testing dataset.

Table 25. Performance metrics for minority classes in UNSW-NB15 multiclass experiments.

Experiment	Class	Precision	Recall	F1 Score	No. of Records
Full features	Analysis	16%	2%	3%	455
	Backdoor	14%	3%	5%	394
	Shellcode	60%	28%	38%	304
	Worms	0%	0%	0%	38
Reduced features	Analysis	0%	0%	0%	455
	Backdoor	0%	0%	0%	394
	Shellcode	49%	24%	33%	304
	Worms	0%	0%	0%	38
Class weights	Analysis	23%	3%	5%	455
	Backdoor	13%	82%	22%	394
	Shellcode	20%	45%	28%	304
	Worms	15%	13%	14%	38

When compared to the first two experiments, the "UNSW-NB15 multiclass—class weights" experiment has seen an improvement in the performance metrics for all these classes. Since Early Stopping was used in order to prevent the model from overfitting, the average number of epochs needed to reach the optimal result while training the model on the UNSW-NB15 dataset was 20. The average time needed to train the network for this dataset was approximately 30 min for each UNSW-NB15 experiment.

7. Conclusions

This research focused on building a deep neural network and training it on two modern datasets for binary and multiclass classification. Despite other works in the literature, the primary goals of our research were to lower the False Negative Rate as much as possible while still keeping the False Positive Rate low and increasing the detection rate of minority classes (classes with a low number of records). We proposed a strategy made of three points: correction of the training and testing subset distribution, feature selection, and usage of class weights. We ran an experimental campaign for two well-established datasets to verify the effectiveness of our strategy in lowering the FNR and increasing the performance of minority classes. In almost all of the experiments, a combination of feature selection and the assignment of correct class weights during the training phase of the neural network gave the best results in lowering the FNR. We observed that the assignment of the class weights needs to be used with caution since it can easily lead to over-fitting and an increase in the FPR. Even when used correctly, it will still give a slight increase in the FPR, as seen from the experiments in this research, but the number is still considered low enough. Only in the case of binary classification for the NSL-KDD dataset the usage of class weights was not the best method for achieving the lowest FNR. A more effective correction was modifying the distribution of the train and test subsets. Regarding multiclass classification, feature selection with class weights proved to be the best method to increase the performance of the minority classes.

Compared to the work surveyed in Section 3, the neural network models constructed in our research incidentally outperform all of them in terms of accuracy, except [35]. This comparison can be seen in the overview given in Table 3. In terms of accuracy, our model reached accuracy values >99% for the NSL-KDD dataset, which is higher than the accuracy achieved by other models on the same dataset, both for binary and multiclass classification. When it comes to the results achieved for the UNSW-NB15 dataset, our proposed model reached the accuracy of >98% for binary classification, and >97% for multiclass classification. Once again, it has overcome most of the other models on the same dataset when it comes to accuracy, with the exception of [35], in which the model has achieved the accuracy of 98.99% for binary classification and the FPR of 0.56%. However, because the datasets used in this research are unbalanced, accuracy is not the best metric to evaluate the performance. Therefore, this research also uses precision, recall, F1 score, and ROC AUC score to assess the performance. The best results for the NSL-KDD dataset show that all of these metrics were >99%, and for the UNSW-NB15, they were >98% for binary classification and >97% for multiclass classification. When it comes to the FPR and FNR, when compared to the models surveyed in Section 3 where the authors focused on calculating these values, the values achieved by our proposed model, once again, outperform most of the surveyed models. The exception is once again [35], when it comes to binary classification for UNSW-NB15 dataset specifically.

The main limitation of our work is that we have limited evidence of the generalization of our strategy. This is because we used only two of the most established datasets to validate our approach. A natural extension of our experiment to other datasets would further confirm the validity of our approach.

We acknowledge that our results for the lowest-represented attack classes are not optimal, and there is still space for increasing the performance. However, the major problem remains: the number of their records is still too low for a deep neural network to learn from it. One of the possible directions could be finding a way to improve these datasets to fix the imbalance and therefore increase the detection rates for minority classes. One idea is to generate and add more records to the minority classes. Another alternative is to use oversampling techniques. Most of the hyperparameter tuning in this research was done manually by doing different experiments. One possible alternative we should consider in future research would be using automatic parameter tuning methods to achieve better performance. Another direction for future work would be to test these models in a live system with actual attacks to see how well they would perform in the real world.

Author Contributions: Conceptualization, J.M. and A.S.; methodology, J.M.; software, J.M.; validation, J.M. and A.S.; formal analysis, J.M.; investigation, J.M.; resources, J.M. and A.S.; data curation, J.M.; writing—original draft preparation, J.M. and A.S.; writing—review and editing, J.M. and A.S.; visualization, J.M.; supervision, A.S.; project administration, J.M. and A.S. All authors have read and agreed to the published version of the manuscript.

Funding: This work was supported in part by the MIUR under grant "Dipartimenti di eccellenza 2018-2022" of the Department of Computer Science of Sapienza University.

Institutional Review Board Statement: Not applicable.

Informed Consent Statement: Not applicable.

Data Availability Statement: Publicly available datasets were analyzed in this study. This data can be found here: https://www.unb.ca/cic/datasets/nsl.html and https://research.unsw.edu.au/projects/unsw-nb15-dataset (accessed on 4 June 2022).

Conflicts of Interest: The authors declare no conflict of interest.

References

1. Liu, H.; Lang, B. Machine Learning and Deep Learning Methods for Intrusion Detection Systems: A Survey. *Appl. Sci.* **2019**, *9*, 4396. [CrossRef]
2. Goeschel, K. Reducing false positives in intrusion detection systems using data-mining techniques utilizing support vector machines, decision trees, and naive Bayes for off-line analysis. In Proceedings of the SoutheastCon, Norfolk, VA, USA, 30 March–3 April 2016; pp. 1–6. [CrossRef]
3. Azeroual, O.; Nikiforova, A. Apache Spark and MLlib-Based Intrusion Detection System or How the Big Data Technologies Can Secure the Data. *Information* **2022**, *13*, 58. [CrossRef]
4. Chahar, R.; Kaur, D. A systematic review of the machine learning algorithms for the computational analysis in different domains. *Int. J. Adv. Technol. Eng. Explor. (IJATEE)* **2020**, *7*, 147–164. [CrossRef]
5. Ahmad, Z.; Shahid Khan, A.; Shiang, C.; Ahmad, F. Network intrusion detection system: A systematic study of machine learning and deep learning approaches. *TRansactions Emerg. Telecommun. Technol.* **2021**, *32*, e4150. [CrossRef]
6. Hodo, E.; Bellekens, X.; Hamilton, A.; Tachtatzis, C.; Atkinson, R. Shallow and deep networks intrusion detection system: A taxonomy and survey. *arXiv* **2017**, arXiv:1701.02145.
7. Al Jallad, K.; Aljnidi, M.; Desouki, M.S. Anomaly detection optimization using big data and deep learning to reduce false-positive. *J. Big Data* **2020**, *7*, 68. [CrossRef]
8. Vijayakumar, D.; Ganapathy, S. Machine Learning Approach to Combat False Alarms in Wireless Intrusion Detection System. *Comput. Inf. Sci.* **2018**, *11*, 67. [CrossRef]
9. Tavallaee, M.; Bagheri, E.; Lu, W.; Ghorbani A. A Detailed Analysis of the KDD CUP 99 Data Set. In Proceedings of the Second IEEE Symposium on Computational Intelligence for Security and Defense Applications (CISDA), Ottawa, ON, Canada, 8–10 July 2009.
10. Moustafa, N.; Slay, J. UNSW-NB15: A comprehensive data set for network intrusion detection systems (UNSW-NB15 network data set). In proceedings of the 2015 Military Communications and Information Systems Conference (MilCIS), Canberra, Australia, 10–12 November 2015; pp. 1–6.
11. Khraisat, A.; Gondal, I.; Vamplew, P.; Kamruzzaman, J. Survey of intrusion detection systems: Techniques, datasets and challenges. *Cybersecurity* **2019**, *2*, 1–22. [CrossRef]
12. Rao, U.H.; Nayak, U. Intrusion Detection and Prevention Systems. In *The InfoSec Handbook*; Apress: Berkeley, CA, USA, 2014; ISBN 978-1-4302-6383-8. [CrossRef]
13. Liao, H.-J.; Lin, C.-H.R.; Lin, Y.-C.; Tung, K.-Y. Intrusion detection system: A comprehensive review. *J. Netw. Comput. Appl.* **2013**, *36*, 16–24. [CrossRef]
14. García-Teodoro, P.; Díaz-Verdejo, J.; Maciá-Fernández, G.; Vázquez, E. Anomaly-based network intrusion detection: Techniques, systems and challenges. *Comput. Secur.* **2009**, *28*, 18–28. [CrossRef]
15. What Is Machine Learning? Available online: https://www.ibm.com/cloud/learn/machine-learning (accessed on 12 June 2022).
16. Bishop, C.M.; Nasrabadi, N.M. *Pattern Recognition and Machine Learning*; Springer: New York, NY, USA, 2006; ISBN 978-1-4939-3843-8.
17. Alzubaidi, L.; Zhang, J.; Humaidi, A.J.; Al-Dujaili, A.; Duan, Y.; Al-Shamma, O.; Farhan, L. Review of deep learning: Concepts, CNN architectures, challenges, applications, future directions. *J. Big Data* **2021**, *8*, 1–74. [CrossRef]
18. Sarker, I.H. Deep Learning: A Comprehensive Overview on Techniques, Taxonomy, Applications and Research Directions. *SN Comput. Sci.* **2021**, *2*, 1–20. [CrossRef] [PubMed]
19. Shone, N.; Ngoc, T.N.; Phai, V.D.; Shi, Q. A Deep Learning Approach to Network Intrusion Detection. *IEEE Trans. Emerg. Top. Comput. Intell.* **2018**, *2*, 41–50. [CrossRef]
20. Kocher, G.; Kumar, G. Machine learning and deep learning methods for intrusion detection systems: Recent developments and challenges. *Soft Comput.* **2021**, *25*, 9731–9763. [CrossRef]
21. Alzaqebah, A.; Aljarah, I.; Al-Kadi, O.; Damaševičius, R. A Modified Grey Wolf Optimization Algorithm for an Intrusion Detection System. *Mathematics* **2022**, *10*, 999. [CrossRef]
22. Ali, M.H.; Jaber, M.M.; Abd, S.K.; Rehman, A.; Awan, M.J.; Damaševičius, R.; Bahaj, S.A. Threat Analysis and Distributed Denial of Service (DDoS) Attack Recognition in the Internet of Things (IoT). *Electronics* **2022**, *11*, 494. [CrossRef]
23. Stolfo, S.J.; Fan, W.; Lee, W.; Prodromidis A.; Chan, P. K. Cost-based modeling for fraud and intrusion detection: Results from the jam project. *DISCEX* **2000** *2*, 1130.

24. Revathi, S.; Malathi, A. A Detailed Analysis on NSL-KDD Dataset Using Various Machine Learning Techniques for Intrusion Detection. *Int. J. Eng. Res. Technol.* **2013**, *2*, 1848–1853.
25. NSL-KDD | Datasets | Research | Canadian Institute for Cybersecurity | UNB. Available online: https://www.unb.ca/cic/datasets/nsl.html (accessed on 4 June 2022).
26. Saporito, G. A Deeper Dive into the NSL-KDD Data Set. Available online: https://towardsdatascience.com/a-deeper-dive-into-the-nsl-kdd-data-set-15c753364657 (accessed on 6 June 2022).
27. Jia, Y.; Wang, M.; Wang, Y. Network intrusion detection algorithm based on deep neural network. *IET Inf. Secur.* **2019**, *13*, 48–53. [CrossRef]
28. Vinayakumar, R.; Alazab, M.; Soman, K.; Poornachandran, P.; Al-Nemrat, A.; Venkatraman, S. Deep Learning Approach for Intelligent Intrusion Detection System. *IEEE Access* **2019**, *7*, 41525–41550. [CrossRef]
29. Yin, C.; Zhu, Y.; Fei, J.; He, X. A Deep Learning Approach for Intrusion Detection Using Recurrent Neural Networks. *IEEE Access* **2017**, *5*, 21954–21961. [CrossRef]
30. Potluri, S.; Diedrich, C. Accelerated deep neural networks for enhanced Intrusion Detection System. In proceedings of the 2016 IEEE 21st International Conference on Emerging Technologies and Factory Automation (ETFA), Berlin, Germany, 6–9 September 2016; pp. 1–8.
31. Kasongo, S.; Sun, Y. A Deep Learning Method With Filter Based Feature Engineering for Wireless Intrusion Detection System. *IEEE Access* **2019**, *7*, 38597–38607. [CrossRef]
32. Fu, Y.; Du, Y.; Cao, Z.; Li, Q.; Xiang, W. A Deep Learning Model for Network Intrusion Detection with Imbalanced Data. *Electronics* **2022**, *11*, 898. [CrossRef]
33. Kanimozhi, V.; Jacob, P. UNSW-NB15 Dataset Feature Selection and Network Intrusion Detection using Deep Learning. *Int. J. Recent Technol. Eng.* **2019**, *7*, 443–446.
34. Mahalakshmi, G.; Uma, E.; Aroosiya, M.; Vinitha, M. Intrusion Detection System Using Convolutional Neural Network on UNSW NB15 Dataset. *Adv. Parallel Comput.* **2021**, *40*, 1–8. [CrossRef]
35. Al-Zewairi, M.; Almajali, S.; Awajan, A. Experimental Evaluation of a Multi-layer Feed-Forward Artificial Neural Network Classifier for Network Intrusion Detection System. In Proceedings of the 2017 International Conference on New Trends in Computing Sciences (ICTCS), Amman, Jordan, 11–13 October 2017; pp. 167–172.
36. Abdulhammed, R.; Musafer, H.; Alessa, A.; Faezipour, M.; Abuzneid, A. Features Dimensionality Reduction Approaches for Machine Learning Based Network Intrusion Detection. *Electronics* **2019**, *8*, 322. [CrossRef]
37. Doshi, S. Various Optimization Algorithms For Training Neural Network. Available online: https://towardsdatascience.com/optimizers-for-training-neural-network-59450d71caf6 (accessed on 8 December 2021).
38. Brownlee, J. Understand the Impact of Learning Rate on Neural Network Performance. Available online: https://machinelearningmastery.com/understand-the-dynamics-of-learning-rate-on-deep-learning-neural-networks/ (accessed on 7 December 2021).
39. McGonagle, J.; Shaikouski, G.; Williams, C.; Hsu, A.; Khim, J.; Miller, A. Backpropagation. Available online: https://brilliant.org/wiki/backpropagation/ (accessed on 16 December 2021).
40. He, K.; Zhang, X.; Ren, S.; Sun, J. Delving Deep into Rectifiers: Surpassing Human-Level Performance on ImageNet Classification. *arXiv* **2015**, arXiv:abs/1502.01852.
41. Wood, T. Softmax Function. Available online: https://deepai.org/machine-learning-glossary-and-terms/softmax-layer (accessed on 18 December 2021).
42. NumPy. Available online: https://numpy.org/ (accessed on 25 June 2022).
43. Pandas. Available online: https://pandas.pydata.org/ (accessed on 25 June 2022).
44. Matplotlib. Available online: https://matplotlib.org/ (accessed on 25 June 2022).
45. Scikit-Learn. Available online: https://scikit-learn.org/ (accessed on 25 June 2022).
46. Tensorflow. Available online: https://www.tensorflow.org/ (accessed on 25 June 2022).
47. Keras. Available online: https://keras.io/ (accessed on 25 June 2022).
48. Jupyter. Available online: https://jupyter.org/ (accessed on 25 June 2022).
49. PyCaret. Available online: https://pycaret.org/ (accessed on 13 June 2022).
50. Feature Selection. Available online: https://pycaret.gitbook.io/docs/get-started/preprocessing/feature-selection (accessed on 16 June 2022).

algorithms

MDPI

Article

Tree-Based Classifier Ensembles for PE Malware Analysis: A Performance Revisit

Maya Hilda Lestari Louk [1] and Bayu Adhi Tama [2,*]

[1] Department of Informatics Engineering, University of Surabaya, Surabaya 60293, Indonesia
[2] Department of Information Systems, University of Maryland, Baltimore County (UMBC), Baltimore, MD 21250, USA
* Correspondence: bayu@umbc.edu

Abstract: Given their escalating number and variety, combating malware is becoming increasingly strenuous. Machine learning techniques are often used in the literature to automatically discover the models and patterns behind such challenges and create solutions that can maintain the rapid pace at which malware evolves. This article compares various tree-based ensemble learning methods that have been proposed in the analysis of PE malware. A tree-based ensemble is an unconventional learning paradigm that constructs and combines a collection of base learners (e.g., decision trees), as opposed to the conventional learning paradigm, which aims to construct individual learners from training data. Several tree-based ensemble techniques, such as random forest, XGBoost, CatBoost, GBM, and LightGBM, are taken into consideration and are appraised using different performance measures, such as accuracy, MCC, precision, recall, AUC, and F1. In addition, the experiment includes many public datasets, such as BODMAS, Kaggle, and CIC-MalMem-2022, to demonstrate the generalizability of the classifiers in a variety of contexts. Based on the test findings, all tree-based ensembles performed well, and performance differences between algorithms are not statistically significant, particularly when their respective hyperparameters are appropriately configured. The proposed tree-based ensemble techniques also outperformed other, similar PE malware detectors that have been published in recent years.

Keywords: portable executable malware; tree-based ensemble; performance comparison; statistical significance test

Citation: Louk, M.H.L.; Tama, B.A. Tree-Based Classifier Ensembles for PE Malware Analysis: A Performance Revisit. *Algorithms* **2022**, *15*, 332. https://doi.org/10.3390/a15090332

Academic Editors: Francesco Bergadano and Giorgio Giacinto

Received: 29 August 2022
Accepted: 14 September 2022
Published: 17 September 2022

Publisher's Note: MDPI stays neutral with regard to jurisdictional claims in published maps and institutional affiliations.

1. Introduction

Malware (e.g., malicious software) is commonly recognized as one of the most potent cyber threats and hazards to modern computer systems [1,2]. It is an overarching word that refers to any code that potentially has a destructive, harmful effect [3]. On the basis of their behavior and execution processes, malicious softwares are categorized as worms, viruses, Trojan horses, rootkits, backdoors, spyware, logic bombs, adware, and ransomware. Computer systems are hacked for a variety of reasons, including the destruction of computer resources, financial gain, the theft of private and confidential information and the use of computing resources, as well as the inaccessibility of system services, to name a few [4].

Malware is recognized using signature-based or behavior-based methods. The signature-based malware detection techniques are quick and effective, but obfuscated malware can quickly circumvent them. In contrast, behavior-based methods are more resistant to obfuscation. Nonetheless, behavior-based methods are relatively time-intensive. Therefore, in addition to the signature-based and behavior-based malware detection techniques, numerous fusion techniques exist that contain the benefits of both [5,6]. The goal of these fusion strategies is to address the shortcomings of signature and behavior-based approaches.

While we work to defend ourselves from malware, cybercriminals continue to create increasingly complex techniques to obtain and steal data and resources. Conventional methods (i.e., rule-based, graph-based, and entropy-based) for analyzing and detecting malware focus on matching known malicious signatures to alleged malicious programs. Such static solutions require a known harmful signature, rendering them unsatisfactory against new (e.g., zero-day) attacks, and depend on end users to maintain system updates. Attackers are aware that these methods may also be vulnerable to obfuscation, such as code obfuscation to avoid detection against known signatures [7]. Hence, it is necessary to update and build malware detection mechanisms that are capable of withstanding significant attacks [8].

Machine learning offers the potential to construct malware detectors that are capable of combating newer versions of malware, and different supervised and unsupervised-algorithm-based machine learning methods have been reported in the literature [9–11]. More specifically, ensemble learning approaches have been utilized and achieved excellent results in malware detection [12–17]. In most cases, ensemble learning algorithms yield superior results as compared to individual classification algorithms, i.e., support vector machine, decision tree, naive Bayes, and neural networks. However, although classifier ensembles demonstrate a significant performance, the majority of these ensembles are deployed in a restricted manner without adequate hyperparameter tuning. Moreover, the performance of classifier ensembles is validated using a single dataset; consequently, no generalizable results are produced.

The tree-based ensemble technique is an ensemble learning paradigm in which a collection of base learners (e.g., decision trees or CART) are constructed and combined from the training data [18]. For instance, random forest [19] is comprised of a large number of individual decision trees that operate as an ensemble. It uses feature randomness to generate an uncorrelated forest of decision trees. In a similar fashion, the gradient boosting decision tree algorithms combine a collection of individual decision trees to form an ensemble. However, unlike random forest, the decision trees in gradient boosting are constructed serially (e.g., additively). Gradient boosting decision tree algorithms have recently been proposed and have demonstrated remarkable results in many domains, such as protein–protein interaction prediction [20], neutronic calculation [21], human activity recognition [22], etc. However, their performance in classifying and detecting malware remains questionable. This motivated us to employ ensembles of tree-based algorithms to classify PE malware. This paper makes the following contributions to the current literature.

(a) Fine-tuned tree-based classifier ensembles, i.e., random forest [19], XGBoost [23], CatBoost [24], GBM [25], and LightGBM [26], to detect PE malware are employed.

(b) The performance differences between classifier ensembles over the most recent datasets, i.e., BODMAS [27], Kaggle, and CIC-MalMem-2022 [28] are benchmarked using statistical significance tests. This study is among the first to utilize the most recent malware BODMAS and CIC-MalMem-2022 datasets. On the BODMAS and CIC-Malmem-2022 datasets, our proposed approaches outperform other baselines with a 99.96% and 100% accuracy rate, respectively.

(c) An in-depth exploratory analysis of each malware dataset is presented to better understand the characteristics of each malware dataset. The analysis includes a feature correlation analysis and t-SNE visualization of pairs of samples' similarities.

The remainder of the paper is structured as follows. An overview of PE malware detection based on classifier ensembles is provided in Section 2. Next, we present the background of tree-based classifier ensembles and datasets in Section 3. Section 4 discusses the experimental results, and in the end, Section 5 concludes the paper.

2. Related Work

Ucci et al. [7], Maniriho et al. [10] provide the machine learning taxonomy for malware analysis, while [11] present an overview of malware analysis in CPS and IoT. Malware analysis can be accomplished via either static or dynamic analysis, or a mix of the two, depending on how the information extraction procedure is carried out. Approaches based on static analysis evaluate the content of samples without necessitating their execution, whereas dynamic analysis examines the behavior of samples by executing them. This study analyzes a static analysis of PE files, since it can yield a plethora of useful information, e.g., the compiler and symbols used.

Meanwhile, machine learning techniques were largely employed in malware detection [29,30]. Malware samples were examined and the extracted features are used to train the classification algorithm. An overview of the machine learning techniques used for the classification of malware is provided in the following. We particularly explore malware detectors that employ at least one ensemble learning technique. Vadrevu et al. [31], Mills et al. [17], Uppal et al. [32], Kwon et al. [33] utilized random forest for malware detection based on PE file characteristics and networks. Furthermore, Mao et al. [34], Wüchner et al. [35], Ahmadi et al. [36] developed a random forest classifier to detect malware using various features, such as system calls, file system, and Windows registry. Amer and Zelinka [13] proposed an ensemble learning strategy to address the shortcomings of the existing commercial signature-based techniques. The proposed technique was able to focus on the most salient features of malware PE files by lowering the dimensionality of the data. Dener et al. [37] and Azmee et al. [38] compared the use of various machine learning algorithms to detect PE malware and showed that XGBoost and logistic regression were the best-performing methods.

Liu et al. [39] employed data visualization and adversarial training on ML-based detectors to effectively detect the various types of malware and their variants in order to address the current issues in malware detection, such as the consideration of attacks from adversarial examples and the massive growth in malware variants. In [40], a deep feature extraction technique for malware analysis was addressed in light of the current progress in deep learning. Deep features were obtained from a CNN and were fed to an SVM classifier for malware classification. Moreover, a CNN ensemble for malware classification was proposed in [15,16]. The proposed architecture was constructed in a stacked fashion, with a machine learning algorithm providing the final classification. A meta-classifier was selected after various machine learning algorithms were analyzed and evaluated. Most recently, Hao et al. [41] proposed a CNN-based feature extraction and a channel-attention module to reduce the information loss in the process of feature image generation of malware samples. Specific deep learning architectures, such as a deep belief network and transformer-based classifier, were also considered when classifying Android [42,43] and PE malware [44], respectively. Table 1 presents a summary of the existing malware detectors described in the literature.

Table 1. Summarization of the existing PE malware detectors.

Study	Algorithm(s)	Data Set	Validation Technique	Best Result
Mills et al. [17]	RF	Private	7-CV	-
Vadrevu et al. [31]	RF	Private	CV and Holdout	TPR: 90%, FPR: 0.1%
Uppal et al. [32]	NB, DT, RF, and SVM	Private	10-CV	Accuracy: 98.5%
Kwon et al. [33]	RF	Private	10-CV	TPR: 98.0%, FPR: 2.00%, F1: 98.0%, AUC: 99.8%
Mao et al. [34]	RF	Private	Repeated hold-out	TPR: 99.88%, FPR: 0.1%
Wüchner et al. [35]	RF	Malicia	10-CV	DR: 98.01%, FPR: 0.48%
Ahmadi et al. [36]	XGBoost	Kaggle	5-CV	Accuracy: 98.62%
Amer and Zelinka [13]	RF and extra trees	Kaggle	Hold-out	Accuracy: 99.8%, FPR: 0.2%
Liu et al. [39]	CNN and autoencoder	MS BIG and Ember	10-CV	Accuracy: 96.25%
Asam et al. [40]	CNN and SVM	MalImg	Hold-out	Accuracy: 98.61%, precision: 96.27%, recall: 96.30%, F1: 96.32%
Azeez et al. [15]	1D CNN and Extra trees	Kaggle	10-CV	Accuracy: 100%, precision: 100%, recall: 100%, F1: 100%
Damaševičius et al. [16]	Stacked CNN	ClaMP	10-CV	Accuracy: 99.9%, precision: 99.9%, recall: 99.8%, F1: 99.9%

Table 1. *Cont.*

Study	Algorithm(s)	Data Set	Validation Technique	Best Result
Hou et al. [42]	DBN	Comodo cloud	10-CV	Accuracy; 96.66%
Hou et al. [43]	DBN and SAEs	Comodo cloud	10-CV	Accuracy: 96.66%
Azmee et al. [38]	XGBoost	Kaggle	10-CV	Accuracy: 98.6%, AUC: 0.99, TPR: 99.0%, FPR: 3.7%
Jingwei et al. [41]	CNN	MS BIG and BODMAS	10-CV	(MS BIG) accuracy: 99.40%, (BODMAS) accuracy: 99.26%
Lu et al. [44]	Transformer	BODMAS and MS BIG	Hold-out	(MS BIG) accuracy: 98.17%, F1: 98.14%, (BODMAS) accuracy:96.96%, F1: 96.96%
Dener et al. [37]	Logistic regression	CIC-MalMem-2022	Repeated hold-out	Accuracy: 99.97%

3. Materials and Methods

This study evaluates the performance of ensembles of tree-based classifiers in detecting PE malware. Figure 1 depicts the stages involved in our comparative analysis. Several tree-based ensemble approaches are trained on three distinct PE malware training datasets in order to generate classification models. The performance of classification models is then determined by validating them on a testing dataset. Finally, a two-step statistical significance test is then utilized to evaluate the performance benchmarks. In the following section, we provide a brief summary of the malware datasets and tree-based classifier ensembles utilized in this study.

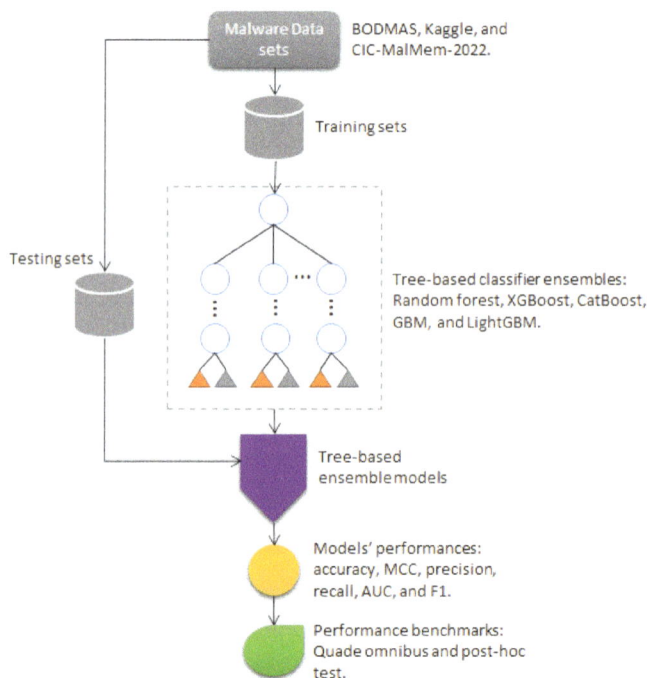

Figure 1. Performance comparison methodology of tree-based ensembles for PE malware detection.

3.1. Datasets

One of the most problematic aspects of using machine learning to solve malware detection problems is producing a realistic feature set from a large variety of unidentified portable executable samples. In essence, the dataset used to train machine learning models determines their level of sophistication. Hence, developing a solid, labeled dataset that represents all analyzed samples is more helpful for malware detection. In light of this, we utilize more recent public datasets that depict the characteristics and attack behaviors of contemporary malware:

(a) BODMAS [27]

The dataset contains 57,293 malicious and 77,142 benign samples (134,435 in total). The malware samples were arbitrarily picked each month from the internal malware database of a security company. The data were collected between 29 August 2019 and 30 September 2020. The benign samples were gathered between 1 January 2007 and 30 September 2020. In order to reflect benign PE binary distribution in real-world traffic, the database of the security company is also processed for benign samples. In addition, SHA-256 hash, the actual PE binary, and a pre-extracted feature vector were given for each malicious sample, whereas only SHA-256 hash and the pre-extracted feature vector were provided for each benign sample. BODMAS is comprised of 2381 input feature vectors and 1 class label feature, of which 0 is labeled as benign and 1 is labeled as malicious.

(b) Kaggle (https://tinyurl.com/22z7u898, access on 25 August 2022)

The dataset was developed using a Python library called *pefile* (https://tinyurl.com/w75zewvr, accessed on 25 August 2022), which is a multi-platform module used to parse and work with PE files. Kaggle dataset contains 14,599 malicious and 5012 benign samples (19,611 in total). The dataset is comprised of 78 input features, denoting PE header files and one class label attribute.

(c) CIC-MalMem-2022 [28]

Unlike the two above-mentioned datasets, CIC-MalMem-2022 is an obfuscated malware dataset that is intended to evaluate memory-based obfuscated malware detection algorithms. The dataset was designed to mimic a realistic scenario as accurately as possible using reowned malware. Obfuscated malware comprises malicious software that conceals itself to escape detection and eradication. The dataset consists of an equal ratio of malicious and benign memory dumps (58,596 samples in total). In addition, CIC-MalMem-2022 is made up of 56 features that serve as inputs for machine learning algorithms.

3.2. Tree-Based Ensemble Learning

The tree-based ensemble is a non-ordinary learning paradigm that constructs and combines a set of base learners (e.g., decision trees or CART) as opposed to the commonplace learning paradigm that attempts to construct individual learners from training data. Normally, an ensemble is formed in two processes, i.e., by first producing the base learners and then integrating them. For a decent ensemble, it is commonly considered that the base learners must be as accurate and diversified as possible [18]. This study considers four tree-based ensemble learning algorithms. It is worth mentioning that tuning the hyperparamters for each algorithm is carried out using *random search* approach [45].

(a) Random forest [19]

As its name implies, a random forest is a tree-based ensemble in which each tree is dependent on a set of random variables. The original formulation of random forest algorithm provided by Breiman [19] is as follows. A random forest employs trees $h_j(\mathcal{X}, \Omega)$ as its base learners. For training data $\mathcal{D} = \{(\mathbf{x}_1, y_1), \dots, (\mathbf{x}_\alpha, y_\alpha)\}$, where $\mathbf{x}_i = (\mathbf{x}_{i,1}, \dots, \mathbf{x}_{i,p})^T$ represents the p predictors and y_i represents the response, and a specific manifestation ω_j of Ω_j, the fitted tree is given as $\hat{h}_j(\mathbf{x}, \omega_j, \mathcal{D})$. More precisely, the steps involved in the random forest algorithm are described in Algorithm 1.

We use a fast random forest implementation called *ranger* [46], available in *R*, which is suitable for high-dimensional data such as ours. The list of random forest's hyperparameters for each malware dataset is provided in Table 2. We set the search space for each hyperparameter tuning is as follows. Number of trees = {50, 100, 250, 500, 750, 1000}, split rule = {'gini','extratrees'}, minimum node size = {1, 2, ..., 10}, *mtry* = number of features × {0.05, 0.15, 0.25, 0.333, 0.4}, sample fraction = {0.5, 0.63, 0.8}, and *replace* = {TRUE, FALSE}.

Algorithm 1: A common procedure of random forest algorithm for classification task.

Training:
Require: Original training set $\mathcal{D} = \{(\mathbf{x}_1, y_1), (\mathbf{x}_2, y_2), \ldots, (\mathbf{x}_\alpha, y_\alpha)\}$,
with $\mathbf{x}_i = (x_{i,1}, \ldots, x_{i,p})^T$
1. **for** $j = 1$ to J
2.   Perform a bootstrap sample $\mathcal{D}_j$ of size α from $\mathcal{D}$.
3.   Using binary recursive partitioning, fit a tree on $\mathcal{D}_j$.
4. **end for**
Testing:
Require: An instance to be classified $\mathbf{x}$.
1. $\hat{f}(\mathbf{x}) = \arg\max_y \sum_{j=1}^{J} I(\hat{h}_j(\mathbf{x}) = y)$
where $\hat{h}_j(\mathbf{x})$ denotes the response variable at $\mathbf{x}$ using the j-th tree.

Table 2. The final learning parameters of random forest used for each dataset after performing a random search.

Hyperparameter	BODMAS	Kaggle	CIC-MalMem-2022
Number of trees	100	1000	500
Split rule	'gini'	'gini'	'extratrees'
Minimum node size	4	8	6
mtry	119	30	18
Sample fraction	0.63	0.80	0.63
replace	FALSE	FALSE	TRUE

(b) Gradient Boosting Decision Trees
In this paper, we also considered various tree-based boosting ensemble approaches for malware detection, such as XGBoost [23], CatBoost [24], GBM [25], and LightGBM [26]. As a rule, GBDT ensembles are a linear additive model, where a tree-based classifier (e.g., CART) was utilized as their base model. Let $\mathcal{D} = \{(\mathbf{x}_i, y_i) | i \in \{1, \ldots, \alpha\}, \mathbf{x}_i \in \mathbb{R}^\eta, y_i \in \mathbb{R}\}$ denote the malware dataset comprising η features and α samples. Considering a collection of j trees, the prediction output $y(\hat{\mathbf{x}})^j$ for an input $\mathbf{x}$ is obtained by calculating the predictions from each tree $y(\hat{\mathbf{x}})^j$, as shown in the following formula.

$$y(\hat{\mathbf{x}})^j = \sum_{i=1}^{j} f_i(\mathbf{x}) \tag{1}$$

where f_i represents the output of the i-th regression tree of the j-tree ensemble. GBDTs minimize a regularized objective function Obj^t in order to create the $(j+1)$-th tree, as follows.

$$\min\{Obj(f)^t\} = \min\{\Omega(f)^t + \Theta(f)^t\} \tag{2}$$

where $\Omega(f)^t$ represents loss function and $\Theta(f)^t$ is a regularization function to control over-fitting. The loss function $\Omega(f)^t$ measures the difference between the prediction $\hat{y}_i$ and the target y_i. On the other hand, the regularization function is defined as $\Theta(f)^t = \gamma T + \frac{1}{2}\lambda \parallel w \parallel^2$, where T and w indicate the number of leaves and leaf weights in the tree, respectively.

(i) XGBoost [23]

XGBoost is a scalable end-to-end tree-boosting strategy that generates a large number of sequentially trained trees. Each succeeding tree corrects the errors made by the preceding one, resulting in an efficient classification model. Through sparsity-aware metrics and multi-threading approaches, XGBoost not only addresses the algorithm's overfitting problem, but also boosts the speed of most real-world computational tasks. This study utilizes two different XGBoost implementations, such as native implementation in *R* [47] and *H2O* [48]. We set the search space of native XGBoost's hyperparameters are as follows. Maximum depth = {2, 3, ..., 24}, eta = {0, 0.1, 0.2, ..., 1.0}, subsample = {0.5, 0.6, 0.7, 0.8}, and column sample by tree = {0.5, 0.6, 0.7, 0.8, 0.9}. Moreover, we set the search space of XGBoost's hyperparameters implemented in *H2O* are as follows. Maximum depth = {1, 3, 5, ..., 29}, sample rate = {0.2, 0.3, ..., 1}, column sample rate = {0.2, 0.21, 0.22, ..., 1}, column sample rate per tree = {0.2, 0.21, 0.22, ...,1}, and minimum rows = {0, 1, ..., $\log_2 \times$ number of rows-1}. The final learning parameters for both XGBoost implementations are presented in Table 3.

Table 3. The final learning parameters of XGBoost used for each dataset after performing a *random search*.

Hyperparameter	BODMAS	Kaggle	CIC-MalMem-2022
Native			
Maximum depth	11	19	19
eta	0.2	0.3	0.1
Subsample	0.6	0.8	0.6
Column sample by tree	0.7	0.5	0.6
H2O			
Maximum depth	24	23	26
Sample rate	0.52	0.94	0.99
Column sample rate	0.42	0.62	0.6
Column sample rate per tree	0.6	0.25	0.5
Minimum rows	2	2	2

(ii) CatBoost [24]

CatBoost is built with symmetric decision trees. It is acknowledged as a classification algorithm that is capable of producing an excellent performance and ten times the prediction speed of methods that do not employ symmetric decision trees. CatBoost, unlike other GBDT algorithms, is able to accommodate gradient bias and prediction shift to increase the accuracy of predictions and generalization ability of large datasets. In addition, CatBoost is comprised of two essential algorithms: ordered boosting, which estimates leaf values during tree structure selection to avoid overfitting, and a unique technique for handling categorical data throughout the training process. An implementation of CatBoost in *R* is employed in this paper, whereas the search space of each hyperparameter is considered as follows. Depth = {1, 2, ..., 10}, learning rate = {0.03, 0.001, 0.01, 0.1, 0.2, 0.3}, l2 leaf regularization = {3, 1, 5, 10, 100}, border count = {32, 5, 10, 20, 50, 100, 200}, and boosting type = {"Ordered", "Plain"}. The final learning parameters of CatBoost for each malware dataset are given in Table 4.

Table 4. The final learning parameters of CatBoost used for each dataset after performing a *random search*.

Hyperparameter	BODMAS	Kaggle	CIC-MalMem-2022
Depth	10	4	2
Learning rate	0.2	0.2	0.2
L2 leaf regularization	5	3	5
Border count	100	100	50
Boosting type	"Plain"	"Plain"	"Ordered"

(iii)   Gradient boosting machine [25]

GBM is the first implementation of GBDT to utilize a forward learning technique. Trees are generated in a sequential manner, with future trees being dependent on the results of the preceding trees. Formally, GBM is achieved by iteratively constructing a collection of functions $f^0, f^1, \ldots, f^t$, given a loss function $\Omega(y_i, f^t)$. We can optimize our estimates of y_i by discovering another function $f^{t+1} = f^t + h^{t+1}(\mathbf{x})$, such that h^{t+1} reduces the estimated value of the loss function. In this study, we adopt GBM implementation in H2O, whereas the hyperparameters' search space is specified as follows. Maximum depth = {1, 3, 5, ..., 29}, sample rate = {0.2, 0.3, ..., 1}, column sample rate per tree = {0.2, 0.21, 0.22, ..., 1}, column sample rate change per level = {0.9, 0.91, ..., 1.1}, number of bins = $2^{\{4,5,\ldots,10\}}$, and minimum rows = {0, 1, ..., $\log_2 \times$ number of rows − 1}. Table 5 shows a list of all the final GBM hyperparameters that were used on each malware dataset.

Table 5. The final learning parameters of GBM used for each dataset after performing *random search*.

Hyperparameter	BODMAS	Kaggle	CIC-MalMem-2022
Maximum depth	24	25	27
Sample rate	0.52	0.44	0.72
Column sample rate per tree	0.42	0.64	0.61
Column sample rate change per level	1.02	1.04	0.92
Number of bins	64	512	1024
Minimum rows	2	2	8

(iv)   LightGBM [26]

LightGBM is an inexpensive gradient boosting tree implementations that employs histogram and leaf-wise techniques to increase both processing power and prediction precision. The histogram method is used to combine features that are incompatible with each another. Before generating a n-width histogram, the core idea is to discretize continuous features into n integers. Based on the discretized values of the histogram, the training data are scanned to locate the decision tree. The histogram method considerably reduces the runtime complexity. In addition, in LightGBM, the leaf with the greatest splitting gain was found and then divided using a leaf-by-leaf strategy. Leaf-wise optimization may result in overfitting and a deeper decision tree. To ensure great efficiency and prevent overfitting, LightGBM includes a maximum depth constraint to leaf-wise. In this study, we employed a LightGBM implementation in R with the following hyperparameter search space; Maximum bin = {100, 255}, maximum depth = {1, 2, ..., 15}, number of leaves = $2^{(\{1,2,\ldots,15\}}$, minimum data in leaf = {100, 200, ..., 1000}, learning rate = {0.01, 0.3, 0.01}, lambda l1 = {0, 10, 20, ..., 100}, lambda l2 = {0, 10, 20, ..., 100}, feature fraction = {0.5, 0.9}, bagging fraction = {0.5, 0.9}, path smooth = {$1 \times 10^{-8}, 1 \times 10^{-3}$}, and minimum gain to split = {0, 1, 2, ..., 15}.

Table 6 contains the list of all final LightGBM hyperparameters used for each malware dataset.

Table 6. The final learning parameters of LightGBM used for each dataset after performing *random search.*

Hyperparameter	BODMAS	Kaggle	CIC-MalMem-2022
Maximum bin	100	100	255
Maximum depth	10	9	3
Number of leaves	8192	8	512
Minimum data in leaf	1000	800	700
Learning rate	0.29	0.27	0.07
Lambda l1	40	0	0
Lambda l2	90	20	90
Feature fraction	0.5	0.9	0.9
Bagging fraction	0.5	0.5	0.5
Path smooth	0.001	1×10^{-8}	0.001
Minimum gain to split	2	15	11

4. Result and Discussion

This section analyzes and discusses the results of the tree-based classifier ensembles applied to malware classification. The results of exploratory analysis are presented first, followed by a performance comparison between the tree-based ensemble models.

4.1. Exploratory Analysis

We first provide a correlation analysis between multiple variables in each malware dataset. Figure 2 shows the correlation coefficient score matrix measured by Pearson correlation. Correlation analysis is useful to understand the relationship between variables in a dataset, since the Good input features of a dataset should have a high correlation with target features, but should be uncorrelated with each other. Figure 2 confirms that both BODMAS and Kaggle datasets have fewer uncorrelated features than CIC-MalMem-2022. Hence, to mitigate the curse of dimesionality, it is strongly recommended to employ feature selection before employing a machine learning method on CIC-MalMem-2022. Highly correlated features have a negligible effect on the output prediction but raise the computational cost.

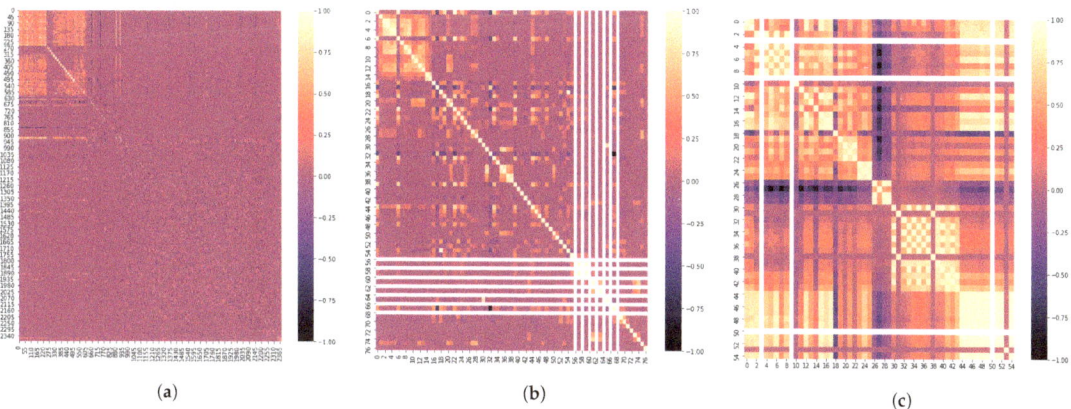

Figure 2. Feature correlation analysis of each malware dataset: (a) BODMAS, (b) Kaggle, and (c) CIC-MalMem-2022.

In addition, we ran a t-SNE algorithm [49] with a learning rate = 5000 and perplexity = 100. The t-SNE is an approach that converts a set of high-dimensional points to two dimensions in such a way that, ideally, close neighbors remain close and far points remain far. Figure 3 provides a spatial representation of the dataset in two dimensions. The t-SNE provides a pliable border between the local and global data structures. It also estimates the size of each datapoint's local neighborhood based on the local density of the data by requiring each conditional probability distribution to have the same perplexity (e.g., Gaussian kernel). Furthermore, Figure 3 demonstrates that both BODMAS and Kaggle datasets are highly imbalanced as compared with CIC-MalMem-2022.

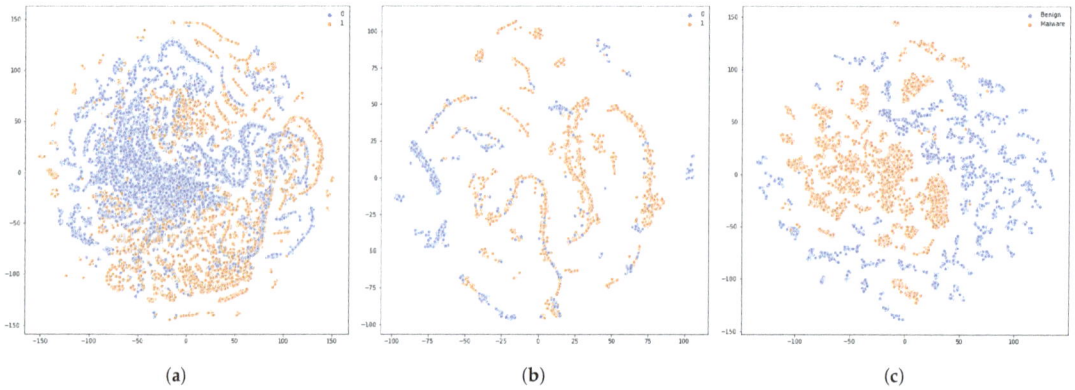

Figure 3. Two-dimensional visualization of instance pairs using t-SNE technique of each malware dataset: (**a**) BODMAS, (**b**) Kaggle, and (**c**) CIC-MalMem-2022.

4.2. Comparison Analysis

In the experiment, we employed a k cross-validation technique ($k = 10$), where the final performance outcome for each tree-ensemble model is the mean of the ten folds. The performance of each model was measured based on six performance metrics, such as accuracy, MCC, precision, recall, AUC, and F1. These metrics are chosen to provide more accurate estimates of the behavior of the classifier ensembles under the experiment. Especially, Chicco et al. [50] have shown that MCC is more informative than accuracy and F1, which yield reliable estimates when used to balanced datasets, but misleading outcomes when applied to imbalanced data sets. For a binary classification problem, the outcome of a tree-based classifier ensemble is typically derived from a contigency matrix, $\mathcal{T} = \begin{pmatrix} TP & FN \\ FP & TN \end{pmatrix}$, where TP is true positive, FN is false negative, FP is false positive, and TN is true negative. Let $\xi_+ = TP + FN$ and $\xi_- = TN + FP$ be the number of samples labeled as malware and non-malware, respectively. Hence, the performance metrics used in this study can be calculated as follows.

$$\text{Accuracy} = \frac{TP + TN}{\xi_+ + \xi_-} \tag{3}$$

$$\text{MCC} = \frac{TP \times TN - FP \times FN}{((TP + FP) \times \xi_- \times (TN + FN) \times \xi_+)^{1/2}} \tag{4}$$

$$\text{Precision} = \frac{TP}{TP + FP} \tag{5}$$

$$\text{Recall} = \frac{TP}{\xi_+} \tag{6}$$

$$AUC = \int_0^1 \text{Recall} \times \frac{FP}{\xi_-} d\frac{FP}{\xi_-} = \int_0^1 \text{Recall} \times \left(\frac{FP}{\xi_-}\right)^{-1} (x)dx \tag{7}$$

$$F1 = \frac{2TP}{2TP + FP + FN} \tag{8}$$

Figure 4 presents the performance score of each algorithm on each dataset. Overall, considering MCC as a performance indicator, LightGBM is the worst-performing algorithm, while XGBoost (native) is the best-performing on BODMAS and Kaggle datasets, followed by GBM (H2O). Interestingly, random forest has also performed well on the remaining dataset. Using accuracy as a performance metric, it is also apparent that there are modest performance disparities amongst algorithms (e.g., all algorithms achieve 100% accuracy). Consequently, our results support the findings stated by [50]. In Table 7, we provide the performance average of each algorithm over various datasets and demonstrate that XGBoost (native) is superior to any competitors on the board in terms of accuracy, MCC, and precision metrics. On the other hand, when recall, AUC, and F1 metrics are utilized, GBM (H2O) shows a superior performance.

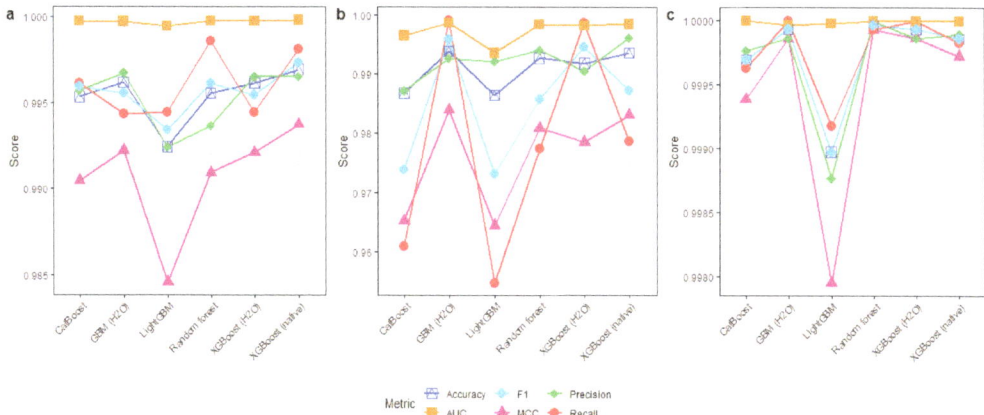

Figure 4. Performance comparison of various tree-based ensemble models on different datasets, i.e., (**a**) BODMAS, (**b**) Kaggle, and (**c**) CIC-MalMem-2022.

Table 7. Performance average of each ensemble technique over various malware datasets.

Ensemble Algorithms	Accuracy	MCC	Precision	Recall	AUC	F1
CatBoost	0.9940	0.9851	0.9943	0.9856	0.9988	0.9899
XGBoost (native)	0.9968	0.9922	0.9975	0.9923	0.9994	0.9949
LightGBM	0.9927	0.9823	0.9945	0.9828	0.9977	0.9885
Random forest	0.9961	0.9906	0.9959	0.9921	0.9994	0.9940
GBM (H2O)	0.9967	0.9920	0.9964	0.9978	0.9995	0.9971
XGBoost (H2O)	0.9960	0.9902	0.9956	0.9977	0.9994	0.9967

This section includes a two-step statistical significance test using Quade omnibus test and Quade post-hoc test [51] to better comprehend the performance difference between tree-based ensemble models. Using a significant threshold $\alpha = 0.05$ and MCC as a performance indicator, the Quade omnibus test demonstrates that at least one classifier performs differently than others (p-value = 0.01725). Since we found significance in the previous test, we then applied the Quade post-hoc test to determine the pairwise performance difference between classifiers. Here, we considered XGBoost (native) as a control classifier for comparison with the remaining algorithms. Table 8 depicts the p-value of the pairwise comparison. It is clear that the performance differences between XGBoost and the other algorithms are not statistically significant (p-value > 0.05).

Table 8. The p-value of Quade post-hoc, in which XGBoost (native) is used as a control algorithm.

CatBoost	XGBoost (Native)	LightGBM	Random Forest	GBM (H2O)	GBM (H2O)
0.250153	-	0.125201	0.7014781	0.6092802	0.8983268

To demonstrate the efficacy of tree-based ensemble models for malware detection, we compared our performance findings to those of previous studies for each dataset. Table 9 denotes the performance comparisons in terms of several performance measures, such as accuracy, precision, recall, and F1. Please note that the comparison is conducted as objectively as possible, given that the prior experiment may have been conducted under different settings, such as validation techniques and the number of training and testing samples. Nevertheless, this study shows that the top-performing tree-based ensemble examined for each dataset outperforms prior research, with a comparable result. More precisely, GBM (H2O), XGBoost (native), and random forest are the best performers on the Kaggle, BODMAS, and CIC-MalMem-2022 datasets, respectively, which also outperform other state-of-the-art malware detection techniques available in the recent literature.

Table 9. Performance comparisons over existing studies. The best performance value on each dataset is shown in bold.

Study	Accuracy (%)	Precision (%)	Recall (%)	F1 (%)
Kaggle				
Hou et al. [42]	93.68	93.96	93.36	93.68
Hou et al. [43]	96.66	96.55	96.76	96.66
Azmee et al. [38]	98.60	96.30	99.00	-
This study (GBM (H2O))	**99.39**	**99.27**	**99.92**	**99.59**
BODMAS				
Jingwei et al. [41]	99.29	98.07	98.26	94.23
Lu et al. [44]	96.96	-	-	96.96
This study (XGBoost (native))	**99.96**	**99.65**	**99.81**	**99.73**
CIC-MalMem-2022				
Dener et al. [37]	99.97	99.98	99.97	99.97
This study (Random forest)	**100.00**	**100.00**	99.99	**100.00**

5. Conclusions

This article examined tree-based ensemble learning algorithms that analyze PE malware. Several tree-based ensemble techniques, including random forest, XGBoost, CatBoost, GBM, and LightGBM, were assessed based on a number of performance criteria, such as accuracy, MCC, precision, recall, AUC, and F1. In addition, we incorporated cutting-edge malware datasets to comprehend the most recent attack trends. This work contributed to the prior research in several ways, including by providing a statistical comparison of fine-tuned tree-based ensemble models utilizing several malware datasets. Furthermore, this article can be expanded in a number of ways, including by looking at the explainability of tree-based ensemble models and signature-based malware classification. Furthermore, a deep neural network model for tabular data, such as TabNet [52], has been underexplored in this application domain, providing a new direction for future research. Finally, it is anticipated that tree-based PE malware detection will be deployed in various real-world settings, such as in host, network, and cloud-based malware detection components.

Author Contributions: Conceptualization, M.H.L.L. and B.A.T.; methodology, B.A.T.; validation, M.H.L.L.; investigation, M.H.L.L.; writing—original draft preparation, M.H.L.L.; writing—review and editing, M.H.L.L. and B.A.T.; visualization, B.A.T.; supervision, B.A.T. All authors have read and agreed to the published version of the manuscript.

Funding: This research received no external funding.

Institutional Review Board Statement: Not applicable.

Informed Consent Statement: Not applicable.

Data Availability Statement: Not applicable.

Conflicts of Interest: The authors declare no conflict of interest. The funders had no role in the design of the study; in the collection, analyses, or interpretation of data; in the writing of the manuscript, or in the decision to publish the results.

List of Acronyms

AUC	Area Under ROC Curve.
CART	Classification and Regression Tree.
CNN	Convolutional Neural Network.
CPS	Cyber-Physical Systems.
CV	Cross Validation.
DBN	Deep Belief Network.
DR	Detection Rate.
DT	Decision Tree.
FPR	False Positive Rate.
GBDT	Gradient Boosting Decision Tree.
GBM	Gradient Boosting Machine.
IoT	Internet of Things.
MCC	Matthews Correlation Coefficient.
NB	Naive Bayes.
PE	Portable Executable.
RF	Random Forest.
SAEs	Stacked AutoEncoders.
SVM	Support Vector Machine.
t-SNE	t-Stochastic Neighbor Embedding.
TPR	True Positive Rate.

References

1. Kleidermacher, D.; Kleidermacher, M. *Embedded Systems Security: Practical Methods for Safe and Secure Software and Systems Development*; Elsevier: Amsterdam, The Netherlands, 2012.
2. Xhafa, F. *Autonomous and Connected Heavy Vehicle Technology*; Academic Press: Cambridge, MA, USA, 2022.
3. Smith, D.J.; Simpson, K.G. *Safety Critical Systems Handbook*; Elsevier: Amsterdam, The Netherlands, 2010.
4. Damodaran, A.; Troia, F.D.; Visaggio, C.A.; Austin, T.H.; Stamp, M. A comparison of static, dynamic, and hybrid analysis for malware detection. *J. Comput. Virol. Hacking Tech.* **2017**, *13*, 1–12. [CrossRef]
5. Mahdavifar, S.; Ghorbani, A.A. Application of deep learning to cybersecurity: A survey. *Neurocomputing* **2019**, *347*, 149–176. [CrossRef]
6. Zhang, W.; Wang, H.; He, H.; Liu, P. DAMBA: Detecting android malware by ORGB analysis. *IEEE Trans. Reliab.* **2020**, *69*, 55–69. [CrossRef]
7. Ucci, D.; Aniello, L.; Baldoni, R. Survey of machine learning techniques for malware analysis. *Comput. Secur.* **2019**, *81*, 123–147. [CrossRef]
8. Milosevic, J.; Malek, M.; Ferrante, A. Time, accuracy and power consumption tradeoff in mobile malware detection systems. *Comput. Secur.* **2019**, *82*, 314–328. [CrossRef]
9. Zhang, H.; Xiao, X.; Mercaldo, F.; Ni, S.; Martinelli, F.; Sangaiah, A.K. Classification of ransomware families with machine learning based onN-gram of opcodes. *Future Gener. Comput. Syst.* **2019**, *90*, 211–221. [CrossRef]
10. Maniriho, P.; Mahmood, A.N.; Chowdhury, M.J.M. A study on malicious software behaviour analysis and detection techniques: Taxonomy, current trends and challenges. *Future Gener. Comput. Syst.* **2022**, *130*, 1–18. [CrossRef]

11. Montes, F.; Bermejo, J.; Sanchez, L.; Bermejo, J.; Sicilia, J.A. Detecting Malware in Cyberphysical Systems Using Machine Learning: A Survey. *KSII Trans. Internet Inf. Syst. (TIIS)* **2021**, *15*, 1119–1139.
12. Singh, J.; Singh, J. Detection of malicious software by analyzing the behavioral artifacts using machine learning algorithms. *Inf. Softw. Technol.* **2020**, *121*, 106273. [CrossRef]
13. Amer, E.; Zelinka, I. An ensemble-based malware detection model using minimum feature set. *Mendel* **2019**, *25*, 1–10.
14. Atluri, V. Malware Classification of Portable Executables using Tree-Based Ensemble Machine Learning. In Proceedings of the 2019 SoutheastCon, Huntsville, AL, USA, 11–14 April 2019; pp. 1–6. [CrossRef]
15. Azeez, N.A.; Odufuwa, O.E.; Misra, S.; Oluranti, J.; Damaševičius, R. Windows PE Malware Detection Using Ensemble Learning. *Informatics* **2021**, *8*, 10. [CrossRef]
16. Damaševičius, R.; Venčkauskas, A.; Toldinas, J.; Grigaliūnas, Š. Ensemble-Based Classification Using Neural Networks and Machine Learning Models for Windows PE Malware Detection. *Electronics* **2021**, *10*, 485. [CrossRef]
17. Mills, A.; Spyridopoulos, T.; Legg, P. Efficient and Interpretable Real-Time Malware Detection Using Random-Forest. In Proceedings of the International Conference on Cyber Situational Awareness, Data Analytics And Assessment (Cyber SA), Oxford, UK, 3–4 June 2019; pp. 1–8. [CrossRef]
18. Zhou, Z.H. *Ensemble Methods: Foundations and Algorithms*; CRC Press: Boca Raton, FL, USA, 2012.
19. Breiman, L. Random forests. *Mach. Learn.* **2001**, *45*, 5–32. [CrossRef]
20. Chen, C.; Zhang, Q.; Ma, Q.; Yu, B. LightGBM-PPI: Predicting protein-protein interactions through LightGBM with multi-information fusion. *Chemom. Intell. Lab. Syst.* **2019**, *191*, 54–64. [CrossRef]
21. Cai, J.; Li, X.; Tan, Z.; Peng, S. An assembly-level neutronic calculation method based on LightGBM algorithm. *Ann. Nucl. Energy* **2021**, *150*, 107871. [CrossRef]
22. Csizmadia, G.; Liszkai-Peres, K.; Ferdinandy, B.; Miklósi, Á.; Konok, V. Human activity recognition of children with wearable devices using LightGBM machine learning. *Sci. Rep.* **2022**, *12*, 5472. [CrossRef]
23. Chen, T.; He, T.; Benesty, M.; Khotilovich, V.; Tang, Y.; Cho, H.; Chen, K. Xgboost: Extreme gradient boosting. In *R Package Version 0.4–2*; R Foundation for Statistical Computing: Vienna, Austria, 2015; Volume 1, pp. 1–4.
24. Dorogush, A.V.; Ershov, V.; Gulin, A. CatBoost: gradient boosting with categorical features support. *arXiv* **2018**, arXiv:1810.11363.
25. Friedman, J.H. Greedy function approximation: A gradient boosting machine. *Ann. Stat.* **2001**, *29*, 1189–1232. [CrossRef]
26. Ke, G.; Meng, Q.; Finley, T.; Wang, T.; Chen, W.; Ma, W.; Ye, Q.; Liu, T.Y. Lightgbm: A highly efficient gradient boosting decision tree. *Adv. Neural Inf. Process. Syst.* **2017**, *30*, 1–9.
27. Yang, L.; Ciptadi, A.; Laziuk, I.; Ahmadzadeh, A.; Wang, G. BODMAS: An Open Dataset for Learning based Temporal Analysis of PE Malware. In Proceedings of the IEEE Security and Privacy Workshops (SPW), San Francisco, CA, USA, 27 May 2021; pp. 78–84. [CrossRef]
28. Carrier, T.; Victor, P.; Tekeoglu, A.; Lashkari, A.H. Detecting Obfuscated Malware using Memory Feature Engineering. In Proceedings of the ICISSP, Online, 9–11 February 2022; pp. 177–188.
29. Singh, J.; Singh, J. A survey on machine learning-based malware detection in executable files. *J. Syst. Archit.* **2021**, *112*, 101861. [CrossRef]
30. Albishry, N.; AlGhamdi, R.; Almalawi, A.; Khan, A.I.; Kshirsagar, P.R. An Attribute Extraction for Automated Malware Attack Classification and Detection Using Soft Computing Techniques. *Comput. Intell. Neurosci.* **2022**, *2022*, 5061059. [CrossRef]
31. Vadrevu, P.; Rahbarinia, B.; Perdisci, R.; Li, K.; Antonakakis, M. Measuring and detecting malware downloads in live network traffic. In Proceedings of the European Symposium on Research in Computer Security, Egham, UK, 9–13 September 2013; Springer: Berlin/Heidelberg, Germany, 2013; pp. 556–573.
32. Uppal, D.; Sinha, R.; Mehra, V.; Jain, V. Malware detection and classification based on extraction of API sequences. In Proceedings of the 2014 International Conference on Advances in Computing, Communications and Informatics (ICACCI), Delhi, India, 34–27 September 2014; pp. 2337–2342.
33. Kwon, B.J.; Mondal, J.; Jang, J.; Bilge, L.; Dumitraş, T. The dropper effect: Insights into malware distribution with downloader graph analytics. In Proceedings of the 22nd ACM SIGSAC Conference on Computer and Communications Security, Denver, CO, USA, 12–16 October 2015; pp. 1118–1129.
34. Mao, W.; Cai, Z.; Towsley, D.; Guan, X. Probabilistic inference on integrity for access behavior based malware detection. In Proceedings of the International Symposium on Recent Advances in Intrusion Detection, Tokyo, Japan, 2–4 November 2015; Springer: Berlin/Heidelberg, Germany, 2015; pp. 155–176.
35. Wüchner, T.; Ochoa, M.; Pretschner, A. Robust and effective malware detection through quantitative data flow graph metrics. In Proceedings of the International Conference on Detection of Intrusions and Malware, and Vulnerability Assessment, Milan, Italy, 9–10 July 2015; Springer: Berlin/Heidelberg, Germany, 2015; pp. 98–118.
36. Ahmadi, M.; Ulyanov, D.; Semenov, S.; Trofimov, M.; Giacinto, G. Novel feature extraction, selection and fusion for effective malware family classification. In Proceedings of the sixth ACM Conference on Data and Application Security and Privacy, New Orleans, LA, USA, 1–9 March 2016; pp. 183–194.
37. Dener, M.; Ok, G.; Orman, A. Malware Detection Using Memory Analysis Data in Big Data Environment. *Appl. Sci.* **2022**, *12*, 8604. [CrossRef]
38. Azmee, A.; Choudhury, P.P.; Alam, M.A.; Dutta, O.; Hossai, M.I. Performance Analysis of Machine Learning Classifiers for Detecting PE Malware. *Int. J. Adv. Comput. Sci. Appl.* **2020**, *11*, 510–517. [CrossRef]

39. Liu, X.; Lin, Y.; Li, H.; Zhang, J. A novel method for malware detection on ML-based visualization technique. *Comput. Secur.* **2020**, *89*, 101682. [CrossRef]
40. Asam, M.; Hussain, S.J.; Mohatram, M.; Khan, S.H.; Jamal, T.; Zafar, A.; Khan, A.; Ali, M.U.; Zahoora, U. Detection of Exceptional Malware Variants Using Deep Boosted Feature Spaces and Machine Learning. *Appl. Sci.* **2021**, *11*, 10464. [CrossRef]
41. Hao, J.; Luo, S.; Pan, L. EII-MBS: Malware Family Classification via Enhanced Instruction-level Behavior Semantic Learning. *Comput. Secur.* **2022**, *112*, 102905.
42. Hou, S.; Saas, A.; Ye, Y.; Chen, L. Droiddelver: An android malware detection system using deep belief network based on api call blocks. In Proceedings of the International Conference on Web-Age Information Management, Nanchang, China, 3–5 June 2016; Springer: Berlin/Heidelberg, Germany, 2016; pp. 54–66.
43. Hou, S.; Saas, A.; Chen, L.; Ye, Y.; Bourlai, T. Deep neural networks for automatic android malware detection. In Proceedings of the 2017 IEEE/ACM International Conference on Advances in Social Networks Analysis and Mining, Sydney, Australia, 31 July–3 August 2017; pp. 803–810.
44. Lu, Q.; Zhang, H.; Kinawi, H.; Niu, D. Self-Attentive Models for Real-Time Malware Classification. *IEEE Access* **2022**, *10*, 95970–95985. [CrossRef]
45. Bergstra, J.; Bengio, Y. Random search for hyper-parameter optimization. *J. Mach. Learn. Res.* **2012**, *13*, 281–305.
46. Wright, M.N.; Ziegler, A. Ranger: A fast implementation of random forests for high dimensional data in C++ and R. *arXiv* **2015**, arXiv:1508.04409.
47. Chen, T.; He, T.; Benesty, M.; Khotilovich, V. Package 'xgboost'. *R Version* **2019**, *90*, 1–66.
48. Cook, D. *Practical Machine Learning with H2O: Powerful, Scalable Techniques for Deep Learning and AI*; O'Reilly Media, Inc.: Sebastopol, CA, USA, 2016.
49. Van der Maaten, L.; Hinton, G. Visualizing data using t-SNE. *J. Mach. Learn. Res.* **2008**, *9*, 2579–2605.
50. Chicco, D.; Tötsch, N.; Jurman, G. The Matthews correlation coefficient (MCC) is more reliable than balanced accuracy, bookmaker informedness, and markedness in two-class confusion matrix evaluation. *BioBata Min.* **2021**, *14*, 13. [CrossRef] [PubMed]
51. Conover, W.J. *Practical Nonparametric Statistics*; John Wiley & Sons: Hoboken, NJ, USA, 1999; Volume 350.
52. Arik, S.Ö.; Pfister, T. Tabnet: Attentive interpretable tabular learning. In Proceedings of the AAAI Conference on Artificial Intelligence, Virtual Conference, 2–9 February 2021; Volume 35, pp. 6679–6687.

algorithms

MDPI

Article

A Momentum-Based Local Face Adversarial Example Generation Algorithm

Dapeng Lang [1,2], Deyun Chen [1,*], Jinjie Huang [1] and Sizhao Li [2]

[1] School of Computer Science and Technology, Harbin University of Science and Technology, Harbin 150080, China
[2] College of Computer Science and Technology, Harbin Engineering University, Harbin 150001, China
* Correspondence: chendeyun@hrbust.edu.cn (D.C.)

Abstract: Small perturbations can make deep models fail. Since deep models are widely used in face recognition systems (FRS) such as surveillance and access control, adversarial examples may introduce more subtle threats to face recognition systems. In this paper, we propose a practical white-box adversarial attack method. The method can automatically form a local area with key semantics on the face. The shape of the local area generated by the algorithm varies according to the environment and light of the character. Since these regions contain major facial features, we generated patch-like adversarial examples based on this region, which can effectively deceive FRS. The algorithm introduced the momentum parameter to stabilize the optimization directions. We accelerated the generation process by increasing the learning rate in segments. Compared with the traditional adversarial algorithm, our algorithms are very inconspicuous, which is very suitable for application in real scenes. The attack was verified on the CASIA WebFace and LFW datasets which were also proved to have good transferability.

Keywords: adversarial examples; face recognition; mask matrix; targeted attack; non-targeted attack

Citation: Lang, D.; Chen, D.; Huang, J.; Li, S. A Momentum-Based Local Face Adversarial Example Generation Algorithm. *Algorithms* **2022**, *15*, 465. https://doi.org/10.3390/a15120465

Academic Editors: Francesco Bergadano and Giorgio Giacinto

Received: 24 October 2022
Accepted: 2 December 2022
Published: 8 December 2022

Publisher's Note: MDPI stays neutral with regard to jurisdictional claims in published maps and institutional affiliations.

1. Introduction

1.1. Introductions

In the field of computer vision, deep learning has become a major technology for applications such as self-driving cars, surveillance, and security. Face verification [1] and face recognition [2] have outperformed humans. The recently proposed ArcFace [3] is an improvement on the previous face recognition model, which uses the loss function in angle space to replace the one in the CosFace [4] model. Earlier, the loss of the Euclidean distance space was used in the FaceNet [5] model. Furthermore, in some face recognition competitions such as the Megaface competition, ArcFace models are comparable to those of Microsoft and Google, and the accuracy rate reached 99.936%. Moreover, many open-source datasets such as LFW [6], CASIA-WebFace [7], etc. are available to researchers.

Despite the extraordinary success of deep neural networks, adversarial attacks against deep models also pose a huge threat in computer vision such as face recognition [8] and person detecton [9]. Szegedy, C. [10] and Goodfellow, I.J. [11] proved from the principle and experiment that the adversarial example is the inherent property of the deep model and proposed a series of classical algorithms. Dong, Y. [12] proposed the momentum algorithm on this basis, which is also one of the research bases of this paper.

However, the fine adversarial noise based on the whole image is not easy to realize, yet adversarial patch is an excellent option. Adversarial patches are covered to an image making it lead to misclassification or undetectable recognition by highlighting salient features of the object classification [13]. In the task of detection and classification, adversarial patches can be on the target or the background, regardless of the location [14]; a sticker on a traffic sign may cause the misclassification of traffic signs [15]; Refs. [16,17] Make it impossible for detctor to detect the wearer by creating wearable adversarial clothing(like a T-shirt or jacket). Ref. [18] is a very powerful attack that uses adversarial glasses to deceive both the digital and physical face recognition system; Based on this idea, researchers turned to the application of adversarial patches in the field of face recognition, and achieved a high success rate [19]. Therefore, adversarial examples are a non-negligible threat in the security field and have received a lot of attention.

1.2. Motivations

There are numerous methods for targeting face recognition models, and many of them have been validated in real scenarios. Ref. [11] proposed that the perturbation direction is the direction of the gradient of the predicted the target category labels; in addition, a GAN-based AGN [1] generates an ordinary eyeglass frame sticker to attack the VGG model. Ref. [3] proposed a new, simple, and replicable method attack the best public Face ID system ArcFace. Adversarial patches generally have a fixed position and visible scale, and also need to consider deformation and spatial mapping [7].

The second idea is rooted in the pixel level, which tricks the FRS with subtle perturbations. As previously described, generating adversarial examples against the full image ignores the semantic information within faces [9]. Such algorithms theoretically validate the feasibility of the attack, but are too restrictive in terms of the environmental requirements, making it difficult to realize. Meanwhile, existing algorithms launch undifferentiated attacks on all the targets in the picture. In real scenes, there are multiple objects in the complex background and foreground, and attacking multiple objects at the same time makes it easy to attract the attention of defenders. To address the above problems, we propose an adversarial example generation algorithm that targets local areas with distinctive facial features.

1.3. Contributions

As shown in Figure 1, our algorithm combines the advantages of adversarial patches and perturbations, generating invisible adversarial examples in the form of a patch. We first extracted a face from the image, and then generated the adversarial example based on the local key features of the face. The adversarial example can be targeted or non-targeted, which can effectively mislead FRS.

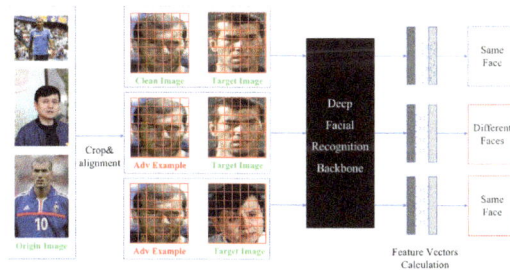

Figure 1. Face-based targeted attack and non-targeted attack diagram. The FRS could identify the first pair of images belonging to the same person but was unable to tell whether the face under attack was the same as the original person.

The work in this paper is as follows.

1. We proposed a white-box adversarial example generation algorithm (AdvLocFace) based on the local face. We circled an area with intensive features on the face to construct an patch-like adversarial example within this range.
2. A momentum optimization module with a dynamic learning rate was proposed. By adopting a dynamic piecewise learning rate, the optimization algorithm can acceler-

ate convergence; the momentum parameter was introduced to avoid the algorithm oscillating near the best point, which improved the attack efficiency.

3. By dynamically calculating the attack threshold, the optimal attack effect parameters were estimated, which reduced the number of modifications to the pixels in the clean images and effectively improved the transferability of the adversarial examples.

4. We compared the algorithm with several traditional algorithms. The experiments showed that the algorithm had a high success rate in the white-box setting, and to also obtain an ideal transferability.

2. Preliminaries

2.1. Deep Model of Face Recognition

DeepFace [1] is the first near-human accuracy model using Labeled Faces in the Wild (LFW) [20] and applies neural networks to face recognition models with nine layers to extract the face vectors. FaceNet [5] computes the Euclidean distance of the feature vectors of face pairs by mapping the face images into the feature space. In addition, they introduced triplet loss as a loss function so that after training, the distance of matched face pairs with the same identity would be much smaller than the distance of unmatched face pairs with different identities [4]. Sphereface [21] uses angular softmax loss to achieve the requirement of "maximum intra-class distance" to be less than "minimum inter-class distance" in the open-set task of face recognition. ArcFace [3] introduces additive angular margin loss, which can effectively obtain face features with high discrimination. The main approach is to add the angle interval m to the θ between x_i and W_{ij} to penalize the angle between the deep features and their corresponding weights in an additive manner. The equation is as follows:

$$L_3 = -\frac{1}{N}\sum_{i=1}^{N} log \frac{e^{s(cos(\theta_{y_i}+m))}}{e^{s(cos(\theta_{y_i}+m))} + \sum_{j=1,j\neq y_i}^{n} e^{s\cdot cos\theta_j}} \tag{1}$$

2.2. Classic Adversarial Attacks Algorithms

Adversarial examples are delicately designed perturbations imperceptible to humans to the input that leads to incorrect classifications [9]. The generation principle is shown in the following equation:

$$X' = X + \epsilon \cdot sign(\nabla_X L(f(x), y)) \tag{2}$$

where ϵ is set empirically, which indicates the learning rate. $L(f(x), y)$ is the linear loss function with the image x and label y. Update the input data by passing back the gradient $\nabla_X L(f(x), y)$, and use the $sign()$ to calculate the update direction.

The fast gradient sign method (FGSM) is a practical algorithm for the fast generation of the adversarial examples proposed by Goodfellow et al. [11]. To improve the transferability of the adversarial examples, Dong et al. [12] proposed the momentum iterative fast gradient sign method by adding the momentum term to the BIM, which prevents the model from entering the local optima and generating overfitting. The C&W [13] attack is a popular white-box attack algorithm that generates adversarial examples with high image quality, and transferability, and is very difficult to defend. Lang et al. [22] proposed the use of the attention mechanism to guide the generation of adversarial examples.

2.3. Adversarial Attacks on Face Recognition

The attack on the face not only needs to deceive the deep model but also requires the semantic expression of the attack method. Ref. [23] studied an off-the-shelf physical attack projected by a video camera, and project the digital adversarial mode onto the face of the adversarial factor in the physical domain, so as to implement the attack on the system. Komkov et al. [19] attached printed colored stickers on hats, called AdvHat, as shown in Figure 2.

Figure 2. AdvHat can launch an attack on facial recognition systems in the form of a hat.

In the context of the COVID-19 epidemic, Zolfi et al. [24] used universal adversarial perturbations to print scrambled patterns on medical masks and deceived face recognition models. Yin et al. [25] proposed the face adversarial attack algorithm of the Adv-Makeup framework, which implemented a black-box attack with imperceptible properties and good mobility. The authors in [26] used a generation model to improve the portability of adversarial patches in face recognition. This method not only realized the digital adversarial example but also achieved success in the physical world. In [27], they generated adversarial patches based on FGSM. The effectiveness of the attack was proven by a series of experiments with different numbers and sizes of patches. However, the patch was still visible and still did not take into account the feature information of the face. The study in [28] introduced adversarial noise in the process of face attribute editing and integrated it into the high-level semantic expression process to make the example more hidden, thus improving the transferability of adversarial attacks.

3. Methodology and Evaluations

3.1. Face Recognition and Evaluation Matrix

We used a uniform evaluation metric to measure whether a face pair matched or not. A positive sample pair is a matched face pair with the same identity; a negative sample pair is a mismatched face pair. To evaluate the performance of the face recognition model, the following concepts are introduced.

The True Positive Rate (TPR) is calculated as follows:

$$TPR = \frac{TP}{TP + FN} \tag{3}$$

where *TP* indicates the matching face pair and is correctly predicted as a matching face pair, and *FN* means a matching face pair and is incorrectly predicted as a mismatched face pair. *TPR* is the probability of correctly predicted positive samples to all positive samples, which is the probability of correctly predicted matched face pairs to all matched face pairs.

The False Positive Rate (FPR) is calculated as follows:

$$FPR = \frac{FP}{FP + TN} \tag{4}$$

where *FP* denotes a face pair whose true label is mismatched and is incorrectly predicted as a matched face pair. *TN* denotes a face pair whose true label is mismatched and is correctly predicted as a mismatched face pair. *FPR* is the probability that the incorrectly predicted negative samples account for all negative samples, and in the face recognition scenario is the probability that the incorrectly predicted mismatched face pairs account for all mismatched face pairs.

Therefore, the accuracy rate (Acc) of the face recognition model is calculated as follows:

$$Acc = \frac{TP + TN}{TP + FN + TN + FP} \tag{5}$$

That is, the accuracy of the face recognition model is the ratio of the number of correctly predicted face pairs to the total number of face pairs. Meanwhile, we chose five face recognition models with different network architectures for validation. These networks are described in the following sections.

3.2. Adversarial Attacks against Faces

The adversarial attacks are classified into non-targeted attacks and targeted attacks. An intuitive way to do this is to set a threshold. When the distance between two faces and this threshold is compared, if the result is less than the threshold, the two faces are from the same person and vice versa. This is obviously more difficult for the FRS to mistake the target face for another designated one [18].

Suppose that for input x, the true label $f(x) = y$ is output by the classification model f. The purpose of the adversarial attack is to generate an adversarial example x^{adv} by adding a small perturbation, and there exists $\left\| x^{adv} - x \right\|_p \leq \varepsilon$, where p can be 0, 1, 2, ∞.

For the non-targeted attack, the generated adversarial example makes $f\left(x^{adv}\right) \neq y$ and the results of the classifier were different from the original label; for the targeted attack, it makes $f\left(x^{adv}\right) \neq y^*$, where $y^* \neq y$, a previously defined specific class.

3.3. Evaluation Indices of Attack

Our goal is to generate adversarial patches to deceive FRS within a small area of the human face. The patch is generated by optimizing the pixels in the area, changing the distance between pairs of faces. The smaller the patch, the less likely it is to be detected by defenders. We explain the process of generating these patches.

1. Cosine Similarity is calculated by the cosine of the angle between two vectors, given as vector X and vector Y, and their cosine similarity is calculated as follows.

$$\cos(X, Y) = \frac{X \cdot Y}{\| X \| \| Y \|} = \frac{\sum_{i=1}^{n} X_i Y_i}{\sqrt{\sum_{i=1}^{n} X_i^2} \sqrt{\sum_{i=1}^{n} Y_i^2}} \tag{6}$$

where X_i and Y_i are the individual elements of vector X and vector Y, respectively. The cosine similarity takes values in the range [–1, 1], and the closer the value is to 1, the closer the orientation of these two vectors (i.e., the more similar the face feature vectors). Cosine similarity can visually measure the similarity between the adversarial example and the clean image.

2. Total variation (TV) [19], as a regular term loss function, reduces the variability of neighboring pixels and makes the perturbation smoother. Additionally, since perturbation smoothness is a prerequisite property for physical realizability against attacks, this lays some groundwork for future physical realizability [18]. Given a perturbation noise r, $r_{i,j}$ is the pixel where the perturbation r is located at coordinate (i, j). The $TV(r)$ value is smaller when the neighboring pixels are closer (i.e., the smoother the perturbation, and vice versa). The TV is calculated as follows:

$$TV(r) = \sum_{i,j} \left(\left(r_{i,j} - r_{i+1,j}\right)^2 + \left(r_{i,j} - r_{i,j+1}\right)^2 \right)^{\frac{1}{2}} \tag{7}$$

3. We used the L_2 constraints to measure the difference between the original image and the adversarial example. L_2 is used as a loss function to control the range of perturbed noise. In the application scenario of attacking, it can be intuitively interpreted as whether the modified pixels will attract human attention.

Given vector X and vector Y, their L_2 distances (i.e., Euclidean distances) can be calculated as follows:

$$\| X, Y \|_2 = \sqrt{\sum_{i}^{n}(x_i - y_i)^2} \tag{8}$$

where x_i and y_i are the elements of the input vector X and the output vector Y, respectively. The larger the L_2 distance between the two vectors, the greater their difference.

4. Our Method

4.1. Configurations for Face Adversarial Attack

After the image preprocessing, we extracted the features from the two face images and calculated their distance. For the input face image x, the face recognition model f extracted the features. For the input face pairs $\{x_1, x_2\}$, the face feature vector $f(x_1)$ and $f(x_2)$ were mapped to 512-dimensional feature vectors, respectively.

Therefore, we compared the distance of $f(x_1)$ and $f(x_2)$ with the specified threshold to determine whether the face pair matched or not. We calculated the angular distance by cosine similarity, which is as follows.

$$Similarity = \cos(f(x_1), \ f(x_2)) \tag{9}$$

$$D(f(x_1), \ f(x_2)) = \frac{\arccos(Similarity)}{\pi} \tag{10}$$

where $\cos(\cdot, \cdot)$ is the cosine similarity of the feature vector of the face pair in the range of $[-1, 1]$. Therefore, $D(f(x_1), \ f(x_2))$, based on the cosine similarity, ranged from $[0, 1]$. The closer the distance is to 0, the more similar the face feature vector and the more likely the face pair is matched, and vice versa. Equation (11) is used to predict the matching result of the face pair.

$$C(x_1, \ x_2) = \mathbb{I}(D(f(x_1), \ f(x_2)) < threshold) \tag{11}$$

where $\mathbb{I}(\cdot)$ is the indicator function; the threshold is the baseline of the detection model that is different depending on the model used. $C(\cdot, \cdot)$ outputs the matching result, if $C = 1$, the face pair is matched; if $C = 0$, the face pair is not matched. A unified attack model is established based on the $\mathbb{I}(\cdot)$ indicator function to implement targeted and non-targeted attacks. The flow of face pair recognition based on the threshold comparison is shown in Figure 3.

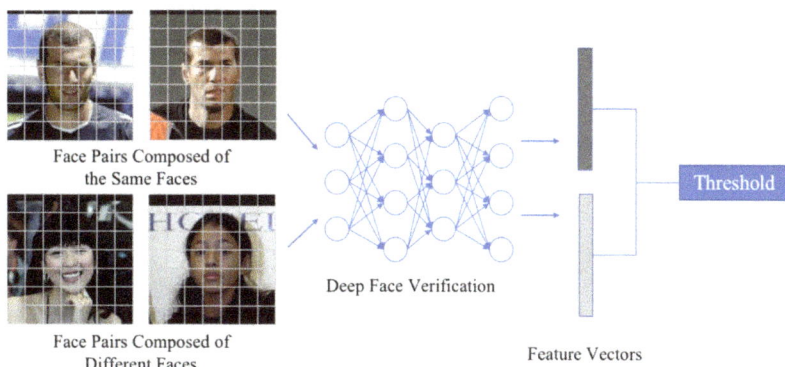

Face Pairs Composed of
the Same Faces

Deep Face Verification

Face Pairs Composed of
Different Faces

Threshold

Feature Vectors

Figure 3. Schematic diagram of the face feature vector and threshold comparison.

4.2. Local Area Mask Generation

The human eye region contains critical semantic information despite its small area [23]. The local region matrix is generated according to the human eye position as the range of constraint against adversarial perturbation. Due to various poses, illumination, and occlusions, we applied a deeply cascaded multitasking framework to integrate face detection and alignment through multitasking learning. First, since images have different sizes, the key points of the extracted face were affined to the unit space using affine transformation to unify the size and coordinate system. The detection and alignment of faces were accomplished by building a multi-level CNN structure containing three stages. Candidate windows will be quickly generated by a shallow CNN. Then, the windows were optimized by more complex CNNs to discard a large number of non-facial windows. Finally, it refines the results. by using a more powerful CNN and outputs the facial marker positions.

The algorithm flow is shown in Figure 4. In Figure 4b, for a given image, we first adjusted it to different scales to construct the image pyramid. In Figure 4c, we referred to the method in [29] to obtain the candidate windows and their bounding box regression vectors. The estimated bounding box regression vectors were then used to calibrate the candidate boxes; in Figure 4d, we used non-maximal suppression (NMS) to merge the highly overlapping candidate objects. In Figure 4e, all candidate frames were used as input to the CNN of the optimization network, which further discarded a large number of incorrect candidate frames, calibrated them using bounding box regression, and merged the NMS candidate frames. Finally, Figure 4f shows the CNN-based classification network that generated a more detailed description of the faces and outputs the critical facial positions.

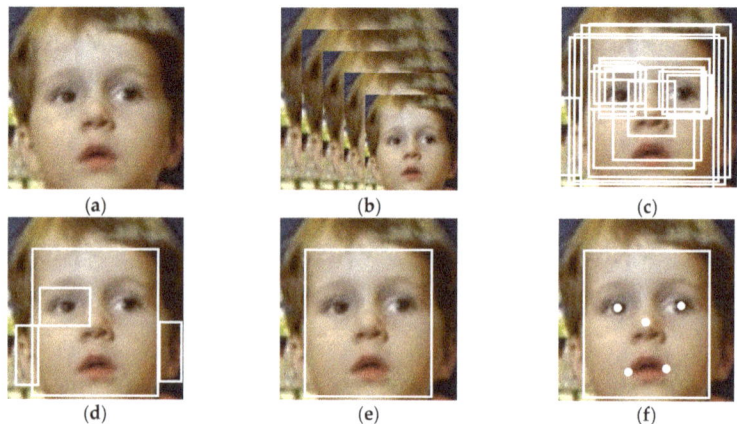

Figure 4. The process of face recognition and face alignment. (**a**) Clean image of a child. (**b**) Image pyramid. (**c**) Bounding box regression. (**d**) Merging the candidate objects. (**e**) Face location. (**f**) Key point location.

Pixels are randomly sampled within the range of key points as the corresponding feature pixel [29]. The feature pixels select the closest initial key point as the anchor and calculate the deviation. The coordinate system of the current pixel after rotation, transformation, and scaling should be close to the initial key point. It acts on the deviation and adds its own position information to obtain the feature pixel of the current key point. Then, we constructed the residual tree and calculated the deviation of the current key point from the target key point. We split the sample and updated the current key point position based on the average residual of the sample. Back to the previous step, it reselected the feature key points, fit the next residual tree, and finally combined the results of all residual trees to obtain the key point locations. According to the default settings, the coordinates of the points 0, 28, 16, 26–17 in the image for the human eye area are shown in Figure 5.

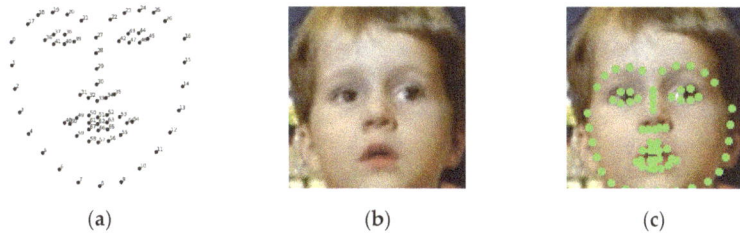

Figure 5. Schematic diagram of key point detection for human face. (**a**) Sixty-eight key points on a human face. (**b**) Face that needs to be matched. (**c**) Key points and face fitting.

We located the human eye region based on the key points of the detected eye in the image and drew the mask against the attacked region. The range of pixel values in the generated mask image was normalized to [0.0, 1.0] to generate a binary-valued mask matrix. This is shown in Figure 6.

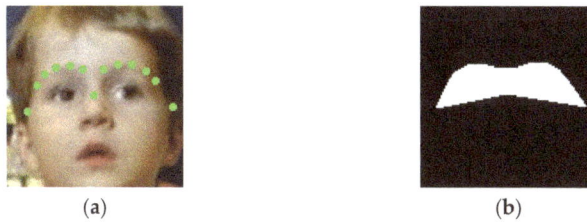

Figure 6. Schematic diagram of eye area matrix generation. (**a**) Locating key areas of the human eye. (**b**) Generating a human eye area mask.

We generated adversarial examples combining the eye region matrix and full face region, respectively, to test the effect of the attacks. Figure 7a shows the clean image used for testing while Figure 7b shows the visualization of the perturbation based on the eye region and the full-face region. Figure 7c is the adversarial example. After testing, both images could successfully deceive the face detector. The adversarial perturbation generated based on the human eye region accounted for 17.8% of the total pixels, while the number of pixels of the adversarial perturbation generated based on the whole face accounted for 81.3% of the image.

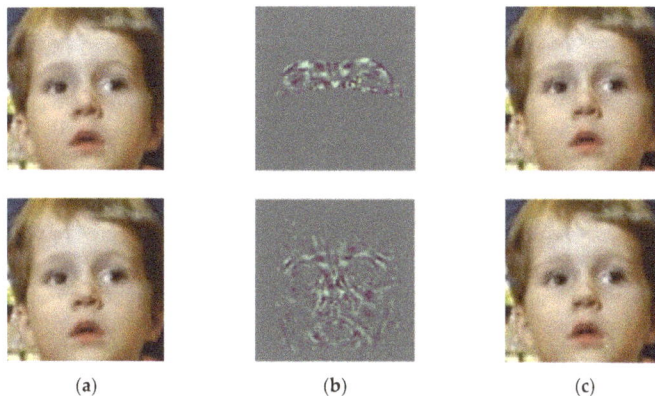

Figure 7. Schematic diagram of key point generation matrix based on eye detection. (**a**) Clean image of a child. (**b**) Adversarial perturbation based on the human eye and full face. (**c**) Adversarial examples based on the human eye and full face.

4.3. Loss Functions

As above-mentioned, this algorithm optimizes the $C\left(x, x^{adv}\right)$ function in the local region. For the targeted attack and non-targeted attack, the relationship between the clean face image x and the three target images x^{tar} and the adversarial example image x^{adv} was compared.

(1) For the non-targeted attack, an adversarial example x^{adv} was generated for the input image x so that the difference between them was as large as possible. When the difference was larger than the threshold value calculated by the deep detection model, the attack was successful; on the other hand, for the targeted attack, the generated adversarial example x^{adv} needed to be as similar as possible to the target image x^{tar}. The loss function $\mathcal{L}oss_1$ is shown as Equation (12).

$$\mathcal{L}oss_1 = \alpha \cdot \cos\left(f(x), f\left(x^{adv}\right)\right) - (1 - \alpha) \cdot \cos\left(f\left(x^{adv}\right), f\left(x^{tar}\right)\right) \tag{12}$$

where $\cos(\cdot, \cdot)$ is the cosine similarity of the feature vector calculated by Equation (10); α takes the value of 0 or 1, representing the non-targeted attack and targeted attack, respectively.

(2) The perturbation size is constrained by the L_2 norm, thus ensuring that the visibility of the perturbation is kept within a certain range when an effective attack is implemented. The loss function in this section constrains the perturbation after the restriction as follows.

$$\mathcal{L}oss_2 = L_2(mask \odot r) \tag{13}$$

where r is the perturbation. The mask is that generated from the first face image of the face pair to restrict the perturbation region. It is a [0, 1] matrix scaled to the same size as the image. The $\odot$ symbol indicates the dot product operation between the elements.

(3) The TV is used to improve the smoothness of the perturbation through Equation (14), and the loss function of this part also deals with the perturbation after restriction, as follows.

$$\mathcal{L}oss_3 = TV(mask \odot r) \tag{14}$$

In summary, for the above targeted and non-targeted attacks, the loss function is minimized by solving the following optimization problem of Equation (15), which can generate the final adversarial perturbation r :

$$\min_r \mathcal{L}oss = \min_r(\lambda_1 \mathcal{L}oss_1 + \lambda_2 \mathcal{L}oss_2 + \lambda_3 \mathcal{L}oss_3) \tag{15}$$

The hyperparameters λ_1, λ_2, and λ_3 are used to control the relative weights of the perturbation losses. The correlation coefficients of the two regular term loss functions $\mathcal{L}oss_2$ and $\mathcal{L}oss_3$ are gradually reduced as the number of iterations increases.

4.4. Momentum-Based Optimization Algorithms

To solve the optimization problem above, the adversarial perturbation is optimized by using an iterative gradient descent method to minimize the objective function. A momentum parameter superimposes in the gradient direction and dynamically stabilizes update directions in each iteration step [12].

In the updating process, due to the different iterations of updating for different scenes, we divided the updating process into several stages, and the learning rate of different stages gradually decreased. The gradient is calculated as follows.Meanwhile, the learning rate $\alpha_{\Delta t}$ is changed according to the number of iterations it and stages st.

$$grad = \beta \cdot m_i + \frac{\nabla_x \mathcal{L}oss(x^{adv}, y)}{||\nabla_x \mathcal{L}oss(x^{adv}, y)||_1}$$
$$m_{i+1} = grad$$
$$r_{t+1} = r_t - \alpha_{\Delta i} * grad \tag{16}$$
$$\alpha_{\Delta i} = \left(\frac{it}{st}\right)^i a_{\Delta(i-1)} + a_{\Delta(i-1)}$$

where $\left|\left|\nabla_x \mathcal{L}oss\left(x^{adv}, y\right)\right|\right|_1$ is the regularized representation of the gradient of $\nabla_x \mathcal{L}oss\left(x^{adv}, y\right)$. The parameter β is the decay factor, adjusting for the influence of momentum on the gradient calculation. r_t is the adversarial perturbation generated in the t iteration. The parameter $\alpha_{\Delta t}$ is dynamically adjusted and is related to the iterations it and the current stages st. If it is high, then * can be set bigger. As it increases, st will become smaller.

$$x_{t+1}^{adv} = clip_{x,\varepsilon}(x + mask \odot r_{t+1}) \tag{17}$$

where $clip_{x,\varepsilon}(\cdot)$ serves to restrict the adversarial examples after superimposed perturbation to a reasonable range (after normalization) of $[-1, 1]$ at the end of each iteration.

The final elaborate perturbation is processed and added to the original face image so that the final adversarial example is generated by restricting the perturbation to a reasonable range of $[0, 255]$ using $clip_{x,\varepsilon}(\cdot)$. The process is shown in Figure 8.

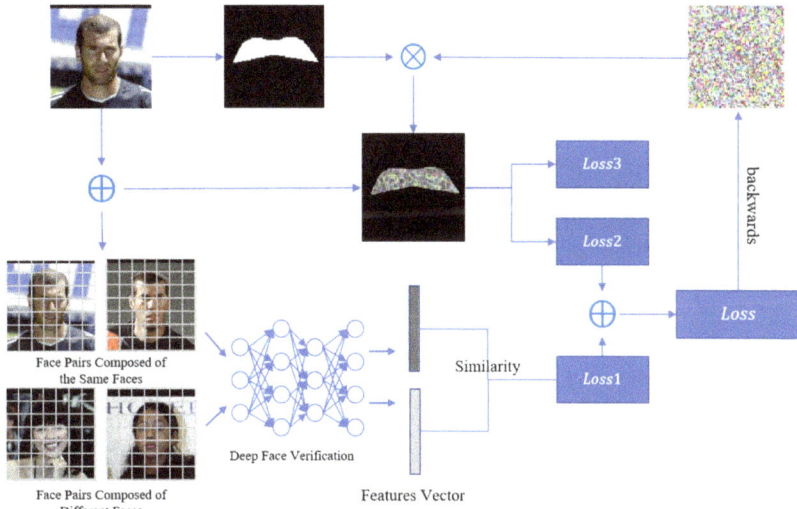

Figure 8. Flowchart of the adversarial attack based on the eye area.

Figure 8 shows that the feature vectors are first extracted from the aligned faces. The attack region is mapped by keypoint detection and the adversarial perturbation information is generated based on this region. The local aggressive perturbation is obtained through the optimization of the loss functions. This perturbation information can effectively mislead FRS.

5. Experiments

5.1. Datasets

In this paper, we used CASIA WebFace [7] as the training dataset. All of the pictures are from movie websites and vary in light and angle. In order to verify the effect of the algorithm on different datasets, we choose LFW [22] as our test dataset. It provides 6000 test face pairs, of which 3000 are matched pairs and 3000 are mismatched pairs.

For face detection and alignment, MTCNN [30–34] was used to uniformly crop the images to 112×112. Experimentally, there were 500 pairs of matched faces with the same identity for non-targeted attacks and 500 pairs of unmatched faces with different identities for targeted attacks.

5.2. Performance Evaluation for Face Recognition Models

Five mainstream pre-trained face recognition models were used for comparison, namely ResNet50-IR (IR50) [31], ResNet101-IR (IR101) [32], SEResNet50-IR (IR-SE50) [33], MobileFaceNet [8], and ResNet50 [33]. In order to show the success rate of adversarial attacks more intuitively, the metrics were the True Accept Rate (TAR) and False Accept Rate (FAR) [20]. Given a face pair (x_1, x_2), let the matched face pair be P_s and the unmatched face pair be P_d. Given a threshold, the calculation of *TAR* and *FAR* is as follows.

$$TA = \left\{ \begin{array}{c} (x_1, x_2) \in P_s, \\ \text{with } D(f(x_1), f(x_2)) < threshold \end{array} \right\} \tag{18}$$

$$FA = \left\{ \begin{array}{c} (x_1, x_2) \in P_d, \\ \text{with } D(f(x_1), f(x_2)) < threshold \end{array} \right\} \tag{19}$$

$$TAR = \frac{|TA|}{|P_s|} \tag{20}$$

$$FAR = \frac{|FA|}{|P_d|} \tag{21}$$

where $|TA|$ is the number of all matched pairs whose distance is less than the threshold; $|P_s|$ is the number of all matched pairs; $|FA|$ is the number of all unmatched pairs whose distance is less than the threshold; and $|P_d|$ is the number of all unmatched face pairs.

Different models will have different thresholds that can objectively reflect the success rate of the attack. Accordingly, the threshold was determined according to different values of *FAR*, and was chosen when $FAR = 1 \times 10^{-2}$ or 1×10^{-3}. We traversed the range of thresholds and used the 10-fold cross-validation method to find the threshold closest to the target *FAR*.

As shown in Figure 9, the TPR of each model (i.e., the proportion of correctly predicted matched face pairs to all unmatched face pairs) was maintained above 96.5% when $FAR = 1 \times 10^{-6}$. When $FAR = 1 \times 10^{-2}$, the TPR reached more than 98.5%. This indicates that the performance of these backbone network models had excellent performance.

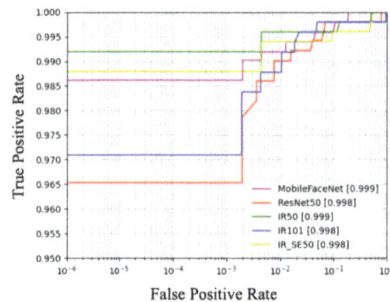

Figure 9. The ROC curve of FPR in the range of 1×10^{-6} to 1.

The test results of the five models on the LFW at a FAR of about 0.01 are shown in Table 1. The value of TAR@FAR = 0.01 (i.e., the probability of correctly identifying matching face pairs when the FAR is close to 0.01) was maintained at more than 98.9%.

Table 1. The TAR and corresponding thresholds for different models.

Models	TAR (%)	FAR	Threshold	Sim-Threshold
IR50	99.596	0.00995	0.43326	0.20814
IR101	98.984	0.01259	0.41311	0.26960
IR-SE50	99.396	0.01025	0.42920	0.22060
ResNet50	99.596	0.00836	0.45207	0.15001
MobileFaceNet	99.196	0.01076	0.43763	0.19469

5.3. Attack Method Evaluation Indicators

The accuracy of a face recognition model intuitively reflects the predictive ability of the model. The attack success rate (ASR) is calculated as follows:

$$ASR = 1 - Acc \tag{22}$$

The higher the ASR, the more vulnerable the model is to adversarial attacks; the lower the ASR, the more robust the model is to adversarial attacks and is able to withstand a certain degree of adversarial attacks.

In order to evaluate the magnitude of the difference between the generated adversarial example and the original face image after the attack, this experiment used the peak signal-to-noise ratio (*PSNR*), and structural similarity (*SSIM*) [34], which are two metrics to measure the image quality of the adversarial example.

The *PSNR* is defined and calculated by the mean squared error (*MSE*). The following equation calculates the *PSNR* for a given image *I*.

$$PSNR = 20 \cdot log_{10}(MAX_I) - 10 \cdot log_{10}(MSE) \tag{23}$$

where MAX_I is the maximum pixel value of the image; *MSE* is the mean square error. The larger the *PSNR*, the less distortion and the better quality of the adversarial example [3].

Considering human intuition, we adopted the evaluation index of structural similarity (*SSIM*), which takes into account the three factors of brightness, contrast, and structure. Given images *x* and *y* with the same dimensions, the structural similarity is calculated as follows.

$$SSIM(x,y) = \frac{(2\mu_x\mu_y + c_1)(2\sigma_{xy} + c_2)}{\left(\mu_x^2 + \mu_y^2 + c_1\right)\left(\sigma_x^2 + \sigma_y^2 + c_2\right)} \tag{24}$$

Among them, μ_x, μ_y are the mean values of image x, y; σ_x^2, σ_y^2 are the variance of image x, y; σ_{xy} is the covariance, and c_1 and c_2 are used to maintain stability. *SSIM* takes values in the range of $[-1, 1]$, and the closer the value is to 1, the higher the structural similarity between the adversarial example and the original image. To a certain extent, it can indicate the more imperceptible the perturbation applied to the adversarial example is to humans.

5.4. Adversarial Attack within Human Eye Area

5.4.1. Non-Targeted Attacks based on Eye Area

A schematic diagram of the non-targeted attack is shown in Figure 10. The first column shows the face pair before the attack. To the human eye, there is no difference between the second image in Figure 10a,b, and the second image in Figure 10b is the adversarial example.

(a)

(b)

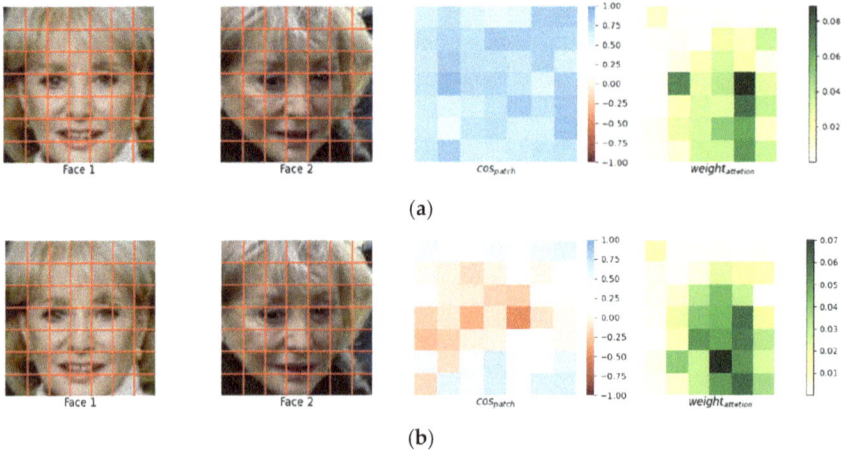

Figure 10. Before and after eye-based non-targeted attack. (**a**) Visualization of facial features of the same person from different angles. (**b**) Visualization of face features after being attacked.

$weight_{attetion}$ in the fourth column indicates the attention of the model, where the darker color indicates that the model paid more attention to the area. It can be seen that there was no significant change in the attention hotspots before and after the attack. In the third column, the xCos [35] module visualizes the face pairs before and after the attack and visualizes the changes in the images from the perspective of the neural network parameters. The bluer color in the similarity plot cos_{patch} indicates that they are more similar, and the redder color indicates that they are less similar. It can be seen that the face pairs changed dramatically after the attack.

5.4.2. Targeted Attacks Based on Eye Area

The purpose of the targeted attack is to deceive the deep detection model into misidentifying another specific face from the original image. As shown in Figure 11a, the similarity graph of the face pair before the attack had a large number of red grids, indicating that this pair was very dissimilar and was a mismatched face pair, while the model's attention focused on the eye area in the middle of the face. The first image in Figure 11b is the generated adversarial example; the second image is the target image. Intuitively, the first images in Figure 11a,b are exactly the same. This is also reflected in the attention map. However, for the face recognition model, the grid of the eye region in cos_{patch} mostly changed to blue, and 43% of the regions changed from yellow to blue. This indicates that the image change affected the classification of deep model.

5.4.3. Quantitative Comparison of Different Attack Models

To verify the effectiveness of the algorithm, we selected 500 faces for targeted and non-targeted attacks. Furthermore, each model has a different threshold for the best performance. The attack success rates of different attack models are shown in Table 2.

Table 2. The accuracy and success rates of different models under the specified thresholds.

Models	ACC (%)	Targeted-ASR (%)	Non-Targeted-ASR (%)	Threshold
IR50	98.2	90.4	99.2	0.43326
IR101	94.7	96.2	98.6	0.41311
IR-SE50	92.6	92.2	98.8	0.42920
ResNet50	94.5	93.8	99.4	0.45207
MobileFaceNet	96.6	91.4	99.2	0.43763

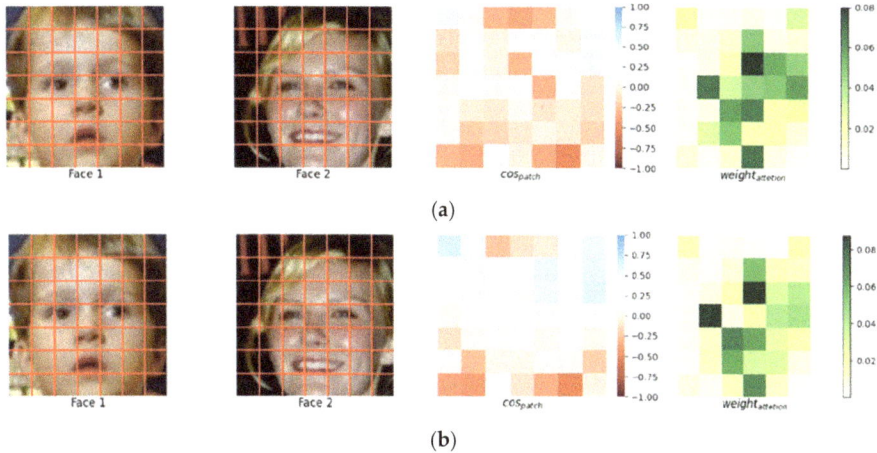

Figure 11. Before and after the eye-based targeted attack. (**a**) Visualization of the features of different faces. (**b**) Feature visualization of different faces after the targeted attack.

Our method was compared with the traditional adversarial algorithms. With a high success rate, we compared the differences between the adversarial examples and the original images, and the evaluated metrics included the image quality of the adversarial examples, the calculated average PSNR and SSIM. The perturbations of the adversarial examples generated by our algorithm for different deep detection models were counted. A PSNR greater than 40 indicates that the image distortion was small; the closer the SSIM takes the value of 1, the closer the adversarial example is to the original image. The comparison results are shown in Table 3.

Table 3. Average PSNR, SSIN, and the perturbed parameters for different models.

Models	Targeted-PSNR	Targeted-SSIM	Targeted-L2	Non-Targeted-PSNR	Non-Targeted-SSIM	Non-Targeted-L2
IR50	43.69864	0.99358	0.71112	39.19224	0.98557	1.29420
IR101	43.68891	0.99379	0.71556	42.37390	0.99232	0.84162
IR-SE50	43.46344	0.99345	0.76280	39.95576	0.98470	1.22338
ResNet50	45.51955	0.99531	0.58352	40.89765	0.98938	1.09996
MobileFaceNet	43.82034	0.99395	0.72392	41.36429	0.98954	1.04512

As shown in Table 3, for different attacks, the PSNR of all models was above 40 dB and the SSIM was above 0.98, indicating that the image distortion was very small and the perturbations were imperceptible to humans; on the other hand, the perturbations generated by the targeted attack was much lower than that of the non-targeted attack. The momentum in this algorithm was updated toward the target image due to the directed output of the image. Therefore, the adversarial example generation algorithm was also guided to optimize in the direction of the specific objects.

5.4.4. Comparison of Adversarial Example Algorithms

In this paper, the validation dataset covered 40 different categories. These categories can be correctly classified by the MobileFaceNet model (Top-1 correct); we also selected ArcFace [3] and SphereFace [21] face recognition models for testing. ArcFace uses IR101 as the network structure and has 99.8% accuracy in the LFW test set; SphereFace's network structure removes the BN module, which differs significantly from the ResNet50 residual network, with 99.5% accuracy in the LFW test set. We selected a variety of typical adversarial example algorithms FGSM [10], I-FGSM [11] algorithms, and the face-specific attack method AdvGlasses [18], AdvHat [19], and our algorithm (AdvLocFace) for cross-testing. The comparison results are shown in Table 4.

Table 4. Accuracy and success rates of different algorithms.

Models	Attack Method	ResNet50	MobileFaceNet	SphereFace	ArcFace
ResNet50	FGSM	77.00%	34.81%	31.88%	30.26%
	I-FGSM	100.00%	24.41%	21.76%	18.82%
	AdvGlasses	100.00%	51.05%	48.02%	40.58%
	AdvHat	97.80%	52.92%	44.87%	44.24%
	AdvLocFace	99.10%	57.15%	51.35%	59.00%
MobileFaceNet	FGSM	39.99%	67.67%	27.83%	29.04%
	I-FGSM	38.86%	100.00%	24.59%	20.75%
	AdvGlasses	69.39%	100.00%	46.06%	45.64%
	AdvHat	77.61%	97.90%	46.29%	37.38%
	AdvLocFace	61.62%	99.20%	52.76%	40.92%
SphereFace	FGSM	38.55%	32.34%	59.20%	28.74%
	I-FGSM	41.94%	33.37%	99.88%	25.91%
	AdvGlasses	76.59%	65.61%	99.58%	53.53%
	AdvHat	68.03%	54.06%	93.91%	64.16%
	AdvLocFace	62.96%	58.90%	96.65%	67.91%
ArcFace	FGSM	37.01%	33.58%	30.76%	75.19%
	I-FGSM	31.56%	25.65%	21.96%	98.67%
	AdvGlasses	57.88%	52.43%	50.20%	97.35%
	AdvHat	68.61%	65.90%	63.19%	100.00%
	AdvLocFace	73.24%	70.64%	70.57%	100.00%

In Table 4, the diagonal lines are white-box attack settings. I-FGSM improved the success rate of white-box attacks by increasing the iterative process, but reduced the mobility of the attack method due to the overfitting of the perturbation. The AdvHat algorithm is an advanced physical attack method that attacks realistic attacks by pasting stickers on the hat, and it is easy to replicate this attack. The optimization process, based on the consideration of pixel smoothing and color printability, limits the effect of mobility in digital attacks. AdvLocFace, with the best threshold for similar models based on the base model of training, obtained a more stable success rate of black-box attacks. For network models with different structures and different training data, the attack success rate decreased significantly.

6. Conclusions

This paper proposed a face adversarial example generation algorithm based on local regions. The algorithm uses the principle of a face recognition system to build a local area containing key features and generates momentum-based adversarial examples. This algorithm is a typical white-box attack method but still achieves good results in the black-box attack scenario.

Compared with the traditional adversarial attack method, the adversarial perturbation generated by our method only needs to cover a small part of the original image. Because the region contains the key features of the face, it can successfully mislead the face recognition system. In addition, the generated adversarial example is patch-like, which is highly similar to the original image and therefore more inconspicuous. Our algorithm can selectively attack any target in the image, so it can be extended to attack other types of images. the experiments show that the proposed algorithm can effectively balance the modified pixel area and attack successfully, achieving good transferability.

Author Contributions: Investigation, J.H.; Methodology, D.L.; Software, S.L.; Writing—Original draft, D.C.; Writing—Review & editing, J.H. All authors have read and agreed to the published version of the manuscript.

Funding: This research was funded by the State Administration of Science, Technology and Industry for National Defense, PRC, Grant No. JCKY2021206B102.

Institutional Review Board Statement: Not applicable.

Informed Consent Statement: Not applicable.

Data Availability Statement: All data and used models during the study are available in a repository or online. The datasets were CASIA WebFace [7], LFW [6], and MTCNN [30].

Conflicts of Interest: The authors declare no conflict of interest.

References

1. Taigman, Y.; Yang, M.; Ranzato, M.A.; Wolf, L. DeepFace: Closing the gap to human-level performance in face verification. In Proceedings of the IEEE Conference on Computer Vision and Pattern Recognition, Columbus, OH, USA, 23–28 June 2014.
2. Taigman, Y.; Yang, M.; Ranzato, M.A.; Wolf, L. Web-scale training for face identification. In Proceedings of the IEEE Conference on Computer Vision and Pattern Recognition, Boston, MA, USA, 7–12 June 2015.
3. Deng, J.; Guo, J.; Xue, N.; Zafeiriou, S. Arcface: Additive angular margin loss for deep face recognition. In Proceedings of the IEEE/CVF Conference on Computer Vision and Pattern Recognition, Long Beach, CA, USA, 16–20 June 2019; pp. 4690–4699.
4. Wang, H.; Wang, Y.; Zhou, Z.; Ji, X.; Gong, D.; Zhou, J.; Li, Z.; Liu, W. Cosface: Large margin cosine loss for deep face recognition. In Proceedings of the IEEE Conference on Computer Vision and Pattern Recognition, Salt Lake City, UT, USA, 18–22 June 2018.
5. Florian, S.; Kalenichenko, D.; Philbin, J. Facenet: A unified embedding for face recognition and clustering. In Proceedings of the IEEE Conference on Computer Vision and Pattern Recognition, Boston, MA, USA, 7–12 June 2015.
6. Tianyue, Z.; Deng, W.; Hu, J. Cross-age lfw: A database for studying cross-age face recognition in unconstrained environments. *arXiv* **2017**, arXiv:1708.08197.
7. Yi, D.; Lei, Z.; Liao, S.; Li, S.Z. Learning Face Representation from Scratch. *arXiv* **2014**, arXiv:1411.7923.
8. Chen, S.; Liu, Y.; Gao, X.; Han, Z. MobileFaceNets: Efficient CNNs for Accurate Real-time Face Verification on Mobile Devices. In Proceedings of the Chinese Conference on Biometric Recognition, Beijing, China, 3–4 December 2018; pp. 428–438.
9. Thys, S.; Van Ranst, W.; Goedeme, T. Fooling automated surveillance cameras: Adversarial patches to attack person detection. In Proceedings of the 2019 IEEE/CVF Conference on Computer Vision and Pattern Recognition Workshops (CVPRW), Long Beach, CA, USA, 16–17 June 2019; pp. 49–55.
10. Szegedy, C.; Zaremba, W.; Sutskever, I.; Bruna, J.; Erhan, D.; Goodfellow, I.; Fergus, R. Intriguing properties of neural networks. *arXiv* **2014**, arXiv:1312.6199.
11. Goodfellow, I.J.; Shlens, J.; Szegedy, C. Explaining and harnessing adversarial examples. *arXiv* **2014**, arXiv:1412.6572. Available online: https://arxiv.org/abs/1412.6572 (accessed on 23 October 2022).
12. Dong, Y.; Liao, F.; Pang, T.; Su, H.; Zhu, J.; Hu, X.; Li, J. Boosting Adversarial Attacks with Momentum. In Proceedings of the IEEE Conference on Computer Vision and Pattern Recognition, Salt Lake City, UT, USA, 18–22 June 2018; pp. 9185–9193.
13. Carlini, N.; Wagner, D.A. Towards Evaluating the Robustness of Neural Networks. In Proceedings of the 2017 IEEE Symposium on Security and Privacy (SP), San Jose, CA, USA, 22–26 May 2017; pp. 39–57.
14. Liu, X.; Yang, H.; Liu, Z.; Song, L.; Chen, Y.; Li, H. DPATCH: An Adversarial Patch Attack on Object Detectors. *arXiv* **2019**, arXiv:1806.02299.
15. Kevin, E.; Ivan, E.; Earlence, F.; Bo, L.; Amir, R.; Florian, T.; Atul, P.; Tadayoshi, K.; Dawn, S. Physical Adversarial Examples for Object Detectors. *arXiv* **2018**, arXiv:1807.07769.
16. Wu, Z.; Lim, S.; Davis, L.; Goldstein, T. Making an Invisibility Cloak: Real World Adversarial Attacks on Object Detectors. In Proceedings of the 16th European Conference on Computer Vision, Glasgow, UK, 23–28 August 2020.
17. Xu, K.; Zhang, G.; Liu, S.; Fan, Q.; Sun, M.; Chen, H.; Chen, P.; Wang, Y.; Lin, X. Evading Real-Time Person Detectors by Adversarial T-shirt. *arXiv* **2019**, arXiv:1910.11099v1.
18. Sharif, M.; Bhagavatula, S.; Bauer, L.; Reiter, M.K. Accessorize to a Crime: Real and Stealthy Attacks on State-of-the-Art Face Recognition. In Proceedings of the CCS'16: 2016 ACM SIGSAC Conference on Computer and Communications Security, Vienna, Austria, 24–28 October 2016.
19. Komkov, S.; Petiushko, A. AdvHat: Real-world adversarial attack on ArcFace Face ID system. In Proceedings of the 2020 25th International Conference on Pattern Recognition (ICPR), Milan, Italy, 10–15 January 2021.
20. Huang, G.B.; Mattar, M.; Berg, T.; Learned-Miller, E. Labeled faces in the wild: A database for studying face recognition in unconstrained environments. In Proceedings of the Workshop on Faces in 'Real-Life' Images: Detection, Alignment, and Recognition, Marseille, France, 17 October 2008; pp. 1–11.
21. Liu, W.; Wen, Y.; Yu, Z.; Li, M.; Raj, B.; Song, L. Sphereface: Deep hypersphere embedding for face recognition. In Proceedings of the IEEE Conference on Computer Vision and Pattern Recognition, Honolulu, HI, USA, 21–26 July 2017; pp. 212–220.

22. Lang, D.; Chen, D.; Shi, R.; He, Y. Attention-Guided Digital Adversarial Patches On Visual Detection. *Secur. Commun. Netw.* **2021**, *2021*, 6637936:1–6637936:11. [CrossRef]
23. Nguyen, D.; Arora, S.S.; Wu, Y.; Yang, H. Adversarial Light Projection Attacks on Face Recognition Systems: A Feasibility Study. Available online: https://arxiv.org/abs/2003.11145 (accessed on 23 October 2022).
24. Ioffe, S.; Szegedy, C. Batch Normalization: Accelerating Deep Network Training by Reducing Internal Covariate Shift. In Proceedings of the International Conference on Machine Learning, Lille, France, 6–11 July 2015.
25. Zolfi, A.; Avidan, S.; Elovici, Y.; Shabtai, A. Adversarial Mask: Real-World Adversarial Attack Against Face Recognition Models. *arXiv* **2021**, arXiv:2111.10759.
26. Yin, B.; Wang, W.; Yao, T.; Guo, J.; Kong, Z.; Ding, S.; Li, J.; Liu, C. Adv-Makeup—A New Imperceptible and Transferable Attack on Face Recognition. In Proceedings of the International Joint Conference on Artificial Intelligence, Montreal, QC, Canada, 19–27 August 2021; pp. 1252–1258.
27. Xiao, Z.; Gao, X.; Fu, C.; Dong, Y.; Gao, W.; Zhang, X.; Zhou, J.; Zhu, J. Improving transferability of adversarial patches on face recognition with generative models. In Proceedings of the IEEE/CVF Conference on Computer Vision and Pattern Recognition, Virtual, 19–25 June 2021; pp. 11845–11854.
28. Parmar, R.; Kuribayashi, M.; Takiwaki, H.; Raval, M.S. On Fooling Facial Recognition Systems using Adversarial Patches. In Proceedings of the 2022 International Joint Conference on Neural Networks, Padova, Italy, 18–23 July 2022; pp. 1–8.
29. Jia, S.; Yin, B.; Yao, T.; Ding, S.; Shen, C.; Yang, X.; Ma, C. Adv-Attribute: Inconspicuous and Transferable Adversarial Attack on Face Recognition. *arXiv* **2022**, arXiv:2210.06871.
30. Zhang, K.; Zhang, Z.; Li, Z.; Qiao, Y. Joint face detection and alignment using multitask cascaded convolutional networks. *IEEE Signal Process. Lett.* **2016**, *23*, 1499–1503. [CrossRef]
31. Farfade, S.S.; Saberian, M.J.; Li, L.J. Multi-view Face Detection Using Deep Convolutional Neural Networks. In Proceedings of the 5th ACM on International Conference on Multimedia Retrieval, Shanghai, China, 23–26 June 2015; pp. 643–650.
32. He, K.; Zhang, X.; Ren, S.; Sun, J. Deep residual learning for image recognition. In Proceedings of the IEEE Conference on Computer Vision and Pattern Recognition, Las Vegas, NV, USA, 27–30 June 2016; pp. 770–778.
33. Hu, J.; Shen, L.; Sun, G. Squeeze-and-excitation networks. In Proceedings of the IEEE Conference on Computer Vision and Pattern Recognition, Salt Lake City, UT, USA, 18–23 June 2018; pp. 7132–7141.
34. Huynh-Thu, Q.; Ghanbari, M. Scope of validity of PSNR in image/video quality assessment. *Electron. Lett.* **2008**, *44*, 800–801. [CrossRef]
35. Lin, Y.S.; Liu, Z.Y.; Chen, Y.A.; Wang, Y.S.; Chang, Y.L.; Hsu, W.H. xCos: An Explainable Cosine Metric for Face Verification Task. ACM Trans. *Multimed. Comput. Commun. Appl.* **2021**, *17*, 1–16.

![algorithms logo] **algorithms**

MDPI

Article

From Iris Image to Embedded Code: System of Methods

Ivan Matveev [1] and Ilia Safonov [2,*]

[1] Federal Research Center "Computer Science and Control" of the Russian Academy of Sciences, Vavilov Street, 44/2, 119333 Moscow, Russia

[2] Computer Science and Control Systems Department, National Research Nuclear University MEPhI, Kashirskoye Highway, 31, 115409 Moscow, Russia

* Correspondence: ilia.safonov@gmail.com

Abstract: Passwords are ubiquitous in today's world, as are forgetting and stealing them. Biometric signs are harder to steal and impossible to forget. This paper presents a complete system of methods that takes a secret key and the iris image of the owner as input and generates a public key, suitable for storing insecurely. It is impossible to obtain source data (i.e., secret key or biometric traits) from the public key without the iris image of the owner, the irises of other persons will not help. At the same time, when the iris image of the same person is presented the secret key is restored. The system has been tested on several iris image databases from public sources. It allows storing 65 bits of the secret key, with zero possibility to unlock it with the impostor's iris and 10.4% probability to reject the owner in one attempt.

Keywords: biometric cryptosystem; iris identification; error-correcting codes

1. Introduction

Nowadays, cryptographic algorithms are widely used for information protection. A large number of them, as well as their applications, have been invented and introduced [1]. These algorithms and systems are mathematically grounded and reliable. The weak link in their implementation and usage, as usual, is human. Cryptography requires keys, i.e., sequences of digits, which should be reproduced precisely. While a human is able to remember and reproduce a personally invented password (though there are difficulties here already), it is practically impossible to memorize a long sequence of pseudorandom symbols, which is created automatically [2]. Meanwhile, humans possess biometric features that are simple to retrieve, difficult to alienate, and contain a significant amount of information. The disadvantage of biometric traits is their variability: it is impossible to exactly replicate the measurement results, we can only say that two sets of traits taken from one person are in some sense closer than the sets obtained from different people. It is of great interest to combine these two approaches, i.e., to develop methods for obtaining unique cryptographic keys from variable biometry data of a given person.

The eye iris is the most suitable biometric modality among all non-invasive ones due to its highest information capacity. The number of degrees of freedom of the iris template was evaluated as 249 [3]. It promises to be almost as good as a strong symmetric cryptography key length of 256 bit, while the net coming fingerprint is reported to have 80 bits [4]. In order to design a practically usable system it is advisable to base it on the iris. Up to now a major focus in developing automated biometric is building an identification system, i.e., the system, which executes a scenario: sample biometric features once, record, take them sometime later and decide whether these samples belong to the same person.

The workflow of the biometric identification system can be combined of the blocks: capture, segmentation, template generation, and template matching, see Figure 1.

Citation: Matveev, I.; Safonov, I. From Iris Image to Embedded Code: System of Methods. *Algorithms* **2023**, *16*, 87. https://doi.org/10.3390/a16020087

Academic Editors: Francesco Bergadano and Giorgio Giacinto

Received: 15 December 2022
Revised: 19 January 2023
Accepted: 3 February 2023
Published: 6 February 2023

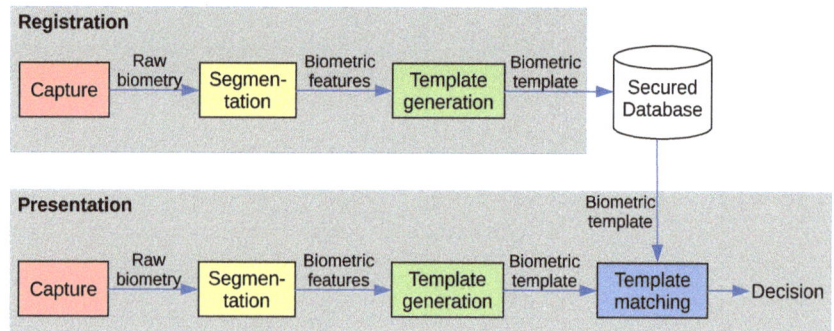

Figure 1. Biometric system workflow.

Note that in this scenario the biometric template should be securely stored and exclude the intruder from obtaining it. Here, a different problem is solved, thus only capture, segmentation and partly template generation blocks are inherited, and matching is replaced by embedding/extracting the cryptographic key into/from the biometric features.

The explanation here goes alongside the data processing: from the source iris image to the embedding of the secret key. The capture process, i.e., obtaining eye images with a camera device, is beyond the scope of this paper. We start from the image segmentation task and present a framework for locating the iris in an eye image. In the next section clue methods of the framework are described. Then feature extraction and matching methods are given. Following is the discussion of the application scenario of embedding the secret key to biometric features. The successful extraction of the embedded key depends on the difference between registered and presented biometric features, the value of this difference is determined based on several databases. In the next section the methods of encoding and decoding the key are presented, and the selection of their optimal parameters is discussed.

The contribution of this work is comprised of the following.

- The system of iris segmentation methods is presented which combines preliminary detection with refinement steps. The first steps use the most common geometric properties of the eye and accept the widest range of image conditions, while the final steps take care of details. The core of the system is a special base radii detection method.
- The cascade of error correction codecs adopted to iris code nature. A novel step of pseudorandom bit shuffling is introduced, accompanied by bit dubbing. This contradicts known methods, which do not use bit dubbing and deliberately avoid bit shuffling.
- The combination of the iris segmentation system and error correction cascade results in a practically applicable method, proven for several databases of variable image quality.

2. Eye Segmentation Framework

Methods, algorithms and applications of iris biometrics have attracted much attention during the last three decades [5–7] and continue developing rapidly in recent years [8]. The main trend of the latest development is switching from heuristic approaches and hand-crafted methods to employing neural networks in various tasks. A wide variety of artificial neural networks has emerged and is applied to iris segmentation, starting from earlier research with fully connected nets [9] to latest applications of attention-driven CNNs [10], U-Nets [11], hybrid deep learning [12]. Another trend comes from the in-born ability of neural networks to *classify* objects (say, pixels, textures, images) rather than *calculate* their positions and other numerical properties. Due to this, most of the works in iris segmentation rely on detecting masks, i.e., pixels belonging to regions of the iris (or pupil, or else) in the image. Positions and sizes of pupil and iris are then derived from these masks. Surely, detecting masks is what one calls segmentation; however, such an approach ignores the clear and simple geometry of the iris and is prone to detecting irises of unnatural shape as is shown in [12]. Some works [13,14] apply a neural network to detect the position of the iris as a number; however, it seems a strained decision.

Here we adopt a "classical" method. The obvious approach to iris region extraction in eye imaging is a chain of operations that starts with the detection of the pupil (the most distinctive area that is dark and has an expressed circular shape). Then outer iris border is presumably located. Finally, the visible iris part is refined by cutting off the areas distorted by reflections, eyelids and eyelashes. Most researchers and developers follow this method. Detection of each iris feature is usually carried out only one time and it is not recalculated any more even after obtaining other features, which can serve for refinement. For instance, in [15–18] full set of iris features is detected; however, pupil parameters are obtained at the first step and are not revised any more.

Only a few papers develop something different from this sequence "first pupil, then iris, once determined, and never reviewed". In [19,20], the position of the iris center is estimated first which makes pupil detection more robust. In [21], pupil parameters are refined using iris size after the iris is located. In [21,22], detection methods run iteratively several times for refinement. In [20,23], a multi-scale approach is used, and methods run in several scales. However, none of these works use various types of methods for detecting any iris feature.

Here we develop a *system of methods* for segmentation of the iris in an eye image. Evaluating of each of parameters is performed at several steps. The main idea of this system is that at first the most general characteristics of objects are defined, which are then sequentially supplemented by more specific and refined ones. Beginning steps do not need to output precise final parameters, used as final. Instead, they should be robust and general and tolerate a wide range of conditions, i.e., detect the object of any quality. Later steps should have the highest possible precision and may reject poor quality data.

Iris region in frontal images is delimited by two nearly concentric nearly circular contours, called inner and outer borders. Hereinafter the contour separating iris and pupil is referred to as *inner border*, *pupil border* or simply *pupil*, and the contour delimiting iris and sclera is called *outer border* or *iris*. In most cases pupil border is wholly visible in the image, but some part of the iris border is frequently overlapped by eyelids and eyelashes.

Since the pupil and iris are almost concentric, one *eye center* point may serve as an approximate center for both contours. It can be considered the most general geometric property of the iris, and the first step of eye detection should be locating this eye center. Note that only the position of the center is to be found, rather than the size of any contour. Excluding size and allowing approximate detection involves both concentric borders in the process. This is especially significant for eyes with poorly visible inner boundaries, where pupil location alone fails frequently. A modification of Hough method [24] is used.

It is very likely that after center detection pupil size should be estimated. To the best of our knowledge, this is carried out in all works where iris segmentation starts from eye center location, as in [19]. However, this method is not stable and universal for a wide range of imaging conditions. Detecting the radius may easily mistake the outer border for the inner, especially for images with poor inner border contrast [25]. Here we decide to employ both correlated contours around the detected center, and detect sizes of them simultaneously. Hereinafter this detection is referred to as *base radii* detection, meaning that it finds approximate (base) radii of inner and outer circles around a given center. The method relies on circular projections of gradient [26]. Base radii detection produces approximate center coordinates and radii of pupil and iris circles, which satisfy some reasonable limitations. Furthermore, the quality of detection is calculated. The quality should be high enough to pass the image to further processing.

Then both boundaries are re-estimated with better precision (refined). Pupil refinement is carried out by a specially developed version of the shortest path method [27]. Iris is refined by the same method as that of base radius. The difference is that the position of the pupil is now fixed and only the iris center and radius are being searched. Iris segmentation here results in detecting two nearly concentric circles, which are approximating the inner and outer borders of the iris ring. Occlusion detection [28] is carried out to ensure the quality of iris data, i.e., to reject strongly occluded irises from further processing, but apart from this the occlusion mask is not used.

Summing up, the segmentation stage of the system employs five steps: center detection, base radii detection, pupil refinement, iris refinement and occlusion detection, see Figure 2.

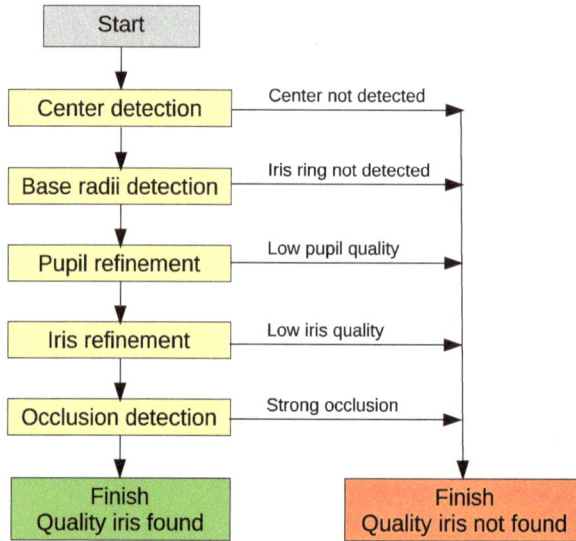

Figure 2. Workflow of iris segmentation methods.

At each stage of segmentation, quality value is estimated and the process is terminated if the quality is below acceptable.

3. Eye Segmentation Methods

Methods of iris segmentation are briefly presented in this section.

3.1. Center Detection

The algorithm finds the coordinates $(x_C, y_C) = \vec{c}$ of eye center in the image $b(\vec{x})$, and does not need to estimate pupil or iris size. There is also no need to find the center position precisely, it is sufficient to locate it somewhere inside the pupil. Thus, pixels of both pupil and iris borders are used in Hough's procedure. Furthermore, the algorithm has low computational complexity since only two parameters are estimated and a two-dimensional Hough accumulator is used.

The algorithm runs the following five steps.

Step 1. Gradient calculation.

Consider rectilinear coordinate system Oxy in the image with the center in the left bottom corner and axes Ox and Oy directed along its borders. Denote brightness $b(\vec{x})$ in image point $\vec{x}$. Brightness gradient $\vec{\nabla} b(\vec{x}) = \vec{g}(\vec{x})$ is estimated by standard Sobel masks [29].

Step 2. Outlining voting pixels.

We need edge pixels to vote. These are selected with the help of a gradient value threshold. Cumulative distribution of brightness gradient values in pixels over the image is calculated, and set Ω_1 of pixels with brightness gradient in the upper 5% of this distribution are selected:

$$H(G) = |\{\vec{z} : \|\vec{g}(\vec{z})\| \leqslant G\}| \,,$$
$$\Omega_1 = \{\vec{x} : H(\|\vec{g}(\vec{x})\|) \geqslant (1 - \tau_s)N\} \,, \tag{1}$$

where $|S|$ is power (count of elements) of set S, N is total number of image pixels, $\tau_s = 0.05$ is the share of points being selected.

Step 3. Voting to accumulator.

Hough methods use *accumulator* function, which is defined over a parameter space. We detect the eye center, which is some point in the image, and its parameters are its coordinates in the image. Thus, the parameter space is 2D vector $\vec{x}$ and the accumulator is $A(\vec{x})$ with the same size as the source image.

Ray from some given point $\vec{x} \in \Omega_1$ in anti-gradient direction $-\vec{\nabla}b(\vec{x})$ is the locus of all possible dark circles with border passing through this point. A set of such rays, drawn in the accumulator, traced from each pixel coordinated selected at step 2 will produce clotting at the center of any roundish dark object. The more circle-like this object is, the more expressed will be its central clotting.

Step 4. Accumulator blurring.

The accumulator $A(\vec{x})$ is subject to a low-pass filter, to suppressed noise such as singular sporadic rays produced by non-circular edges in the image. Denote the result as $A_B(\vec{x})$.

Step 5. Maximum location.

Maximum position

$$\vec{c} = arg \max_{\vec{x}} A_B(\vec{x}) \tag{2}$$

in blurred accumulator corresponds to the center of the best round-shaped object in the image. It is the most probable eye center. However, local maxima exist in any image due to noise. In order to decide whether there is a noticeable circular object, one can compare the value of local maxima against the values produced by noise. Since τ_s pixels of the image are voting and for each point voting procedure draws a segment of approximately 0.5W pixels, where W is a linear size of the image, the average brightness level is near $0.5\tau_s W$. Selecting desirable signal to noise ratio P_{SNR}, one can write the condition of accepting the located maximum (2) for eye center:

$$Q_C = \max_{\vec{x}} A_B(\vec{x}) > \frac{1}{2} P_{SNR} \tau_s W \,. \tag{3}$$

If condition (3) does not hold, the decision is made that there is no eye in the image $b(\vec{x})$.

3.2. Base Radii Detection

The algorithm simultaneously locates two iris boundaries as circle approximations: inner (pupil) (x_P, y_P, r_P) and outer (iris) (x_I, y_I, r_I) starting from the center $\vec{c}$ (2). In this section, we set $(x_C, y_C) = \vec{c}$ as coordinate origin. Anti-gradient vector at the boundary of the dark circle and the direction to the circle center coincide or form a small angle. As the pupil and iris are both dark circles on the brighter background, one can state the following condition for pixels $\vec{x}$ of their boundaries:

$$\phi(\vec{x}) = \arccos \frac{\vec{g}(\vec{x}) \cdot \vec{x}}{\|\vec{g}(\vec{x})\| \, \|\vec{x}\|} < \tau_\phi \,. \tag{4}$$

We use a threshold value $\tau_\phi = 45^0$.

Furthermore, the condition for gradient value (1) is applicable. Pixel $\vec{x}$ satisfying the conditions (1), (4) probably belongs to the inner or outer boundary. Call it *candidate*. Define the set of candidate pixels as Ω_2:

$$\Omega_2 = \left\{ \vec{x} : \phi(\vec{x}) < \tau_\phi, H(\|\vec{g}(\vec{x})\|) \geqslant (1 - \tau_s)N \right\} . \tag{5}$$

For each radius r a ratio of candidate count at this radius to the count of all pixels at this radius is estimated:

$$\Pi(r) = \frac{|\{\vec{x} : \|\vec{x}\| \in [r - 0.5, r + 0.5), \vec{x} \in \Omega_2\}|}{|\{\vec{x} : \|\vec{x}\| \in [r - 0.5, r + 0.5)\}|} . \tag{6}$$

If there is a dark circle of some radius ρ with the center near the coordinate origin its border pixels are likely to belong to the set Ω_2, and are likely to have distance ρ to the coordinate origin. Thus, $\Pi(\rho)$ will be big, i.e., have local maximum. Other contours will not vote to the same radius of circular projection and will not form local maxima therein.

The image plane is divided into four quadrants, left, right, top and bottom by the lines $y = x$ and $y = -x$. In each quadrant, a *sub-projection* is calculated separately according to (6). Positions of local maxima on the right, left, top, and bottom sub-projections are:

$$\mu_\alpha(n) = \arg\operatorname*{loc\,max}_{n\ \ r} \Pi_\alpha(r), \quad \alpha = \{R, L, T, B\} . \tag{7}$$

The quality of maxima is simply the value of histogram at the point

$$q_\alpha(n) = \Pi_\alpha(\mu_\alpha(n)) . \tag{8}$$

If not occluded, each of the two circular contours (inner and outer borders) gives a local maximum in each sub-projection. Other maxima may arise due to occlusions such as eyelashes and eyelids or due to other details in eye images, including patterns of the iris itself. Combining local maxima positions (7) gives set of hypothetical pupils:

$$
\begin{aligned}
x_P^{i,j} &= \frac{1}{2}(\mu_R(i) - \mu_L(j)), \quad i = \overline{1, n_R}, \ j = \overline{1, n_L}, \\
y_P^{k,l} &= \frac{1}{2}(\mu_T(k) - \mu_B(l)), \quad k = \overline{1, n_T}, \ l = \overline{1, n_B}, \\
r_P^{i,j,k,l} &= \frac{1}{4}(\mu_R(i) + \mu_L(j) + \mu_T(k) + \mu_B(l)) .
\end{aligned}
\tag{9}
$$

Qualities of combinations are also defined from values (8):

$$q_P^{i,j,k,l} = \frac{1}{4}(q_R(i) + q_L(j) + q_T(k) + q_B(l)) . \tag{10}$$

Irises are estimated by just the same formulas:

$$
\begin{aligned}
x_I^{i,j} &= \frac{1}{2}(\mu_R(i) - \mu_L(j)), \quad i = \overline{1, n_R}, \ j = \overline{1, n_L}, \\
y_I^{k,l} &= \frac{1}{2}(\mu_T(k) - \mu_B(l)), \quad k = \overline{1, n_T}, \ l = \overline{1, n_B}, \\
r_I^{i,j,k,l} &= \frac{1}{4}(\mu_R(i) + \mu_L(j) + \mu_T(k) + \mu_B(l)), \\
q_I^{i,j,k,l} &= \frac{1}{4}(q_R(i) + q_L(j) + q_T(k) + q_B(l)) .
\end{aligned}
\tag{11}
$$

The nature of the pupil and iris imposes certain limitations on their locations and sizes. We use the following four inequalities: pupil size is not less than 15% of iris size and not more than 75% of iris size; center of the iris is inside pupil; pupil cannot be displaced too much from iris center. This can be written as:

$$r_P > 0.15r_I , \quad r_P < 0.75r_I , \quad d < r_P , \quad 2(r_I - r_P - d) > r_I - r_P + d , \tag{12}$$

where $\vec{c}_P = (x_P, y_P)$, $\vec{c}_I = (x_I, y_I)$ are centres of pupil and iris, $d = \|\vec{c}_P - \vec{c}_I\|$ is a distance between these centres.

From all possible variants of pupil-iris pair given by (9)–(11) we select those satisfying conditions (12). The quality of combination is a sum of pupil and iris qualities (10) and a weighted quality of fitting to conditions (12):

$$Q = q_P + q_I + \gamma q_{fit} ,$$
$$q_{fit} = \min\left\{ \frac{r_P - 0.15r_I}{r_P}, \frac{0.75r_I - r_P}{r_P}, \frac{r_P - d}{r_P}, \frac{r_I - r_P - 3d}{r_I - r_P} \right\} . \tag{13}$$

The combination with the best quality is selected. If Q is below the given threshold, it is supposed that the eye in the image is squinted and upper and lower eyelids cover a big share of the iris border. In this case, the variant with absent top and bottom iris local maxima is tested. The formulas (9) and (10) are modified accordingly, iris center vertical position is taken equal to that of the pupil: $y_I \equiv y_P$. If Q is below the threshold again, it is decided that there is no feasible iris ring and in the image. Other types of occlusion are not treated, the iris images are considered too bad for processing in this case. Thresholds for Q and value of γ in (13) are estimated experimentally so as to reject the biggest share of erroneously detected irises while preserving good outcomes. So, the method runs in six steps:

Step 1. Gradient calculation.

This step is common with center detection.

Step 2. Candidates selection.

This step is similar to Step 2 of center detection. In addition to gradient value condition (1) angular condition (4) is imposed.

Step 3. Circular projecting.

Calculating circular projections (6) in four quadrants.

Step 4. Enumeration of maxima.

Finding local maxima (7) in projections. Prior to this the projections are smoothed with a Gaussian filter to suppress redundant local maxima originating from noise.

Step 5. Enumerations of hypothetical irises.

Finding coordinates and radii of inner and outer circles from combinations of maxima (9), which hold conditions (12).

Step 6. Selecting the best iris.

Pair of circles is selected according to the qualities (8), (10), (13).
If no feasible iris is detected in step 5, the result is "no eye detected".
A sample of the projection combination is presented in Figure 3. Real positions of pupil and iris borders, taken from expert marking are depicted by arrows. There is no local maxima corresponding to the iris border in the top projection $\Pi_T(r)$ since the iris border is occluded. Such minor obstacles do not prevent choosing correct combination.

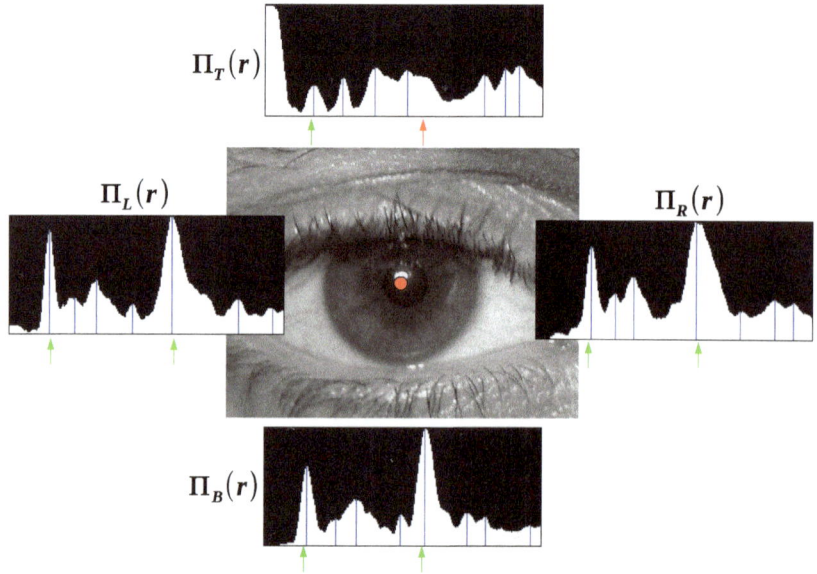

Figure 3. Four circular projections, their maxima positions and correct position of borders.

3.3. Pupil Refinement

Circular shortest path method constructs a closed contour in a circular ring [30]. The ring is centered at a given point and has inner and outer radii. CSP method is a kind of optimal path method, i.e., it optimized the functional, which is the cost of the path. We take the ring concentric to the approximate pupil circle and spread 30% of its radius inside and outside.

In order to ease calculations polar transformation is carried out. The ring shape in the source image is unwrapped to a rectilinear raster. Radial and angular coordinates of the ring are mapped to abscissa and ordinate. Thus, the problem of locating the circular shortest path is reduced to a problem of detecting the optimal path from the left to the right side of the rectangle such that terminal points of the path have the same ordinate. Contour is represented as a function $\rho(\phi)$, $\phi \in [0; 2\pi]$, $\rho(0) = \rho(2\pi)$ with limited derivative $d\rho/d\phi < 1$. In a discrete rectilinear raster of size $W \times H$ the contour is turns to a sequence of points: $\{(n, \rho_n)\}$, $n \in [0; W-1]$. Limitations to the derivative transforms to $|\rho_{n+1} - \rho_n| \leqslant 1$, edge condition is set as $|\rho_{W-1} - \rho_0| \leqslant 1$.

Consider points (n, ρ') and $(n+1, \rho'')$ from adjacent columns of the raster. Denote the cost of passing between them as

$$C((n, \rho'), (n+1, \rho'')) \equiv C_n(\rho', \rho'') = C_n^{(I)}(\rho', \rho'') + C_n^{(O)}(\rho', \rho'') . \tag{14}$$

This cost is a sum of inner and outer parts.

Inner cost is a function of contour shape, designed in a way to promote its smoothness:

$$C_n^{(I)}(\rho', \rho'') = \begin{cases} 0 , & \rho' = \rho'' , \\ \tau_i , & |\rho' - \rho''| = 1 , \\ \infty , & |\rho' - \rho''| > 1 . \end{cases} \tag{15}$$

Value of $\tau_i > 0$ is a parameter defining the magnitude of a "force", which pulls the contour towards a straight line. Optimizing the inner part alone would give horizontal lines in polar raster, i.e., ideal circles with the given center in source image.

Outer cost is designed to make the contour pass through border pixels. So it is low in boundary points (the gradient vector is big and perpendicular to the local direction of the contour) and is high otherwise. The outer part is the cost of passing the point (n, ρ'):

$$C^{(O)}(n,\rho) = \begin{cases} 0, & (x,y) \in \Omega_3, \\ \tau_o;, & (x,y) \notin \Omega_3, \end{cases} \tag{16}$$

where Ω_3 is the set of points defined by (5), x and y are the coordinates of the source image point, which was mapped to (n, ρ).

Optimal contour $S = \{\rho_n\}_{n=1}^{W}$ is the one minimizing the total cost:

$$S^* = \arg\min_S \sum_{n=1}^{W} C_n(\rho_n, \rho_{n+1}). \tag{17}$$

This discrete optimization problem can be solved by various methods. Here the method works in quite a narrow ring and the exhaustive search is faster due to small overhead.

Denote sum in (17) as Σ. In the best case $\Sigma = 0$, in the worst case $\Sigma = W(\tau_i + \tau_o)$. Mapping this into the range $[0; 1]$ where value 1 stands for best we obtain the quality

$$Q_{ref} = 1 - \frac{\Sigma}{W(\tau_i + \tau_o)}. \tag{18}$$

The contour is considered acceptable if $Q_{ref} > 0.5$, otherwise the decision is made that the pupil border cannot be detected with the required precision and the segmentation is terminated.

The algorithm runs in five steps.

Step 1. Candidates selection.

The same gradient calculation as in the first step of previous methods is used. Then the conditions (1), (4) are imposed as in Step 2 of base radii detection. However, a smaller angular threshold $\tau_\phi = 30^0$ is set since the center position is known with better precision.

Step 2. Polar transform.

The transform creates an image (rectangular raster) $P(\phi, \rho)$, $\phi \in [0, W-1]$, $\rho \in [0; H-1]$ by calculating a brightness value in each of its pixels (ϕ, ρ). This brightness is taken from source image $b(x,y)$ where its coordinates are estimated as

$$x(\phi, \rho) = \left(r_1 + \frac{r_2 - r_1}{H}\rho\right)\cos\left(\frac{2\pi\phi}{W}\right),$$
$$y(\phi, \rho) = \left(r_1 + \frac{r_2 - r_1}{H}\rho\right)\sin\left(\frac{2\pi\phi}{W}\right), \tag{19}$$

where r_1 and r_2 are the inner and outer radii of the ring in the source image, and the coordinate origin of the source image is placed at the center of the ring. The brightness of the point of the polar image is obtained by bilinear interpolation:

$$N(\rho, \phi) =$$
$$(1-\{x\})(1-\{y\})b(\lfloor x\rfloor, \lfloor y\rfloor)+$$
$$\{x\}(1-\{y\})b(\lfloor x\rfloor+1, \lfloor y\rfloor)+$$
$$(1-\{x\})\{y\}b(\lfloor x\rfloor, \lfloor y\rfloor+1)+$$
$$\{x\}\{y\}b(\lfloor x\rfloor+1, \lfloor y\rfloor+1), \tag{20}$$

where $\lfloor a\rfloor$ and $\{a\}$ define integer and fractional parts of a.

Step 3. Optimal path tracking.

Finding S^* according to (14)–(17).

Step 4. Transforming to original coordinates.

Restore the coordinates of the optimal path from $O\rho\phi$ polar system back to the source image Oxy system.

Step 5. Estimating equivalent circle.

Pupil border contour is not a circle precisely; however, we can define an *equivalent circle*, with area and center of mass same as those of the figure enclosed into the pupil border contour. The center and radius of the equivalent circle are:

$$x_{eq} = \frac{M_x}{M}, \quad y_{eq} = \frac{M_y}{M}, \quad r_{eq} = \sqrt{\frac{M}{\pi}},$$
$$M = |\Omega_4|, \quad M_x = \sum_{(x,y)\in\Omega_4} x, \quad M_y = \sum_{(x,y)\in\Omega_4} y,$$

(21)

where Ω_4 is the area inside contour S^* in source image. This equivalent circle is further used as the pupil border, and it happens to be a better model due to its stability [31].

4. Experiments with Iris Segmentation

Iris segmentation here results in detecting two nearly concentric circles, which are approximating inner and outer borders of iris ring.

Assessment of the iris segmentation quality can be carried out in the following ways:

- Matching against manual segmentation by a human.
- Matching against rivals disclosed in the literature.
- Applying obtained segmentation further to iris recognition. Under the assumption that more precise detection reduces the number of classification errors this will indirectly estimate segmentation quality.

In order to compare the results of the proposed system with the known analogs, the following publicly available iris image databases were used: CASA-3-Lamp and CASIA-4-Thousand [32] (totally 54,434 images), BATH [33] (31,988 images), NDIRIS [34] (64,980 images), UBIRIS-1 [35] (1207 images).

4.1. Matching against Manual Segmentation

All images were processed by a human expert, who marked two circles approximating iris borders in each of them or rejected if the iris was not visible or of poor quality. (In fact, there were very few, less than a hundred altogether, images rejected at this stage.) We assume that the expert did it accurately; therefore this segmentation is taking for ground truth. Denote the manually marked circles as $(x, y, r)_P^*$ for pupil and $(x, y, r)_I^*$ for the iris. Values of absolute and relative errors of eye center detection averaged in databases

$$\varepsilon_{C,abs} = \langle\Delta\rangle, \quad \varepsilon_{C,rel} = \left\langle\frac{\Delta}{r_P^*}\right\rangle, \quad \Delta = \left((x_C - x_P^*)^2 + (y_C - y_P^*)^2\right)^{1/2}$$

(22)

are given in Table 1.

It can be seen that for all databases except for the small bases MMU and UBIRIS, which contain low-resolution images, and UBIRIS, which contain images with small pupil size, the mean absolute deviation of the detected eye center from the true center of the pupil does not exceed four pixels and the relative deviation does not exceed one-tenth of the radius.

Table 1. Errors of eye center detection.

Database	$\varepsilon_{C,abs}$, Pixels	$\varepsilon_{C,rel}$, %
BATH	2.85	5.27
CASIA	2.94	7.47
MMU	3.22	15.5
NDIRIS	3.12	6.30
UBIRIS	8.29	22.3

The next method of the system is base radius detection. Table 2 presents the average deviations of the detected centers and radii of the pupil and the iris from those marked by human experts.

$$\varepsilon_{P,abs} = \left\langle \left((x_P - x_P^*)^2 + (y_P - y_P^*)^2 \right)^{1/2} \right\rangle, \ \varepsilon_{rP,abs} = \langle r_P - r_P^* \rangle,$$

$$\varepsilon_{I,abs} = \left\langle \left((x_I - x_I^*)^2 + (y_I - y_I^*)^2 \right)^{1/2} \right\rangle, \ \varepsilon_{rI,abs} = \langle r_I - r_I^* \rangle, \tag{23}$$

Table 2. Errors of base radii detection, pixels.

Database	$\varepsilon_{P,abs}$	$\varepsilon_{rP,abs}$	$\varepsilon_{I,abs}$	$\varepsilon_{rI,abs}$
BATH	2.15	1.66	6.23	2.05
CASIA	2.58	1.60	17.41	4.21
MMU	3.31	2.05	13.01	5.11
NDIRIS	2.33	2.73	5.34	2.34
UBIRIS	6.03	5.78	7.68	11.35

It is seen that the mean error in detecting the pupil center is reduced compared with the first column of Table 1.

Table 3 shows the errors for the final circles of the pupil and the iris obtained by the system, calculated according to (23).

Table 3. Errors of final iris parameters detection, pixels.

Database	$\varepsilon_{P,abs}$	$\varepsilon_{rP,abs}$	$\varepsilon_{I,abs}$	$\varepsilon_{rI,abs}$
BATH	0.52	1.42	2.48	1.71
CASIA	1.05	1.13	2.44	1.62
MMU	0.97	1.77	1.92	4.35
NDIRIS	0.84	1.14	1.97	2.26
UBIRIS	2.27	5.37	3.25	5.82

4.2. Matching against Other Methods

Table 4 compares the computation time and errors in determining the pupil parameters for the presented system and its analogs. The comparison was carried out with the methods described in [3,36–39].

The third method to assess the algorithm of iris segmentation, i.e., applying its results to iris recognition is disclosed further.

Table 4. Matching against other methods.

Database	Error, Pixels	Method					
		Wildes	Daugman	Masek	Ma et al.	Daugman-2	Presented
BATH	$\varepsilon_{P,abs}$	3.44	3.73	5.32	4.29	3.27	0.52
	$\varepsilon_{rP,abs}$	4.38	4.54	6.72	4.65	3.19	1.42
CASIA	$\varepsilon_{P,abs}$	5.37	2.15	3.67	4.79	1.19	2.44
	$\varepsilon_{rP,abs}$	6.12	4.39	5.15	5.39	3.02	1.62
MMU	$\varepsilon_{P,abs}$	3.15	2.61	4.98	3.92	1.14	0.97
	$\varepsilon_{rP,abs}$	3.96	4.18	5.78	4.67	3.76	1.77
NDIRIS	$\varepsilon_{P,abs}$	6.37	2.13	5.59	5.92	1.79	0.83
	$\varepsilon_{rP,abs}$	7.51	3.53	7.23	7.38	3.11	1.14

5. Feature Extraction and Matching

We use the standard approach [3] here, which first transforms the iris ring to a so-called *normalized* image. This image is a rectangular raster, it is obtained from the iris ring by the polar transformation, analogous to (19), (20), where r_1 and r_2 are set to the radius of pupil and iris, respectively. In fact, more elaborate version of (19) is used:

$$x(\phi,\rho) = \left(1 - \frac{\rho}{H}\right)x_1\left(\frac{2\pi\phi}{W}\right) + \frac{\rho}{H}x_2\left(\frac{2\pi\phi}{W}\right),$$

$$y(\phi,\rho) = \left(1 - \frac{\rho}{H}\right)y_1\left(\frac{2\pi\phi}{W}\right) + \frac{\rho}{H}y_2\left(\frac{2\pi\phi}{W}\right),$$

$$x_1(\phi) = x_P + r_P\cos(\phi), y_1(\phi) = y_P + r_P\sin(\phi),$$

$$x_2(\phi) = x_I + r_I\cos(\phi), y_2(\phi) = y_I + r_I\sin(\phi),$$

(24)

where x_P, y_P, r_P are the position and radius of pupil and x_I, y_I, r_I are the position and radius of iris. In comparison to (19) this variant accounts for the difference of pupil and iris centres.

The key idea of standard iris feature extraction is to convolve the normalized iris image with a filter, calculating the most informative features of the texture. Earlier Gabor wavelet was used for feature extraction. In one-dimensional space, it is represented as

$$g_{\sigma\lambda}(x) = \exp\left(-\frac{x^2}{2\sigma^2}\right)\exp\left(-i\frac{x}{\lambda}\right), \quad G_{\sigma\lambda}(u) = \exp\left(-\frac{(u-\lambda^{-1})^2\sigma^2}{2}\right),$$

(25)

where σ defines the width of the wavelet in the spatial domain, λ is the wavelength of modulation of the Gaussian by a harmonic function. By introducing inverse values $S = 1/\sigma$ and $W = 1/\lambda$, a simplified representation in the frequency domain can be obtained:

$$G_{SW}(u) = \exp\left(-\frac{(u-W)^2}{2S^2}\right).$$

(26)

It turned out that the modification of the Gabor wavelet called Log-Gabor function is better for feature extraction. Log-Gabor is given in the frequency domain as:

$$G_{SW}(u) = \exp\left(\frac{-\log^2(u/W)}{2\log^2(S/W)}\right) = \exp\left(-\frac{(\log u - \log W)^2}{2\log^2 L}\right).$$

(27)

This is equivalent to (26), in which each variable is replaced by its logarithmic counterpart. $L = S/W = \lambda/\sigma$ represents the ratio of the modulation wavelength to the width of the Gaussian. Research has shown that Log-Gabor wavelets are most likely optimal for the template generation problem. Therefore, we use this type of filter. The parameter λ is essentially the characteristic size of the objects in the image extracted by this filter, and L

is the number of periods of the harmonic function in Equation (25) which have sufficient amplitude and influence to the result. Optimal values of λ and L are selected according to [40].

Iris features $V(\phi,\rho)$ are calculated by convolution of the normalized image (20) with a Gabor or Log-Gabor filter, the transformation is performed in the spectral domain:

$$
\begin{aligned}
V(\phi,\rho) &= N(\phi,\rho) * g_{\sigma\lambda}(\phi) = \\
&= \mathcal{F}^{-1}\{\mathcal{F}\{N(\phi,\rho)\}\mathcal{F}\{g_{\sigma\lambda}(\phi)\}\} = \\
&= \mathcal{F}^{-1}\{\mathcal{F}\{N(\phi,\rho)\}G_{\lambda L}(u)\} .
\end{aligned}
\tag{28}
$$

where σ and λ define the width of the wavelet along the angular axis and the modulation frequency, s is the width along the radial axis, $\mathcal{F}$ is the Fourier transform. The features used to form the patterns are computed as binary values of real and imaginary parts of the array $V(\phi,\rho)$:

$$
\begin{aligned}
T_{Re}(\phi,\rho) &= \mathcal{H}[\mathfrak{R}(V(\phi,\rho))] , \\
T_{Im}(\phi,\rho) &= \mathcal{H}[\mathfrak{I}(V(\phi,\rho))] ,
\end{aligned}
\tag{29}
$$

where $\mathcal{H}[\cdot]$ is the Heavyside function. So, eye image $b(x,y)$ produces a template $T(\phi,\rho)$, and each element of the template contains two bits.

We use features raster of 13 pixels in radial direction r and 256 pixels in tangential direction ϕ. Since each pixel produces two bits in (29) the total size of the template is 6656 bit [40].

Although here we do not build a classification system, which calculates a distance between templates and compares it against a classification threshold, template matching is implicitly present, as it will be shown below. Thus, we need to describe the matching method.

In a standard iris recognition approach templates T_1 and T_2 are matched by normalized Hamming distance:

$$
\rho_0(T_1, T_2) = \frac{1}{|\Omega|}|\{T_1(\phi,\rho) \neq T_2(\phi,\rho), (\phi,\rho) \in \Omega\}| ,
\tag{30}
$$

where $\Omega = M_1 \cap M_2$ is the intersection of the visible areas (presenting true data) of the two irises. Because of the uncertainty of the iris rotation angle, a more complex distance formula is used. The rotation of the original image of the eye is equivalent to a cyclic shift of the normalized image along the $O\phi$ axis. Therefore, one of the templates (together with the mask) is subjected to several shift and compare operations:

$$
\begin{aligned}
\rho(T_1, T_2) &= \min_{\psi} \rho_\psi(T_1, T_2) , \\
\rho_\psi(T_1, T_2) &= \frac{1}{\Omega(\psi)}|\{T_1(\phi + \psi, \rho) \neq T_2(\phi,\rho), (\phi,\rho) \in \Omega(\psi)\}| \\
\Omega(\psi) &= M_1(\phi + \psi) \cap M_2(\phi) ,
\end{aligned}
\tag{31}
$$

where $\psi \in [-S; S]$ is the image rotation angle.

Here things may be simplified. For the embedding method, only irises with low occlusion levels are acceptable. Thus, it is supposed that masks M_1 and M_2 cover all of the iris area, and Ω set spans all templates. Omitting mask, rewriting $|\{T_1 \neq T_2\}|$ as $\sum T_1 \oplus T_2$ and using single order index i instead of coordinates (ϕ,ρ) put (30) as:

$$
\rho_0(T_1, T_2) = \frac{1}{N}\sum_{i=1}^{N} T_1(i) \oplus T_2(i) ,
\tag{32}
$$

where $T(i)$ is the i-th bit of the template, operation $\oplus$ is the sum modulo 2, N is the size of the template. Furthermore, (31) transforms to

$$\rho(T_1, T_2) = \min_{\psi} \rho_{\psi}(T_1, T_2) \,,$$

$$\rho_{\psi}(T_1, T_2) = \frac{1}{N} \sum_{i=1}^{N} T_1(i(\psi)) \oplus T_2(i) \,, \tag{33}$$

where $i(\psi)$ index is recalculated accordingly.

The recognition system is designed to supply the following conditions with the lowest possible errors:

$$
\begin{aligned}
T_1, T_2 \text{ taken from same person} &\implies \rho(T_1, T_2) \leqslant \theta \,, \\
T_1, T_2 \text{ taken from different persons} &\implies \rho(T_1, T_2) > \theta \,.
\end{aligned}
\tag{34}
$$

Violation of the first condition in (34) is called *false reject* and its probability is referred to as *false reject rate* (FRR). FRR of the system is estimated in tests as the ratio of the number of false rejects to the number of all matches of biometric traits of the same persons. Analogously, violation of the second condition in (34) is called *false accept* and its probability is named *false accept rate* (FAR). The threshold θ is chosen from a trade-off between FRR and FAR.

6. Selecting the Embedding Method

There are many works, where biometry is used in combination in combination with other security measures such as usual secured passwords, for instance [41,42]. Here, we intend to develop a system that uses only data transmitted insecurely—the only protection is the iris of the owner.

We also limit ourselves to the case of symmetric encryption. During encoding the *message M* and the secret *key K* are combined into the *code* by the *encoder* function Φ: $C = \Phi(M, K)$, and during decoding the message is reconstructed from code and key by *decoder* functions Ψ: $M = \Psi(C, K)$. If key K is not present, it is impossible to obtain M from C, thus the code C can be made public. Symmetric encryption requires that K is repeated exactly. Not a single bit of it can be changed.

The central problem in automatic biometry systems can be put as developing the optimal classifier. The classifier consists of a distance function between two biometric data samples $\rho(D_1, D_2)$ and a threshold θ (34). The function ρ can be treated as a superposition of two sub-functions. The first one is the calculation of the biometric template T from source data $T = T(D)$, Second sub-function is the calculation of the distance itself $\rho(T_1, T_2)$. Features should be selected, which are stably close for the same person and stably far for different persons with respect to function ρ. As a rule, the elements of biometric templates are highly correlated. On the contrary, cryptographic keys are deliberately developed so as to have uncorrelated bits. However, the entropy (information amount) of an iris template is comparable to that of currently used cryptographic keys [43]. This suggests that it is possible to implement a cryptographic key in biometrics without reducing its robustness.

It should be noted that most of the works presenting the application of cryptographic methods to biometrics, develop the scenario of *cancelable biometrics* [44]. Its essence is producing such biometric templates that source biometric data cannot be extracted or guessed anyhow from any number of templates. Cancelable biometrics is nothing but a kind of fuzzy hashing [45]. Formally, an additional step is introduced in the calculation of the distance function ρ. Distance $\rho(S_1, S_2)$ is calculated, $S = S(T)$ is the hash function. Obviously, the recognition problem is still being solved here. Thus, cancelable biometrics is just a remake of identification and cannot be used for our purposes.

There are two approaches to how to process volatile biometrics, leading them to an unchanging cryptographic key. The first approach employs already computed biometric features constituting the template T, which are supplemented with error correction using different variants of redundant coding. This approach is used here. In the second approach [46] biometric features are not obtained in explicit form. Instead, a neural network is trained, which directly produces a target key from raw biometric data D. The advantage of this approach is said to be less coding redundancy by using continuous data at all stages and quantization only at the end. Disadvantages are the unpredictability of neural network training, lack of guaranteed quality of performance, including uncertainty in retaining quality in a wider set of data than that used in training.

The task of reproducing a cryptographic key is accomplished by *biometric cryptosystems* (BC) [45,47], also called *biometric encryption* [48]. There are two classes of BCs, which implement different approaches: *key generation* and *key binding*.

Methods of key generation, i.e., direct production of the key from raw biometry or template without using supplementary code are studied in [49–51]. Biometric template T is mapped into the space of cryptographic keys (usually bit strings) by a special function: $K(T) : T \to \{0,1\}^n$, where n is the length of the key. One function is used for registration and recognition. The conditions must hold

$$
\begin{aligned}
T_1, T_2 \text{ taken from one person} &\implies K_1 = K_2 \,, \\
T_1, T_2 \text{ taken from different persons} &\implies K_1 \neq K_2 \,.
\end{aligned}
\tag{35}
$$

These conditions are closely related to (34); however, in (35) the task is to reproduce the sequence of bits. The results of the methods without supplementary data are not very hopeful for practical applications. Error level is generally very high in this approach. In [50] the authors report FRR = 24% at FAR = 0.07% even for homogeneous high-quality images [32]. In [51], the idea is based on assumption that two iris codes can be mapped to some "closest" prime number and this number will be the same for the codes from one person. Considering the variability of iris codes even for ideal conditions this is unlikely to happen. The authors do not report the study of recognition errors.

Scenario with *helper code* demonstrates much better performance. During registration the encoder takes the template T_1, computes the key $K_1 = K(T_1)$, encrypts the message M with K_1 and additionally outputs some helper code $h = \Phi(T_1)$. Immediately after this the original T_1, M, and K_1 are destroyed, leaving only encoded message M' and helper code h. The original template T_1 or key T_1 cannot be recovered from M' and h. During presentation another set of biometric traits T_2 is obtained and the key $K_2 = \Psi(T_2, h)$ is calculated. Functions Φ and Ψ are designed so as to satisfy (35). Thus, by providing biometrics and helper code, the registered user can obtain the original key $K_2 = K_1$, and hence the message M. At the same time, the intruder, even knowing h, will not be able to obtain K_1 [52], so the helper code h can be made non-secret.

The biometric data itself may be used as a key: $K \equiv T$. In this case, at the stage of presentation, original biometrics T_1 is restored from presented T_2. This scenario is called *secure sketch* [53]. However, the works based on secure sketches and available in the literature show rather modest results. For example, the system [54] is workable under the assumption that intraclass variability of features is below 10%. In practice, the variability is more than 20%. This conditions the inoperability of the proposed method.

The *key binding* scheme in the above terms looks like a pair of encoder function $C = \Phi(K_1, T_1)$ and decoder function $K_2 = \Psi(C, T_2)$, which holds the (35) condition. The advantage is that K_1 is set externally, rather than created by the developed algorithm. From this point of view, K_1 can be perceived as a message M, which is external to the encryption system. This immediately simplifies the biometric cryptosystem to a symmetric encryption scenario. The difference is that the secret key K must be the same in encoding and decoding in symmetric encryption, whereas the biometric features (also secret) differ: $T_1 \neq T_2$. This scenario is called *fuzzy extractor* [53].

If Ψ is an inverse of Φ and biometric data are composed of real numbers the scenario is referred to as *shielding functions* [55]. So-called *fuzzy vault* [43] is another popular method of key embedding. It is founded on Shamir's secret separation scheme [56]. Here rather low, practically meaningful error values are obtained: [57] reports FRR = 0.78% and [58] reports FRR = 4.8% at zero FAR. However, both results are shown using a single small image database (less than 1000 samples).

The most promising for use in iris biometry is *fuzzy commitment* scenario [59]. In [60], a simple algorithm is proposed. The basic idea is to use employ the *error correcting coding* (ECC) [61]. ECC is widely used in data transmission over noisy channels. Generally, data transmission involves a pair of functions also called encoder and decoder. The encoder $R = \Phi_p(K)$ maps the transmitted message K into a larger redundant code R. Then R is passed through the transmission channel, which alters each of its symbols independently with the probability q, and the altered code R' is received at the other side. The decoder $K = \Psi_p(R')$ is able to restore K back from R' under the condition that no more than a p share of values were altered. Call p as *tolerated error probability*. Thus, if $q < p$ then the message is restored with a probability close to 1. Otherwise, the probability to restore K is close to 0. One can design Φ and Ψ for a wide range of transition error probabilities $p \in [0; 0.5)$. Redundancy grows as p approaches to 0.5, for $p = 0.5$ it becomes infinite.

Here ECC is used as follows. The encoder and decoder are constructed so as to have a tolerated error probability equal to the classification threshold of the biometric recognition system: $p = \theta$. Upon registration, a password K_1 is constructed and the user's template T_1 is obtained. The code $R_1 = \Phi_p(K_1)$ (it generally looks like pseudorandom numbers) is bitwise summed modulo 2 (exclusive or) to the iris template yielding the public code $C = R_1 \oplus T_1$. After C is calculated, template T_1, message K_1, and redundant code R_1 are destroyed. None of them can be extracted from C alone. Thus, it is possible to expose C publicly and transmit it through unprotected channels. Upon presentation iris of a person is registered once more and a new template T_2 is formed. Of course, it is not equal to the original one. Since $R_2 = C \oplus T_2 = (R_1 \oplus T_1) \oplus T_2$, then $R_1 \oplus R_2 = T_1 \oplus T_2$. If the templates T_1 and T_2 are taken from one person, the distance is very likely to be less than the classification threshold: $\rho(T_1, T_2) \leqslant \theta$, so $\rho(R_1, R_2) \leqslant \theta$. By the nature of (32) it means that less than p share of bits differ in R_1 and R_2 and the original secret key $K_1 = K_2 = \Psi_p(R_2)$ will be recovered. On the other hand, if the templates are classified as belonging to different persons $\rho(T_1, T_2) \geqslant \theta$, the probability of restoring the original K_1 is close to zero. The scenario of operation is shown in Figure 4.

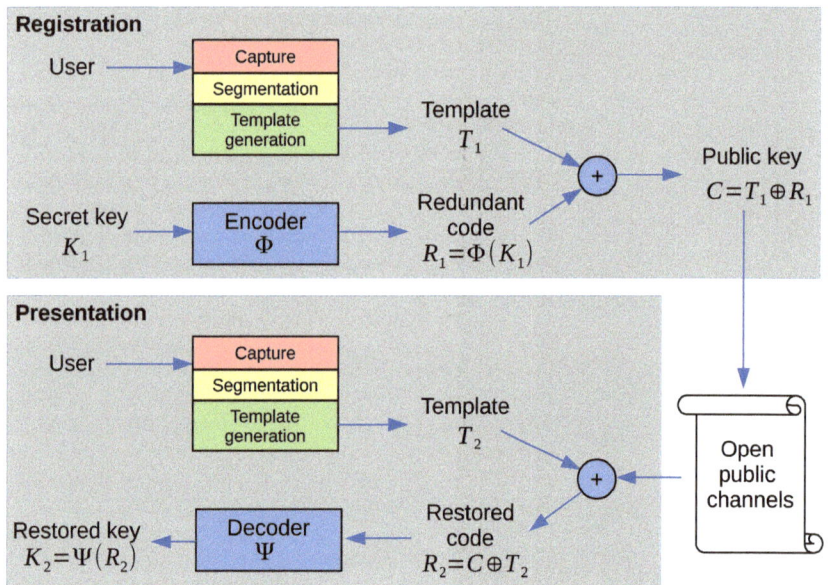

Figure 4. Scenario of the method [60].

Work [60] proposes a cascade of two ECC algorithms: Reed–Solomon [62] and Hadamard [61]. Reed–Solomon coding handles an entire block of data of length L, processing it as a set of L/s s-bit symbols. Any arbitrary symbols (not bits!) can be different as long as their number is not greater than pL. In [60], this coding is aimed to combat group errors appearing from various occlusions (eyelashes, eyelids), which cover significant areas of the iris. Hadamard coding processes small chunks of data (few bits), and corrects no more than 25% of the errors in each chunk. For Hadamard code to be most successful in error correction, the errors (deviations of T' from T) should be evenly scattered across the template with a density of no more than 25%. This coding is designed to deal with single pixel deviations arising from camera noise. The key K is encoded by Reed–Solomon code, the result is processed by Hadamard code.

This cascade performs well if the share of altered bits in one person's templates does not exceed 25%. However, in practical databases and applications this share is bigger which leads to an unacceptably high (more than 50%) false reject probability. To overcome this difficulty, it is proposed [42] to introduce additional template masking: every fourth bit of the iris templates is set to zero. Due to this, the proportion of altering bits in the templates of one person is reduced below 20%. This easy solution ruins the very idea of security: if some bits of the template are fixed, then appropriate bits of redundant code are made known to code crackers and can be used to attack the code. A critique of this method in terms of resistance to cracking is given in [46]. The attack is carried out by gradually restoring the original template.

Here we attempt to refine the fuzzy extractor [60] in a more feasible method and build a practically applicable key embedding method. Based on the iris feature extraction system, experiments against several publicly available iris databases are carried out. Two steps are added to the encoder tail (and hence, decoder head): majority coding of single bits and pseudorandom bit mixing. Three of these four steps have parameters, which affect their properties, including error tolerance and size. Optimal values of these parameters are selected to fit the redundant code size into the iris template size, keep the error tolerance near the target level, and maximize the size of encoded key.

7. Determining the Threshold

So, if the registration template T and the presentation template T' are at Hamming distance (32) below the threshold θ, then the encrypted message M is recovered with high confidence; otherwise, the probability to recover it is extremely low (order of random guess). Thus, the threshold θ separates "genuine" and "intruder" templates T' with respect to T.

It is necessary to determine the value of the threshold p, which will be used for separating "genuines" and "impostors". With this value redundant coder Φ_p and decoder Ψ_p will be devised, capable to restore message for "genuine" template and making it impossible for "impostor" template.

The following publicly available databases were used for the experiments: CASIA-4-Thousand [32], BATH [33], ICE subset of NDIRIS [34], UBIRIS-1 [35].

Table 5 gives a list of databases used with the obtained thresholds.

Table 5. Database characteristics and thresholds.

Database	Number of Eyes	Images	θ at FAR$=10^{-4}$	FAR at $\theta=0.35$, $\times 10^{-4}$	FRR at $\theta=0.35$, $\times 10^{-2}$
BATH	1600	31988	0.402	0.03	4.46
CASIA	2000	20000	0.351	0.97	6.71
ICE	242	2953	0.395	0.011	7.13
UBIRIS	240	1207	0.401	0.001	5.18

For each database the following numbers are given:

- The number of individual eyes. It does not match the number of participating persons as some persons in the database supplied images from both eyes.
- The number of eye images. Each eye produced from one to several hundred images, depending on the database collection scenario.
- Value of the threshold θ, at which the false accept rate (FAR) is 10^{-4}. This value is the probability of a random guess of a four-digit pin code. It is used for reference in developing biometric recognition systems so that they would have the same or less probability of being defeated as pin-code-based systems.
- False accept rate at $\theta = 0.35$.
- False reject rate at $\theta = 0.35$.

Since $FAR(\theta)$ is a monotonous growing function, we select a minimal $\theta(FAR = 10^{-4})$ from the fourth column of the table. It is $\theta = 0.351$ for the CASIA database. Other databases have even smaller FAR with this value of θ.

So, the value of $\theta = 0.35$ is the tolerated error probability p for constructing the ECC. Table 5 shows the values of false accept and false reject rates for this threshold. The maximum false reject rate does not exceed 8%.

8. Error Correction Cascade

We describe the applied methods in the sequence of their execution by the decoder, which is also the method "from simple to complex". In the beginning data unit is a single bit, at the end it is the whole message. The problem is to devise an error correction method, which encodes a message into a block of data with redundancy, and then is able to reconstruct the message if no more than $p \leqslant 0.35$ share of these bits is altered. Popular Walsh-Hadamard and Reed-Muller [63] methods can be used only for $p < 0.25$, thus they are not directly applicable. Furthermore, the errors of neighboring elements of the template are strongly correlated, whereas almost all ECC methods show the best performance against uncorrelated errors.

8.1. Decorrelation by Shuffling

It is more difficult to design methods usable against correlated errors, and their performance is worse compared to the case of uncorrelated errors. Much of the effort in this case is directed precisely at decorrelation. Luckily, the whole block of data is available in our task (rather than sequentially feeding with symbols as in many transmission channel systems) and a simple method of decorrelation can be applied which is the quasi-random shuffling of iris template bits. A bit of the template array T is placed from position i into position $ij \mod N$:

$$\tilde{T}(ij \mod N) = T(i), \quad i = \overline{0, N-1}, \tag{36}$$

where j is a number, relatively prime to total number N of bits in array. Relatively prime condition guarantees $ij \mod N$ number being unique for $i = \overline{0, N-1}$ and the shuffling $T \rightarrow \tilde{T}$ being reversible. After shuffling the neighboring bits of $\tilde{T}$ are taken from bits, which were far away from each other in T and their errors are uncorrelated. If one rule of shuffling is always applied then bits in all templates change their position in the same method, and calculation (32) affects the same pairs of bits. Thus, Hamming distance is unchanged and all developments from it are preserved.

This method does not change the size of the code.

8.2. Bit-Majority Coding

The error rate $p = 0.35$ is too big for most error correction codes. Practically, the only possibility here is the majority coding of single bits. It is applicable for $p < 0.5$. At the coding stage, the bit is repeated n times. At the decoding stage, the sum of n successive bits is counted. If it is below $n/2$, zero bit value is decoded, otherwise a unit. It is easy to see that odd values of n are preferable. If p is the error probability of a single bit and bits are altered independently the error probability of decoded bit is

$$p_D(p) = 1 - \sum_{l=0}^{(n-1)/2} \binom{n}{l} p^{n-l}(1-p)^l = 1 - (1-p)^n \sum_{l=0}^{(n-1)/2} \binom{n}{l} \left(\frac{p}{1-p}\right)^l. \quad (37)$$

If the error probability of one bit of the code is $p = 0.35$, then bit majority coding with $n = 7$ will transmit bits with error probability $p_D = 0.2$. This value is below 0.25 and allows the use of Hadamard codes. Majority coding with $n = 15$ will give $p_D = 0.12$ for $p = 0.35$. A larger duplication is possible, but results in a larger code size.

The parameter of this method, affecting its size and error probabilities is the bit repetition rate n.

8.3. Block Coding

Denote the set of all bit strings of length n as $\mathbb{B}^n$. This set can be viewed as a set of vertices of n-dimensional binary cube. Consider the string of length k called *message* here: $M \in \mathbb{B}^k$. There can be 2^k different messages. Consider the set of 2^k strings of length $n > k$ called *codes*. There is a one-to-one correspondence of messages and codes. Since the code length is greater than the message length, the coding is redundant and it is possible to alter some bits of the code but still be able to restore the corresponding message. The idea of block coding is to select such 2^k codes out of their total amount of 2^n, that the probability of restoration error is minimal. The set of selected codes is called *code table* $\mathbb{C}$. In terms of the n-dimensional binary cube, this means selecting 2^k vertices so as to maximize minimal Hamming distance between selected vertices:

$$\mathbb{C}^* = \arg\max_{\mathbb{C}} \min_{\substack{u,vs.\in\mathbb{C} \\ u\neq v}} \rho(u,v), \quad \rho^* = \min_{\substack{u,vs.\in\mathbb{C}^* \\ u\neq v}} \rho(u,v), \quad (38)$$

where ρ is the distance (32).

Hadamard coding is based on a Hadamard matrix, which is constructed iteratively:

$$H_0 = (0), \quad H_{n+1} = \begin{pmatrix} H_n & H_n \\ H_n & \overline{H_n} \end{pmatrix}, \quad (39)$$

where $\overline{H}$ is a bit inversion of all elements of H. Hadamard matrix H_k is a square matrix with 2^k rows. It gives the coding table naturally: each row number is the message and the row contents is the code. It can be proven that for Hadamard codes $\rho^* = 2^{k-1}$. We use so-called augmented Hadamard codes, where another 2^k strings are added to the code table. These strings are bitwise inverted versions of the strings obtained from (39). For this code table $\rho^* = 2^{k-2}$.

There is a simple and well-known estimation (called *Hamming boundary*) of the probability of block coding error, which for augmented Hadamard codes is:

$$p_H \leqslant 1 - P_{corr} = 1 - (1-p)^n \sum_{l=0}^{(n-1)/4} \binom{n}{l} \left(\frac{p}{1-p}\right)^l, \quad (40)$$

where p is the probability of bit inversion. Since this stage inputs the output of bit majority decoding, the value of p here is the value of p_D from (37). Let us redefine $p_D \to p$ in this section for simplicity. Furthermore, one can note that (40) is the same as (37) except for the upper summation limit.

However, this is a rather rough estimate, which grows worse with the increase in n. For small n exact calculations performed by simple exhaustive search, the results are given below. The message is decoded assuming that the original code is distorted minimally, i.e., for the code C we will look for the closest code $C^* \in \mathbb{C}$. Let us call it *attractor*. There can be several attractors (several codes can have the same minimal distance to C). If there are several attractors, a random one is chosen. Let us denote the set of attractors for C as $A(C)$.

The probability of decoding the correct message is that of choosing the correct attractor

$$P_{corr} = \sum_M P(M) \sum_C p(C^*|C) p(C|C^*) , \tag{41}$$

where $P(M)$ is the probability to obtain message M as input. Message M is encoded by code $C^* \in \mathbb{C}$. Then $p(C|C^*)$ is the probability to obtain distorted code C while transmitting C^*, $p(C^*|C)$ is probability to recover C^* (hence, correct M) from distorted code C. Suppose the probability of all messages is the same. Then the sum by M is reduced and

$$P_{corr} = \sum_C p(C^*|C) p(C|C^*) . \tag{42}$$

Without loss of generality, due to the symmetry of Hadamard codes [61], we can assume C^* to be a zero code, i.e., a string of zero bits (as is in standard code). Then the probability of obtaining a certain code C from zero code is $p^{b(C)}(1-p)^{n-\beta(C)}$, where $\beta(C) \equiv \rho(0, C)$ is number of unit bits in C, and

$$P_{corr} = \sum_C p(0|C) p^{\beta(C)} (1-p)^{n-\beta(C)} , \tag{43}$$

where $p(0|C)$ is the likelihood to obtain zero code from C. Define the set of attractors for string C as code table entries with minimal distance to the code:

$$A(C) = \{C' \in \mathbb{C} : \rho(C', C) = \rho_{min}(C)\} ,$$
$$\rho_{min}(C) = \min_{C' \in \mathbb{C}} \rho(C, C') . \tag{44}$$

Define the cardinality of this set as $\alpha = |A(C)|$ and state that $\alpha = 0$ if there is another code table entry more close to C then the correct code: $\exists C' \in \mathbb{C}, C' \neq C^* : \rho(C', C) < \rho(C^*, C)$. Then we can write

$$p(0|C) = \begin{cases} 0, & \alpha = 0, \\ 1/\alpha, & \alpha \neq 0 . \end{cases} \tag{45}$$

For small values of n all points of code space $\mathbb{B}^n$ can be enumerated and their distribution by distance to zero β and attractor number α can be estimated:

$$H(\tilde{\beta}, \tilde{\alpha}) = |\{C : 0 \in A(C), |\alpha(C)| = \tilde{\alpha} , \ \beta(C) = \tilde{\beta}\}| ,$$
$$\tilde{\beta} \in [0; n] , \ \tilde{\alpha} \in [1; 2^k] . \tag{46}$$

Substituting to (43) we get:

$$P_{corr} = \sum_{\alpha \neq 0} \sum_\beta \frac{H(\beta, \alpha)}{\alpha} p^\beta (1-p)^{n-\beta} = \sum_\beta p^\beta (1-p)^{n-\beta} \sum_{\alpha \neq 0} \frac{H(\beta, \alpha)}{\alpha} \tag{47}$$

and decoding error

$$p_H = 1 - P_{corr} = 1 - (1-p)^n \sum_\beta h(\beta) \left(\frac{p}{1-p}\right)^\beta , \ h(\beta) = \sum_{\alpha \neq 0} \frac{H(\beta, \alpha)}{\alpha} . \tag{48}$$

The formula is the same as (40) except for the coefficients and summation limits. The values of $h(\beta)$ for the augmented Hadamard code of order 5 ($n = 2^5 - 1 = 31$, $k = 5 + 1 = 6$) are given in Table 6. All meaningful values are given, and values for other α and β are zero. For instance, if the code is distorted in 12 bits or more, it will be never recovered to the correct value, since there will be another valid code from the code table, closer to distorted value.

Table 6. Values $h(\beta)$ and $p_H(\beta)$ for Hadamard code $k = 6$, $n = 31$.

β	$h(\beta)$	$p_H(\beta)$
0	1	0.
1	31	0.
2	465	0.
3	4495	0.
4	31,465	0.
5	169,911	0.
6	736,281	0.
7	2,629,575	0.
8	7,490,220	0.026
9	13,798,100	0.164
10	8,265,964	0.508
11	427,924	0.810

Up to values $\beta = 7$ only one attractor is chosen, i.e., at this or less divergence the message is definitely recovered. This corresponds to Hamming boundary (40). However, even with larger divergences, up to $\beta = 11$ there is a significant probability of correct recovery. This plays a big role since the majority of distorted codes fall outside of Hamming boundary but still have the significant probability to restore the message correctly. Thus, the probability (40) is overestimated. For example, for the considered code and error $p = 0.250$, the formula (40) gives $p_H = 0.527$, which would seem to prevent using such a code. However, the calculation using the formula (48) gives $p_H = 0.261$, which is fairly suitable for use in the next step of coder.

The Hadamard coding parameter is only the word length k. The size of the codeword n is dependent: $n = 2^{k-1}$ for the augmented variant.

8.4. Reed–Solomon Message Coding

The unit of encoding for Reed–Solomon's algorithm is the entire message, which is divided into codewords of fixed size, s bits each. A stream of L bits is cut into $k = \lceil L/s \rceil$ words. Then additional words can be added up to the total count of n by the coding algorithms. It turns out that if no more than $t = \lfloor (n - k)/2 \rfloor$ codewords are altered then it is possible to recover the message. So, Reed–Solomon code corrects no more than t of errors, where t is half the number of redundant words. Denoting $p = t/n$, we get

$$p \leqslant \frac{n - k}{2n}. \tag{49}$$

This number is an estimate of the tolerated error probability of a codeword. The Reed–Solomon method also imposes a limitation to a codeword count in the whole message:

$$n \leqslant 2^s - 1. \tag{50}$$

The error probability in (49) is determined by the previous step: $p = p_H$. Hence, possible Reed–Solomon codes here are determined by codeword length s and message length L.

9. Selection of Code Parameters

Four ECC methods are organized in a chain. Encoder runs Reed–Solomon, Hadamard, bit majority and shuffling to obtain the redundant code. The decoder executes this chain in reverse order. The encoder should obtain the code of size no more than the size of the iris template, i.e., $N = 6656$ bits for the presented system. The code size cannot be larger, duplication and masking are unacceptable, as they make it trivial to break such a code.

Furthermore, of course, it is desired to embed the message of reasonable size. This is a discrete constrained optimization problem.

ECC methods used here have the following parameters affecting their characteristics: (1) decorrelation has no parameters; (2) majority coding is governed by the bit duplication count n; (3) Hadamard coding depends on word size k, Reed–Solomon coding is parameterized by word size s and message length L. Combinations of (n, k, s, L) values yield different encoding with specific code length $C(n, k, s, L)$ and error probabilities. The errors are the aforementioned FRR and FAR. False rejection is a failure to recover the embedded key after presenting the same person's biometrics. False acceptance is recovering the person's key with another person's biometric. The errors depend on ECC parameters: $FRR(n, k, s, L)$ and $FAR(n, k, s, L)$. Furthermore, the formal statement of the problem is

$$FRR(n, k, s, L) \to \min ,$$
$$s.t. \ FAR(n, k, s, L) \leqslant 10^{-4} , \ C(n, k, s, L) \leqslant N. \tag{51}$$

10. Results and Comparison with Literature

The solution of (51) was found: $L = 65$, $n = 13$, $k = 5$, $s = 5$, $FRR = 10.4\%$. Message size $L = 65$ bits is considered satisfactory for "common" user keys. For bigger message sizes there reasonable solution was not discovered. It should be noted that without bit shuffling (not explicitly involved in (51)) the problem is not solved even for $L = 65$.

Table 7 contains the results reported in the literature in contrast with those presented in this work.

Table 7. The results of iris biometric cryptosystems

Authors	FRR/FAR	DataSet	Keybits
Wu et al. [64]	5.45/0.73	CASIA v1	1024
Rathgeb & Uhl [50]	4.92/0.0	CASIA v3	128
Hao et al. [60]	0.42/0.0	70 persons	140
Bringer et al. [65]	5.62/0.0	ICE 2005	40
Kanade08 et al. [42]	2.48/0.0008	ICE 2005	234
Presented	10.4/0.0	mixed	65

The presented system may seem not very successful against its rivals with respect to error level and key length. However, one should note that each of these systems was tested with a single database. Both CASIA databases have images of one eye taken from adjacent video frames that results in the extremely high similarity of iris codes, inaccessible in practice. The same issue concerns [60], they use a small laboratory database and their results cannot be extended to real applications. ICE 2005 database is much closer to the real world, it contains images of varying quality, time and conditions of registration. However, both works [42,65] based on it use interleaving bits. If the bit sequence is fixed and known, this ruins the cryptographic strength. If it is made secret, then it turns to just another secret key, which should be passed securely: the very thing we try to avoid. Although the presented system has the highest FRR, it is practically applicable and has no obvious holes in security.

11. Conclusions

The set of methods allowing to build biometric cryptosystem based on iris images is presented. It contains three main parts: iris segmentation, biometric template generation, and the method of embedding/extracting the cryptographic key to/from biometric features. The original system of iris segmentation methods is described. Its distinction is estimating iris parameters at several steps (initial rough calculation, then refinement), by algorithms of a different kind. The sequence of detection of iris parameters is different from commonly employed as well. The template creation method is a de facto standard Daugman-style convolution. The method for introducing a cryptographic key into iris biometrics is constructed using a fuzzy extractor paradigm. A key of size up to 65 bits can be embedded, for larger sizes no solution has been obtained. The challenge of high variance of biometric features has been overcome by introducing bit majority coding. A high local correlation of errors was removed by quasi-random shuffling. The system was tested on several databases of iris images. A study of cryptography stability is still required.

Author Contributions: Conceptualization, I.M.; methodology, I.M.; software, I.M.; validation, I.M. and I.S.; investigation, I.M.; data curation, I.M.; writing—original draft preparation, I.M.; writing—review and editing, I.S.; supervision, I.M.; project administration, I.M. All authors have read and agreed to the published version of the manuscript.

Funding: This research received no external funding.

Data Availability Statement: Not applicable.

Conflicts of Interest: The authors declare no conflict of interest.

References

1. Bertram, L.A.; van Gunther, D. (Eds.) *Nomenclatura: Encyclopedia of Modern Cryptography and Internet Security—From AutoCrypt and Exponential Encryption to Zero-Knowledge-Proof Keys*; Books on Demand: Paris, France, 2019; ISBN 978-3-7460-6668-4. Available online: https://www.amazon.com/Nomenclatura-Encyclopedia-Cryptography-Internet-Security/dp/3746066689 (accessed on 2 February 2023).
2. Chmora, A.L. Key Masking Using Biometry. *Probl. Inf. Transm.* **2011**, *47*, 201–215. [CrossRef]
3. Daugman, J. How Iris Recognition Works. *IEEE Trans. Circuits Syst. Video Technol.* **2004**, *14*, 21–30. [CrossRef]
4. Shekar, B.H.; Bharathi, R.K.; Kittler, J.; Vizilter, Y.V.; Mestestskiy, L. Grid structured morphological pattern spectrum for off-line signature verification. In Proceedings of the International Conference on Biometrics, Phuket, Thailand, 19–22 May 2015; Volume 8, pp. 430–435.
5. Bowyer, K.; Hollingsworth, K.; Flynn, P. Image understanding for iris biometrics: A survey. *Comput. Vis. Image Underst.* **2008**, *110*, 281–307. [CrossRef]
6. Radman, A.; Jumari, K.; Zainal, N. Iris Segmentation: A Review and Research Issues. In *Software Engineering and Computer Systems. ICSECS 2011. Communications in Computer and Information Science*; Mohamad Zain, J., Wan Mohd, W.M.B., El-Qawasmeh, E., Eds.; Springer: Berlin/Heidelberg, Germany, 2011; Volume 179.
7. Bowyer, K.; Hollingsworth, K.; Flynn, P. A survey of iris biometrics research: (2008–2010). In *Handbook of Iris Recognition*; Burge, M., Bowyer, K., Eds.; Springer: London, UK; Heidelberg, Germany; New York, NY, USA; Dordrecht, The Netherlands, 2012.
8. Malgheet, J.R.; Manshor, N.B.; Affendey, L.S. Iris Recognition Development Techniques: A Comprehensive Review. *Complexity* **2021**, *2012*, 6641247. [CrossRef]
9. Liu, N.; Li, H.; Zhang, M.; Liu, J.; Sun, Z.; Tan, T. Accurate iris segmentation in non-cooperative environments using fully convolutional networks. In Proceedings of the 2016 International Conference on Biometrics (ICB), Halmstad, Sweden, 13–16 June 2016; pp. 1–8.
10. Wang, C.; Muhammad, J.; Wang, Y.; He, Z.; Sun, Z. Towards Complete and Accurate Iris Segmentation Using Deep Multi-Task Attention Network for Non-Cooperative Iris Recognition. *IEEE Trans. Inf. Forensics Secur.* **2020**, *15*, 2944–2959. [CrossRef]
11. Huo, G.; Lin, D.; Yuan, M.; Yang, Z.; Niu, Y. Heterogeneous iris segmentation method based on modified U-Net. *J. Electron. Imaging.* **2021**, *30*, 063015. [CrossRef]
12. Meng, Y.; Bao, T. Towards More Accurate and Complete Heterogeneous Iris Segmentation Using a Hybrid Deep Learning Approach. *J. Imaging.* **2022**, *8*, 246. [CrossRef]
13. Korobkin, M.; Odinokikh, G.; Efimov, I.; Solomatin, I.; Matveev, I. Iris Segmentation in Challenging Conditions. *Pattern Recognit. Image Anal.* **2018**, *28*, 817–822. [CrossRef]
14. Hofbauer, H.; Jalilian, E.; Uhl, A. Exploiting superior cnn-based iris segmentation for better recognition accuracy. *Pattern Recognit. Lett.* **2019**, *120*, 17–23. [CrossRef]
15. Cui, J.; Wang, Y.; Tan, T.; Ma, L.; Sun, Z. A Fast and Robust Iris Localization Method Based on Texture Segmentation. In *Biometric Authentication and Testing, National Laboratory of Pattern Recognition*; Chinese Academy of Sciences: Beijing, China, 2004; pp. 401–408.
16. Liu, X.; Bowyer, K.W.; Flynn, P.J. Experiments with an Improved Iris Segmentation Algorithm. In Proceedings of the 4th IEEE Workshop on Automatic Identification Advanced Technologies, New York, NY, USA, 17–18 October 2005; pp. 118–123.
17. Dey, S.; Samanta, D. A Novel Approach to Iris Localization for Iris Biometric Processing. *Intern. J. Biol. Life Sci.* **2007**, *3*, 180–191.
18. Ling, L.L.; de Brito, D.F. Fast and Efficient Iris Image Segmentation. *J. Med. Biol. Eng.* **2010**, *30*, 381–392. [CrossRef]
19. Yuan, W.; Lin, Z.; Xu, L. A Rapid Iris Location Method Based on the Structure of Human Eyes. In Proceedings of the 27th Annual Conf. Engineering in Medicine and Biology, Shanghai, China, 1–4 September 2005; pp. 3020–3023.

20. Pan, L.; Xie, M.; Ma, Z. Iris Localization Based on Multiresolution Analysis. In Proceedings of the 19th International Conference on Pattern Recognition, Tampa, FL, USA, 8–11 December 2008; pp. 1–4.
21. He, Z.; Tan, T.; Sun, Z.; Qiu, X. Toward Accurate and Fast Iris Segmentation for Iris Biometrics. *IEEE PAMI* **2009**, *31*, 1670–1684.
22. Maenpaa, T. An Iterative Algorithm for Fast Iris Detection. In Proceedings of the International Workshop on Biometric Recognition Systems, Beijing, China, 22–23 October 2005; p. 127.
23. Nabti, M.; Ghouti, L.; Bouridane, A. An Effective and Fast Iris Recognition System Based on a Combined Multiscale Feature Extraction Technique. *Pattern Recognit.* **2008**, *41*, 868–879. [CrossRef]
24. Matveev, I. Iris Center Location Using Hough Transform with Two-dimensional Parameter Space. *J. Comput. Syst. Sci. Int.* **2012**, *51*, 785–791. [CrossRef]
25. Proenca, H.; Alexandre, L.A. Iris Segmentation Methodology for Non-cooperative Recognition. *IEEE Proc. Vision Image Signal Process.* **2006**, *153*, 199–205. [CrossRef]
26. Matveev, I. Detection of Iris in Image by Corresponding Maxima of Gradient Projections. In Proceedings of the Computer Graphics, Visualization, Computer Vision and Image Processing 2010, Freiburg, Germany, 27–29 July 2010; pp. 17–21.
27. Novik, V.; Matveev, I.A.; Litvinchev, I. Enhancing Iris Template Matching with the Optimal Path Method. *Wirel. Netw.* **2020**, *26*, 4861–4868. [CrossRef]
28. Solomatin, I.A.; Matveev, I.A.; Novik, V.P. Locating the Visible Part of the Iris with a Texture Classifier with a Support Set. *Autom. Remote Control* **2018**, *79*, 492–505. [CrossRef]
29. Pratt, W.K. *Digital Image Processing: PIKS Scientific Inside*, 4th ed.; Wiley–Interscience: New York, NY, USA, 2007.
30. Sun, C.; Pallottino, S. Circular Shortest Path in Images. *Pattern Recognit.* **2003**, *36*, 709–719. [CrossRef]
31. Gankin, K.A.; Gneushev, A.N.; Matveev, I.A Iris Image Segmentation Based on Approximate Methods with Subsequent Refinements. *J. Comput. Syst. Sci. Int.* **2014**, *53*, 224–238. [CrossRef]
32. CASIA. Iris Image Database, Institute of Automation, Chinese Academy of Sciences. 2010. Available online: http://biometrics.idealtest.org/findTotalDbByMode.do?mode=Iris (accessed on 2 February 2023).
33. Woodard, D.L.; Ricanek, K. Iris Databases. In *Encyclopedia of Biometrics*; Li, S.Z., Jain, A., Eds.; Springer: Boston, MA, USA, 2009.
34. Phillips, P.J.; Scruggs, W.T.; O'Toole, A.J.; Flynn, P.J.; Bowyer, K.W.; Schott, C.L.; Sharpe, M. Frvt2006 and Ice2006 Large-scale Experimental Results. *IEEE PAMI* **2010**, *5*, 831–846. [CrossRef]
35. Proenca, H.; Alexandre, L. UBIRIS: A Noisy Iris Image Database. In Proceedings of the 13th International Conference Image Analysis and Processing, Cagliari, Italy, 6–8 September 2005; pp. 970–977.
36. Wildes, R.P. Iris Recognition: An Emerging Biometric Technology. *Proc. IEEE* **1997**, *85*, 1348–1363. [CrossRef]
37. Daugman, J. New Methods in Iris Recognition. *IEEE Trans. Syst. Man-Cybernatics-Part B Cybernatics* **2007**, *37*, 1167–1175. [CrossRef]
38. Masek, L. Recognition of Human Iris Patterns for Biometric Identification. 2003. Available online: https://www.peterkovesi.com/studentprojects/libor/ (accessed on 2 February 2023).
39. Ma, L.; Tan, T.; Wang, Y.; Zhang, D. Local Intensity Variation Analysis for Iris Recognition. *Pattern Recognit.* **2004**, *37*, 1287–1298. [CrossRef]
40. Matveev, I.A.; Novik, V.; Litvinchev, I. Influence of Degrading Factors on the Optimal Spatial and Spectral Features of Biometric Templates. *J. Comput. Sci.* **2018**, *25*, 419–424. [CrossRef]
41. Kumar, M.; Prasad, M.; Raju, U.S.N. BMIAE: Blockchain-based Multi-instance Iris Authentication using Additive ElGamal Homomorphic Encryption. *IET Biom.* **2020**, *9*, 165–177. [CrossRef]
42. Kanade, S.; Camara, D.; Krichen, E.; Petrovska-Delacrétaz, D.; Dorizzi, B. Three Factor Scheme for Biometric-based Cryptographic Key Regeneration Using Iris. In Proceedings of the Biometrics Symposium, Tampa, FL, USA, 23–25 September 2008; pp. 59–64.
43. Juels, A.; Sudan, M. A Fuzzy Vault Scheme. *Des. Codes Cryptogr.* **2006**, *38*, 237–257. [CrossRef]
44. Patel, V.M.; Ratha, N.K.; Chellappa, R. Cancelable Biometrics: A review. *IEEE Signal Process. Mag.* **2015**, *32*, 54–65. [CrossRef]
45. Rathgeb, C.; Uhl, A. A Survey on Biometric Cryptosystems and Cancelable Biometrics. *EURASIP J. Inf. Secur.* **2011**, *3*, 1–25. [CrossRef]
46. Akhmetov, B.S.; Ivanov, A.I.; Alimseitova, Z.K. Training of Neural Network Biometry-Code Converters. *Izv. NAS RK Ser. Geol. Tech. Sci.* **2018**, *1*, 61.
47. Itkis, G.; Chandar, V.; Fuller, B.W.; Campbell, J.P.; Cunningham, R.K. Iris Biometric Security Challenges and Possible Solution. *IEEE Signal Process. Mag.* **2015**, *32*, 42–53. [CrossRef]
48. Cavoukian, A.; Stoianov, A. Encryption, Biometric. In *Encyclopedia of Biometrics*; Li, S.Z., Jain, A.K., Eds.; Springer: Boston, MA, USA, 2015.
49. Sutcu, Y.; Sencar, H.T.; Memon, N.A. Secure Biometric Authentication Scheme Based on Robust Hashing. In Proceedings of the 7th Workshop Multimedia and Security, New York, NY, USA, 1–2 August 2005; pp. 111–116.
50. Rathgeb, C.; Uhl, A. Privacy preserving key generation for iris biometrics. In *Communications and Multimedia Security*; De Decker, B., Schaumueller-Bichl, I., Eds.; Springer: Berlin/Heidelberg, Germany, 2010; pp. 191–200.
51. Therar, H.M.; Mohammed, E.A.; Ali, A.J. Biometric Signature based Public Key Security System. In Proceedings of the International Conference Advanced Science and Engineering, Duhok, Iraq, 23–24 December 2020; pp. 1–6.

52. Davida, G.I.; Frankel, Y.; Matt, B.; Peralta, R. On the Relation of Error Correction and Cryptography to an Offline Biometric Based Identification Scheme. In Proceedings of the Workshop on Coding and Cryptography, Paris, France, 11–14 January 1999; pp. 129–138.
53. Dodis, Y.; Reyzin, L.; Smith, A. Fuzzy Extractors: How to Generate Strong Keys from Biometrics and Other Noisy Data. *SIAM J. Comput.* **2008**, *38*, 97–139. [CrossRef]
54. Yang, S.; Verbauwhede, I. Secure Iris Verification. In Proceedings of the 2007 IEEE International Conference on Acoustics, Speech and Signal Processing—ICASSP'07, Honolulu, HI, USA, 15–20 April 2007; Volume 2, pp. 33–136.
55. Linnartz, J.-P.; Tuyls, P. New Shielding Functions to Enhance Privacy and Prevent Misuse of Biometric Templates. In Proceedings of the 4th International Conference Audio- and Video-Based Biometric Person Authentication, Guildford, UK, 9–11 June 2003; pp. 393–402.
56. Shamir, A. How to Share a Secret. *Commun. ACM* **1979**, *22*, 612–613. [CrossRef]
57. Lee, Y.J.; Bae, K.; Lee, S.J.; Park, K.R.; Kim, J. Biometric Key Binding: Fuzzy Vault Based on Iris Images. In Proceedings of the 2nd International Conference Biometrics, Seoul, Republic of Korea, 27–29 August 2007; pp. 800–808.
58. Wu, X.; Qi, N.; Wang, K.; Zhang, D. An Iris Cryptosystem for Information Security. In Proceedings of the International Conference Intelligent Information Hiding and Multimedia Signal Processing, Harbin, China, 15–17 August 2008; pp. 1533–1536.
59. Juels, A.; Wattenberg, M. A Fuzzy Commitment Scheme. In Proceedings of the 6th ACM Conference Computer and Communications Security, Singapore, 1–4 November 1999; pp. 28–36.
60. Hao, F.; Anderson, R.; Daugman, J. Combining Crypto with Biometrics Effectively. *IEEE Trans. Comput.* **2006**, *55*, 1081–1088.
61. Morelos-Zaragoza, R.H. *The Art of Error Correcting Coding*; John Wiley and Sons: Hoboken, NJ, USA, 2006.
62. Reed, I.S.; Solomon, G. Polynomial Codes over Certain Finite Fields. *J. Soc. Ind. Appl. Math.* **1960**, *8*, 300–304. [CrossRef]
63. Reed, I.S. A Class of Multiple-error-correcting Codes and the Decoding Scheme. *Trans. Ire Prof. Group Inf. Theory* **1954**, *4*, 38–49. [CrossRef]
64. Wu, X.; Qi, N.; Wang, K.; Zhang, D. A Novel Cryptosystem based on Iris Key Generation. In Proceedings of the 2008 Fourth International Conference on Natural Computation, Jinan, China, 18–20 October 2008; pp. 53–56.
65. Bringer, J.; Chabanne, H.; Cohen, G.; Kindarji, B.; Zemor, G. Optimal iris fuzzy sketches. In Proceedings of the 2007 First IEEE International Conference on Biometrics: Theory, Applications, and Systems, Crystal City, VA, USA, 27–29 September 2007; pp. 1–6.

algorithms

MDPI

Article

CVE2ATT&CK: BERT-Based Mapping of CVEs to MITRE ATT&CK Techniques

Octavian Grigorescu [1], Andreea Nica [1], Mihai Dascalu [1,2,*] and Razvan Rughinis [1,2]

[1] Computer Science & Engineering Department, University Politehnica of Bucharest, 313 Splaiul Independentei, 060042 Bucharest, Romania
[2] Academy of Romanian Scientists, Str. Ilfov, Nr. 3, 050044 Bucharest, Romania
* Correspondence: mihai.dascalu@upb.ro

Citation: Grigorescu, O.; Nica, A.; Dascalu, M.; Rughinis, R. CVE2ATT&CK: BERT-Based Mapping of CVEs to MITRE ATT&CK Techniques. *Algorithms* **2022**, *15*, 314. https://doi.org/10.3390/a15090314

Academic Editor: Frank Werner

Received: 10 August 2022
Accepted: 27 August 2022
Published: 31 August 2022

Abstract: Since cyber-attacks are ever-increasing in number, intensity, and variety, a strong need for a global, standardized cyber-security knowledge database has emerged as a means to prevent and fight cybercrime. Attempts already exist in this regard. The Common Vulnerabilities and Exposures (CVE) list documents numerous reported software and hardware vulnerabilities, thus building a community-based dictionary of existing threats. The MITRE ATT&CK Framework describes adversary behavior and offers mitigation strategies for each reported attack pattern. While extremely powerful on their own, the tremendous extra benefit gained when linking these tools cannot be overlooked. This paper introduces a dataset of 1813 CVEs annotated with all corresponding MITRE ATT&CK techniques and proposes models to automatically link a CVE to one or more techniques based on the text description from the CVE metadata. We establish a strong baseline that considers classical machine learning models and state-of-the-art pre-trained BERT-based language models while counteracting the highly imbalanced training set with data augmentation strategies based on the TextAttack framework. We obtain promising results, as the best model achieved an F1-score of 47.84%. In addition, we perform a qualitative analysis that uses Lime explanations to point out limitations and potential inconsistencies in CVE descriptions. Our model plays a critical role in finding kill chain scenarios inside complex infrastructures and enables the prioritization of CVE patching by the threat level. We publicly release our code together with the dataset of annotated CVEs.

Keywords: MITRE ATT&CK Matrix; techniques classification; BERT-based multi-labeling

1. Introduction

Cyberspace has become a fundamental component of everyday activities, being the core of most economic, commercial, cultural, social, and governmental interactions [1]. As a result, the ever-growing threat of cyber-attacks not only implies a financial loss, but also jeopardizes the performance and survival of companies, organizations, and governmental entities [2]. It is vital to recognize the increasing pace of cybercrime as the estimated monetary cost of cybercrime skyrocketed from approximately $600 billion in 2018 to over $1 trillion in 2020 [3]. This effect has increased even further due to the COVID-19 pandemic [4].

In this context, the necessity for better cyber information sources and a standardized cybersecurity knowledge database is of paramount importance, as a means to identify and combat the emerging cyber-threats [5]. Efforts to build such globally accessible knowledge bases already exist. MITRE Corporation set up two powerful public sources of cyber threat and vulnerability information, namely the Common Vulnerabilities and Exposures list and the MITRE ATT&CK Enterprise Matrix.

The *Common Vulnerabilities and Exposures* list is a community-based dictionary of standardized names for publicly known cybersecurity vulnerabilities. Its effort converges toward making the process of identifying, finding, and fixing software vulnerabilities more efficient, by providing a unified naming system [6]. Despite their benefits and widespread usage, CVE entries offer little to no information regarding mitigation techniques or existing defense strategies that could be employed to address a specific vulnerability. Moreover, the meta-information of a CVE does not include sufficient classification qualities, resulting in sub-optimal usage of this database. Better classification would translate to mitigating a larger set of vulnerabilities since they can be grouped and addressed together [7].

The *MITRE ATT&CK Enterprise Matrix* links techniques to tangible configurations, tools, and processes that can be used to prevent a technique from having a malicious outcome [8]. By associating an ATT&CK technique to a given CVE, more context and valuable information for the CVE can be extracted, since CVEs and MITRE ATT&CK techniques have complementary value. Furthermore, security analysts could discover and deploy the necessary measures and controls to monitor and avert the intrusions pointed out by the CVE and cluster the CVEs by technique [9].

Even though linking CVEs to the MITRE ATT&CK Enterprise Matrix would add massive value to the cybersecurity community, these two powerful tools are currently separated. However, manually mapping all 189,171 [10] CVEs currently recorded to one or more of the 192 different techniques in the MITRE ATT&CK Enterprise Matrix is a non-trivial task and the need for automated models emerges to map all existing entries to corresponding techniques. In addition, even if new CVEs would be manually labeled, an initial pre-labeling using a machine learning model before expert validation would be time effective and beneficial. Moreover, the model would provide technique labeling for zero-day vulnerabilities, which would be extremely helpful for security teams.

The ATT&CK matrix supports a better understanding of vulnerabilities and what an attacker could achieve by exploiting a certain vulnerability. ATT&CK technique details, such as detection and mitigation, are useful for system administrators, SecOps, or DevSec-Ops teams to obtain an assessment risk report in a short period of time while generating a remediation plan for discovered vulnerabilities. The Center for Threat-Informed Defense team has created a very useful methodology [11] that helps the community build a more powerful threat intelligence database. The organization's defender team has to understand how important it is to bridge vulnerability and threat management with the adoption of this methodology as more reliable and consistent risk assessment reports will be obtained [12].

Baker [12] highlights the importance of combining CVEs with the ATT&CK framework to achieve threat intelligence. Years ago, it was considerably harder for security teams to understand the attack surface, thus reducing their capacity to protect the organization against cyber attacks. With the emergence of the ATT&CK project, the security teams have a better overview of the CVEs based on known attack techniques, tactics, and procedures.

Vulnerability management can be divided into three categories, namely: the "Find and fix" game, the "Vulnerability risk" game, and the "Threat vector" game. The first one is a traditional approach where the vulnerabilities are prioritized by CVSS Score; this is applicable for small organizations with less dynamic assets. The second category consists of risk-based vulnerability management where organizational context and threat intelligence (such as CVE exploited in the wild properties) are considered; this applies to organizations that have security teams, but the number of CVEs is too large. The "Threat Vector" game includes the understanding of how the hackers might exploit the vulnerabilities while accounting for the MITRE ATT&CK framework mappings between CVEs and techniques, tactics, and procedures. The third category is the most efficient model of threat intelligence, with inputs delivered to the vulnerability risk management process from cyber attacks that have occurred and are trending. As such, security teams should take into account risks for building the vulnerability management program, but also threat intelligence to have a better understanding of vulnerabilities and to discover the attack chains within the network [13].

The aim of this paper is to develop a model that leverages the textual description found in CVE metadata to create strong correlations with the MITRE ATT&CK Enterprise Matrix techniques. To achieve this goal, a data collection methodology is developed to build our manually labeled CVE corpus containing more than 18,100 entries. Moreover, state-of-the-art Natural Language Processing (NLP) techniques that consider BERT-based architectures are employed to create robust models. We also target addressing the problem of a severely imbalanced dataset by developing an oversampling method based on adversarial attacks.

Efforts have been already undertaken to interconnect CVEs to the MITRE ATT&CK Framework. However, we identified limitations of existing solutions based on the research gap in the literature regarding the identification of correspondences between CVEs to the corresponding techniques from the MITRE ATT&CK Enterprise Matrix. The following subsections details existing state-of-the-art techniques relevant for our task.

1.1. BRON

BRON [9] is a bi-directional aggregated data graph which allows relational path tracing between MITRE ATT&CK Enterprise Matrix tactics and techniques, Common Weakness Enumerations (CWE), Common Vulnerabilities and Exposures (CVE), and Common Attack Pattern Enumeration and Classification list (CAPEC). BRON creates a graph framework that unifies all scattered data through inquiries performed of the resulted graph representation by data-mining the relational links between all these cyber-security knowledge sources. In this manner, it connects the CVE list to MITRE ATT&CK by traversing the relational links in the resulted graph.

Each information source has a specific node type, interconnected by external linkages as edges. MITRE ATT&CK techniques are linked to Attack Patterns. Attack Patterns are connected to CWE Weaknesses, which have relational links to a CVE entry. Thus, BRON can respond to several different queries, including linking the CVE list to the MITRE ATT&CK Framework.

However, the model falls short as it does not connect new CVEs to MITRE ATT&CK Enterprise Matrix techniques, but it uses already existing information and links to create a more holistic overview of the already available knowledge. It does not solve our problem, since the main aim is to correctly label new emergent samples.

1.2. CVE Transformer (CVET)

The CVE Transformer (CVET) [14] is a model that combines the benefits of using the pre-trained language model RoBERTa with a self-knowledge distillation design used for fine-tuning. Its main aim is to correctly associate a CVE with one of 10 tactics from the MITRE ATT&CK Enterprise Matrix. Although the CVET approach obtains increased performance in F1-score, it is unable to identify all 14 tactics from the MITRE ATT&CK Matrix on the training knowledge base.

Moreover, the problem of technique labeling is much more complex than tactic mapping, since the number of available techniques is ten times higher (i.e., there are 14 tactics and 192 different techniques in the MITRE ATT&CK Enterprise Matrix). Additionally, tactic labeling can be viewed as a subproblem of our main goal given the correlation between tactics and techniques. Overall, technique labeling is out of scope for the CVE Transformer project.

1.3. Unsupervised Labeling Technique of CVEs

The unsupervised labeling technique introduced by Kuppa et al. [15] considers a multi-head deep embedding neural network model that learns the association between CVEs and MITRE ATT&CK techniques. The proposed representation identifies specific regular expressions from the existing threat reports and then uses the cosine distance to measure the similarity between ATT&CK technique vectors and the text description provided in the CVE metadata. This technique manages to map only 17 techniques out of the existing 192. As such, multiple techniques are not covered by the proposed model. Thus, a supervised approach for technique labeling might improve the recognition rate among techniques.

1.4. Automated Mapping to ATT&CK: The Threat Report ATT&CK Mapper (TRAM) Tool

Threat Report ATT&CK Mapping (TRAM) [16] is an open-source tool developed by *The Center for Threat-Informed Defense* that automates the process of mapping MITRE ATT&CK techniques on cyber-threat reports. TRAM utilizes classical pre-processing techniques (i.e., tokenization, stop-words removal, lemmatization) [17] and applies Logistic Regression on the bag-of-words representations. Since the tool maps any textual input on MITRE ATT&CK techniques, it could, in theory, be adapted to link the CVE list to the MITRE ATT&CK Framework by simply using it on the CVE textual description. However, due to its simplicity, the tool has serious limitations when it comes to its capacity to learn the right association between text descriptions and techniques. In addition, TRAM labels each sentence individually, failing to capture dependencies in textual passages. In this way, the overall meaning of the text is lost.

The main contributions of this paper are as follows:

- Introducing a new publicly available dataset of 1813 CVEs annotated with all corresponding MITRE ATT&CK techniques;
- Experiments with classical machine learning and Transformer-based models, coupled with data augmentation techniques, to establish a strong baseline for the multi-label classification task;
- A qualitative analysis of the best performing model, coupled with error analysis that considers Lime explanations [18] to point out limitations and future research directions.

We open-source our dataset on TagTog [19] and the code on GitHub [20].

2. Method

This section provides an overview of our proposed methodology, focusing on: (1) data collection and building the corpus needed for training the models; and (2) exploring various neural architecture for mapping CVEs to ATT&CK techniques.

2.1. Our Labeled CVE Corpus

2.1.1. Data Collection

Since no public datasets exist that map a CVE to all corresponding ATT&CK techniques, the first step consisted of building our own labeled corpus of 1813 CVEs, which was obtained using two different methods.

First, we manually created a knowledge base of 993 labeled CVEs by individually mapping each CVE to tactics and techniques from MITRE ATT&CK Enterprise Matrix. We extracted CVEs that were published between 2020 to 2022 for relevance. The labeling process was performed by 4 experts to ensure consistency, following the standardized approach proposed by the *Mapping MITRE ATT&CK to CVEs for Impact* methodology [11] and a set of common general guidelines.

The *Mapping MITRE ATT&CK to CVEs for Impact* methodology consists of three steps. The first one is to identify the type of vulnerability (e.g., cross-site scripting, buffer overflow, SQL injection) based on the vulnerability type mappings. The next step is to find the functionality to which the attacker gains access by exploiting the CVE. The final step refers to determining the exploitation technique using the provided tips that offer details about the necessary steps to exploit a vulnerability. Our methodology started from these steps and added other common general guidelines before labeling the tactics and techniques, such as searching for more details about a CVE on security blogs to obtain more relevant insights, or analyzing databases (e.g., the Vulnerability Database [21] and the Exploit Database— Exploits for Penetration Testers, Researchers, and Ethical Hackers [22]) for useful inputs about CVEs.

The labeling was performed by three 4th year undergraduate students in Computer Science with background courses in security, networking, and operating systems, and one Ph.D. student in Computer Science with 5+ years of experience in information security in the industry who provided guidance and helped reach consensus. The entire annotation process was overseen by a professor in cyber security. The dataset can be found on TagTog [19] and is split into the following collections:

1. *Inter-rater*—A collection of 24 CVEs evaluated by all experts to ensure high agreement and consistent annotations; this collection was used for training the raters until perfect consensus was achieved;
2. *Double-rater*—A collection of 295 CVE evaluated by pairs of two raters; this collection was created after some experience was accumulated and consensus among raters was achieved using direct discussions;
3. *Individual*—A collection of 674 CVE evaluated by only one rater; this collection was annotated after the initial training phase was complete and raters gained experience.

Second, besides the manual labeling process, we automatically extracted 820 already labeled CVEs provided by *Mapping MITRE ATT&CK to CVEs for Impact* [11] and imported them in our TagTog project. The provided CVEs date from 2014 to 2019; thus, there is no overlap with the manually annotated CVEs.

Each CVE entry has associated the corresponding ID, the rich text description, and 14 labels denoting the possible tactics found in the MITRE ATT&CK Enterprise Matrix where the corresponding techniques are annotated. Extracting the data from TagTog can be performed automatically, using the TagTog API [23].

2.1.2. Data Analysis

The size of our corpus can be argued by the increased difficulty when annotating a CVE and the impossibility to find other previously build repositories consisting of CVEs mapped on MITRE ATT&CK Enterprise Matrix both tactics and techniques. As discussed previously, more than 189,171 CVEs currently exist and our dataset only captures a fraction of them. Moreover, the distribution of CVEs based on technique is highly imbalanced (see Figure 1) because the CVEs were collected based on their release date, without any other further considerations. About 77% of the collected CVEs cover 5 techniques (*Exploit Public-Facing Application, Exploitation for Client Execution, Command and Scripting Interpreter, Endpoint Denial of Service* and *Exploitation for Privilege Escalation*).

Figure 1 also shows that a large number of techniques contain a far too small number of examples for effective learning. As such, a threshold of a minimum of 15 examples per technique was imposed. In this manner, out of the 192 different techniques from the MITRE ATT&CK Enterprise Matrix, only 31 were considered in follow-up experiments. The CVEs that are not mapped to any of the 31 considered techniques were also discarded, leaving a total of 1665 annotated examples in the dataset. Figure 2 depicts the new distribution of CVEs based on technique after applying the threshold.

2.1.3. Data Augmentation

The severe data imbalance which characterizes our CVE dataset can potentially degrade the performance of many machine learning models since few techniques have high prevalence, while the others have low or very low frequencies [24].

One scheme for dealing with class imbalance is oversampling [24]. This data-level approach consists of randomly oversampling duplicate examples from low-frequency classes to rebalance the class distribution. However, this can result in overfitting and we opted to use the TextAttack Framework [25] for generating adversarial examples. TextAttack is a Python framework designed for adversarial attacks, data augmentation, and adversarial training in NLP. The adversarial attack finds a sequence of transformations to perform on an input text such that the perturbations adhere to a set of grammar and semantic constraints and the attack is successful [26]. These transformations performed can be reused to expand the training dataset by producing perturbed versions of the existing samples. As such, TextAttack Framework offers various pre-packaged recipes for data augmentation [27].

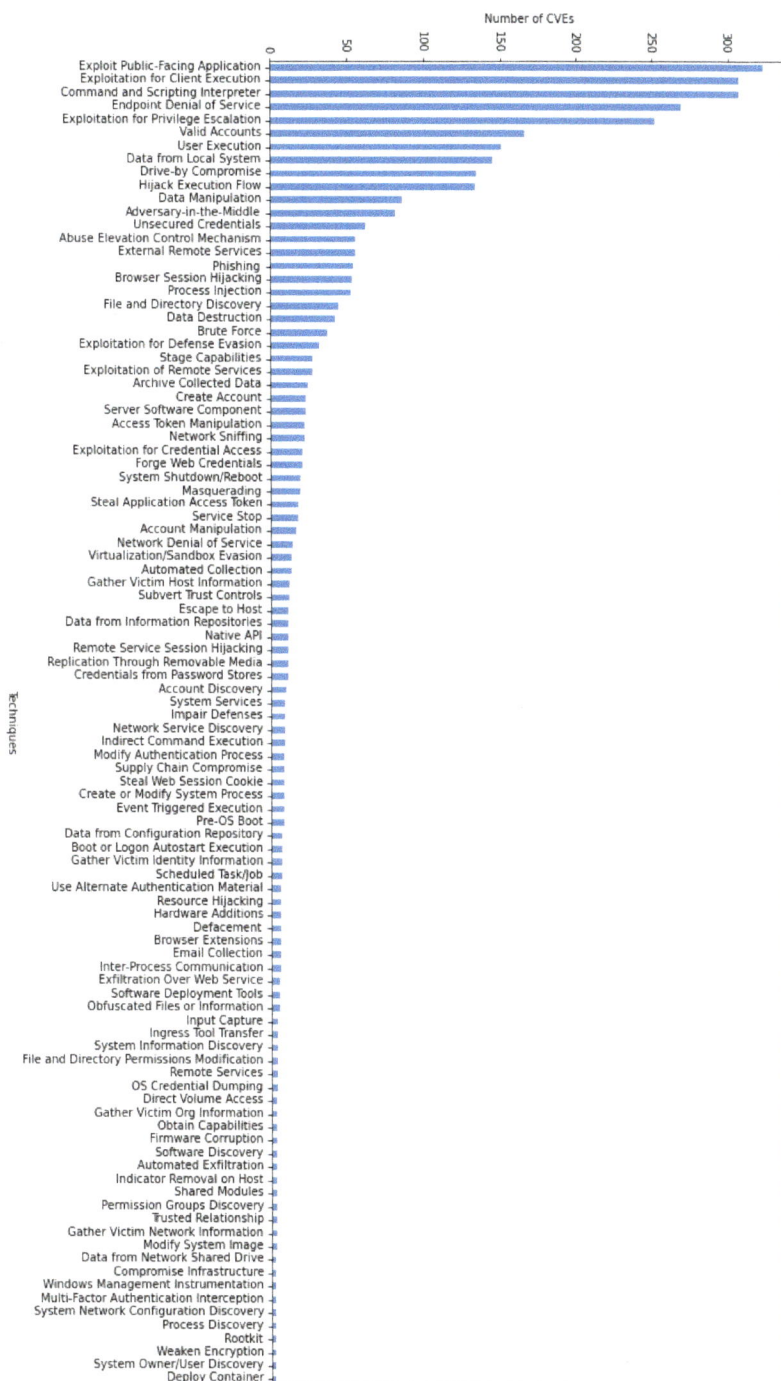

Figure 1. The distribution of CVEs among techniques.

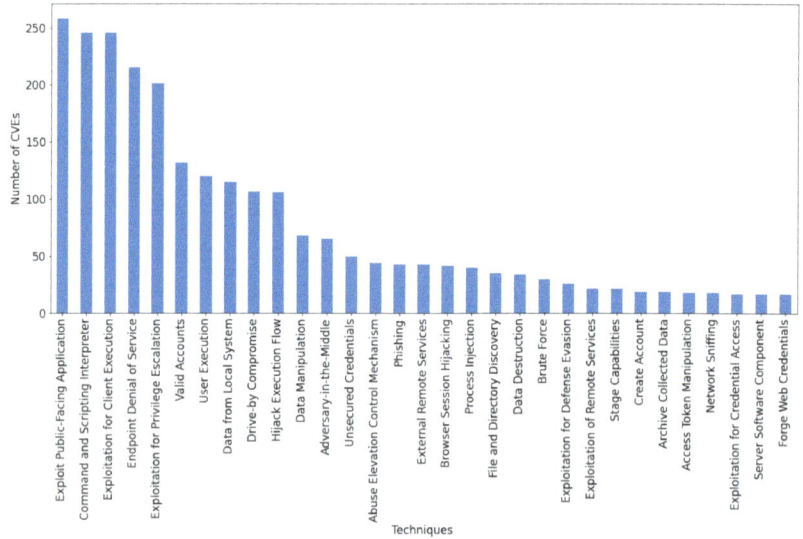

Figure 2. The distribution of CVEs among the 31 considered techniques after applying the threshold.

We chose the EasyDataAugmenter (EDA) for augmenting the CVE dataset, which performs four simple but powerful operations on the input texts: synonym replacement, random insertion, random swap, and random deletion. EDA significantly boosts performance and shows particularly strong results for smaller datasets [28], which makes it the perfect candidate for oversampling our labeled CVE corpus. Moreover, EDA does not perform major alterations of the content and is not as computationally expensive as other recipes, such as CLAREAugmenter, while providing satisfactory results on our CVE corpus.

Since one CVE can be mapped to multiple techniques at the same time, rare techniques among the dataset are usually found in combination with highly prevalent techniques. Using all CVEs that are mapped to a specific technique for augmentation would only preserve the class imbalance, generating new samples for both low-frequency and high-frequency techniques. To counter this undesired effect, EasyDataAugumenter was fed only with CVEs that were particular to only one technique and were mapped to that technique only, thus producing new samples only for the desired class.

Figure 3 displays the distribution of CVEs per technique after performing the data augmentation. The initial severe imbalance among techniques was scaled down, but still exists, due to the reduced number of particular CVEs for low-frequency techniques.

2.2. Machine Learning and Neural Architectures

Our main goal is to create a model that can accurately predict all the techniques that can be mapped to a specific CVE while using its text description. We tacked this task as a multi-label learning problem as each CVE may be assigned to a subset of techniques. Given the challenging nature of the multi-label paradigm [29], we experimented with multiple state-of-the-art machine learning models to find the most predictive architecture.

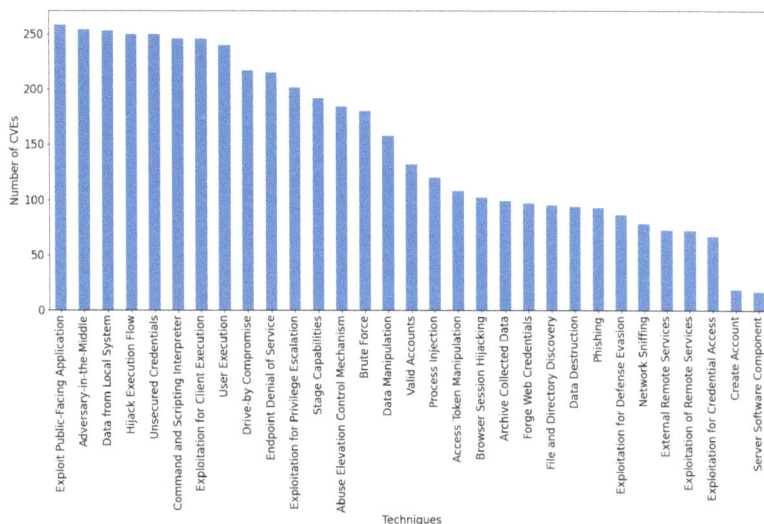

Figure 3. The distribution of CVEs among the 31 considered techniques after data augmentation.

2.2.1. Classical Machine Learning

In order to establish a strong baseline we also considered classical machine learning algorithms applied on bag-of-words representations. All CVE descriptions were pre-processed to remove noise and retain only the relevant words. The pipeline from the spaCy [30] NLP open-source library was employed which included the following steps: text tokenization, removal of stopwords, punctuation, and numbers, followed by lemmatization of remaining tokens. The tokens are afterward converted to bag-of-words representations using Term Frequency-Inverse Document Frequency (TF-IDF).

Multi-Label Learning

The aim of problem transformation methods is to reduce the complexity of the multi-label learning by converting the multi-label problem into one or more single-label classification tasks [31].

Given that the interconnection between techniques is worth taking into account when labelling a CVE since it can provide further insights on general adversarial patterns, we experimented with different problem transformation methods to find the one that captures best the relations between labels:

- *One versus Rest*. This method splits the multi-label problem into multiple binary classification tasks, one for each label, treated independently. The *N* different binary classifiers are separately trained to distinguish the examples of a single class from all the examples from the other labels [32];
- *Label Powerset*. This method considers every unique combination of labels as a single class, reducing the multi-label problem to a multi-class classification problem [29]. The real advantage of this strategy is that correlations between labels are exploited for a more accurate labelling process;
- *Binary Relevance*. This linear strategy groups all positive and negative examples within a label into a set, later training a classifier for each resulted set. The final prediction is then computed by merging all the intermediary predictions of the trained classifiers [29]. An advantage of this strategy consists of the possibility to perform parallel executions;
- *RaKEL(Random k-Labelsets)*. This state-of-the-art approach builds an ensemble of Label Powerset classifiers trained on a different subset of the labels [33].

Naive Bayes Classifiers

The Naive Bayes classifier makes the simplifying assumption that features are conditionally independent, given a class. Even though the assumption of independence is generally unrealistic, Naive Bayes performs well in practice, competing with more sophisticated classifiers models especially for text classification [34]. We chose to experiment with a Naive Bayes variant for multinomial distributed data because of the model's simplicity and relatively good results.

Support Vector Machines

A Support Vector Machine (SVM) searches for the maximum margin hyperplane that separates two classes of examples. Because SVMs have shown efficiency to capture high dimensional spaces and performed successfully on a number of distinctive classification tasks [35], we decided to use it in our experiments for CVE technique labelling. We performed an exhaustive search over specified parameters values using GridSearchCV [36] to determine the optimum configuration of parameters.

2.2.2. Convolutional Neural Network (CNN) with Word2Vec

Convolutional Neural Networks (CNNs) consist of multiple layers designed to extract local features in the form of a feature map. Since CNN uses back-propagation to update its weights in the convolutional layers, the CNN feature extractors are self-determined through continuous tuning of the model [37]. In the field of NLP, CNNs have proved to be extremely effective in several tasks, such as semantic parsing [38] and sentence modeling [39]. This intuition pointed in the direction to experiment with CNN for our model since CNNs with Word2Vec embeddings are robust even on small datasets. In addition, we considered SecVuln_WE [40] that includes word representation especially designed for the cybersecurity vulnerability domain. SecVuln_WE was trained on security-related sources such as Vulners, English Wikipedia (Security category), Information Security Stack Exchange Q&As, Common Weakness Enumeration (CWE) and Stack Overflow.

Figure 4 presents the architecture in which the pre-trained SecVuln_WE embeddings are passed through the convolutional layer containing 100 filters with a kernel size of 4. In this way, each convolution will consider a window of 4 word embeddings. Afterward, we perform batch normalization of the activations of the previous layer at each batch. Next comes the MaxPool and the Dropout layers, followed by a dense layer with sigmoid activation. Since we are dealing with a multi-label classification problem, the output layer has a designated node for each technique and each output indicates the binary probability to have a specific technique mapped to the considered CVE.

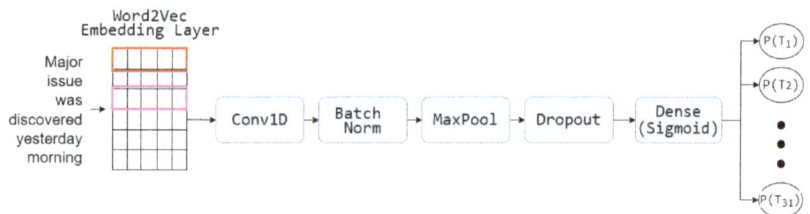

Figure 4. Architecture of the CNN with Word2Vec embeddings.

2.2.3. BERT-Based Architecture with Multiple Output Layers

Reducing the considerable complexity of the multi-label problem was first among our considerations when designing this architecture. Converting our multi-labeling problem into multiple binary classification tasks following the *One versus Rest* method has the advantage of conceptual simplicity; yet, having a distinct BERT layer for contextualized embeddings for each one of the 31 techniques was redundant.

The proposed architecture from Figure 5 considers a pre-trained BERT encoder, a Dropout layer, and an individual dense layer for each technique, which outputs the probability that a particular CVE points to that particular technique. The model is consistent with the considerations of the *One VS Rest* method, while also taking advantage of the shared embeddings layer.

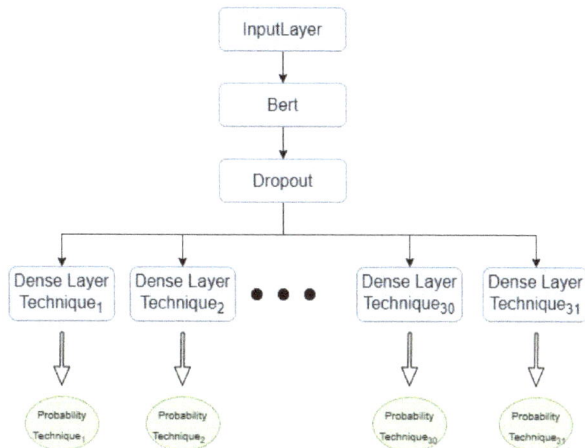

Figure 5. BERT-based architecture with multiple output layers.

2.2.4. BERT-Based Architecture Adapted for Multi-Labeling

Analyzing each label separately might overlook the strong correlation between techniques. This correspondence has multiple roots, as techniques in a given tactic are connected through their attack behavior pattern, whereas techniques across multiple tactics are connected through the attack vector of the vulnerability. Thus, we explored creating a model capable of exploiting the link between multiple techniques.

The specific architectural decision taken for this last design was to have only one output layer, with one individual node for each technique. In this manner, we aim to capture the specifics for each technique, while also considering how subsets of techniques are interconnected.

Figure 6 details the proposed model which considers 768-dimensional contextual embeddings from various BERT-based models (i.e., BERT [41], SciBERT [42], and SecBERT [43]) passed through a Dropout layer. The Dropout layer output goes through a Linear layer with 768 input features and 31 output nodes, one for each technique. We considered BCEWithLogitsLoss [44] (the combination of a Sigmoid layer and the BCELoss) as a loss function, the most commonly used for multi-label classification tasks, because each output node reveals the probability of a technique to be tagged for a specific CVE (i.e., the probabilities need to be treated independently).

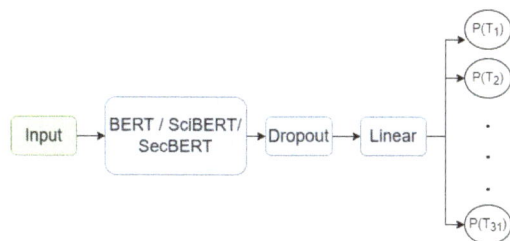

Figure 6. The design of the multi-labeling BERT-based architecture.

2.3. Performance Assessment

For a predicted technique, we wanted to make sure that our mapping was correct (i.e., high precision—P) and we wanted to correctly classify as many examples as possible for a given class (i.e., high recall—R). Thus, we considered the F1-score as a performance metric for all models, defined as the harmonic mean of the P and R per class. Moreover, we used the weighted version of the F1-score given the imbalance between classes, which calculated a general F1-score per model by proportionally combining the F1-scores obtained for each label separately. We also computed the weighted precision and recall for the tested models.

3. Results

This section analyses the results of the empirical experiments performed using the previously detailed models. First, it compares the performance of various models. Second, it assesses the impact of data augmentation on performance and investigates the metrics obtained by the best model.

Multiple observations can be made based on the results of our experiments shown in Table 1. From the classical machine learning models, LabelPowerset is the best multi-label strategy and SVC with a linear kernel and C = 32 has the higher F1-score, competing even with our deep-learning models. The SecBERT model has the highest F1-score (42.34%) among all considered models, proving to be the most powerful solution to labeling a CVE. An important observation is that the CNN + Word2Vec architecture obtained better results than those using simple BERT. Thus, domain-related pre-training on large security databases leads to increased performance by providing better contextualization and partially compensating for the scarce training set.

Table 1. Results for the proposed models (italics marks the best multi-label strategy for classical ML, while bold marks the best model).

Model Type	Model	Multi-Label Strategy	Weighed P	Weighed R	Weighed F1-Score
Classical ML	Naive Bayes	OneVsRestClassifier	57.35%	9.18%	14.47%
		LabelPowerset	31.40%	24.59%	24.76%
		BinaryRelevance	57.35%	9.18%	14.47%
		RakelD	53.71%	9.83%	15.31%
	SVC	OneVsRestClassifier	31.97%	35.57%	33.32%
		LabelPowerset	*46.73%*	*34.75%*	*37.98%*
		BinaryRelevance	33.45%	34.91%	33.75%
		RakelD	36.20%	33.77%	34.50%
Deep Learning	CNN + Word2Vec	-	48.32%	**35.40%**	39.39%
	Multi-Output BERT	-	46.85%	31.47%	35.92%
	Multi-label BERT	-	55.25%	30.98%	37.43%
	Multi-label SciBERT	-	**59.26%**	34.42%	41.87%
	Multi-label SecBERT	-	57.66%	**35.40%**	**42.34%**

Table 2 points out the appropriateness of employing data augmentation techniques on our dataset for deep learning models (approximately 6% performance gain). Only the best multi-label strategy for classical machine learning algorithms was considered. The F1-score falls considerably by 10% for Naive Bayes, in particular, since Naive Bayes places great importance on the number of appearances of a word in a document; however, swapping a relevant word with synonyms and performing random insertions or deletions (i.e., the strategies employed by the EasyDataAugmenter [28]) only confuse the model. The SVC model had a similar performance, whereas the BERT-based models take advantage of the increased sample size/the decreased class imbalance, and generalize better. Not only is performance increased, but the models also tend to learn faster (see faster convergence in Figure 7 in terms of training loss for each output layer associated with a technique in the multi-output BERT model). Moreover, Figure 7 denotes which techniques are more easily learned by the model.

Table 2. Side-by-side comparison of performance with and without data augmentation (bold denotes the best model).

Model	Data Augmentation	Weighted P	Weighted R	Weighted F1-Score
Naive Bayes (LabelPowerset)	No	31.40%	24.59%	24.76%
	Yes	29.40%	14.42%	14.42%
SVC (LabelPowerset)	No	46.73%	34.75%	37.98%
	Yes	45.90%	34.09%	36.79%
CNN + Word2Vec	No	48.32%	35.40%	39.39%
	Yes	50.48%	35.59%	41.59%
Multi-Output BERT	No	46.85%	31.47%	35.92%
	Yes	49.81%	35.57%	39.66%
Multi-label SciBERT	No	**59.26%**	34.42%	41.87%
	Yes	52.52%	**45.90%**	**47.84%**
Multi-label SecBERT	No	57.66%	35.40%	42.34%
	Yes	54.70%	42.45%	46.54%

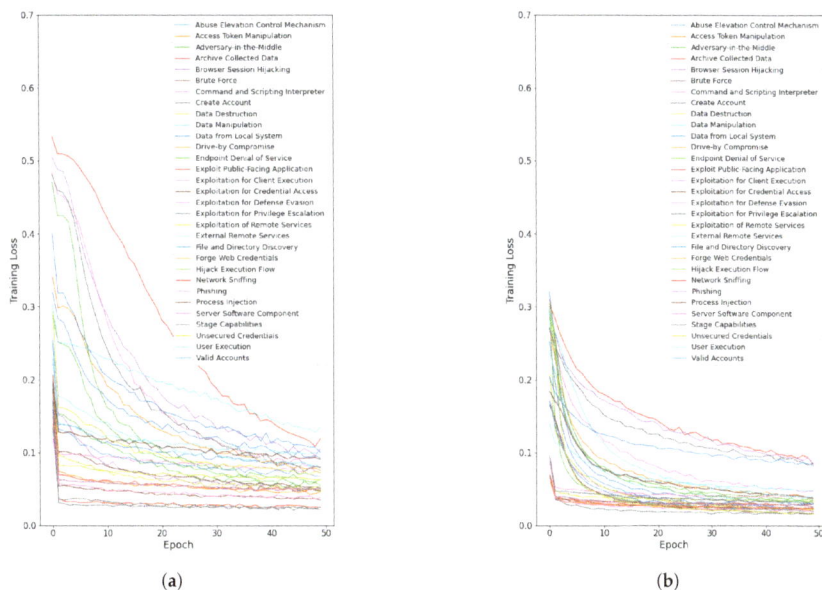

(a)   (b)

Figure 7. Comparison of training loss for the multi-output BERT architecture. (**a**) Without data augmentation; (**b**) With data augmentation.

Since Table 2 only provides a global overview of the average performance of the SciBERT model trained on the augmented data, exploring the particular difference between how the model handles different techniques provides additional insights into our model's behavior. Figure 8 plots the F1-score obtained for each individual technique, for both the original model and the one trained on the augmented dataset. Apart from four exceptions (*Data from Local System*, *Hijack Execution Flow*, *User Execution* and *File and Directory Discovery*), the model obtains considerably higher or at least equal scores for all the other 27 techniques. Moreover, the difference between models is minimal (close to 0) for the techniques where the initial model obtains a better F1-score.

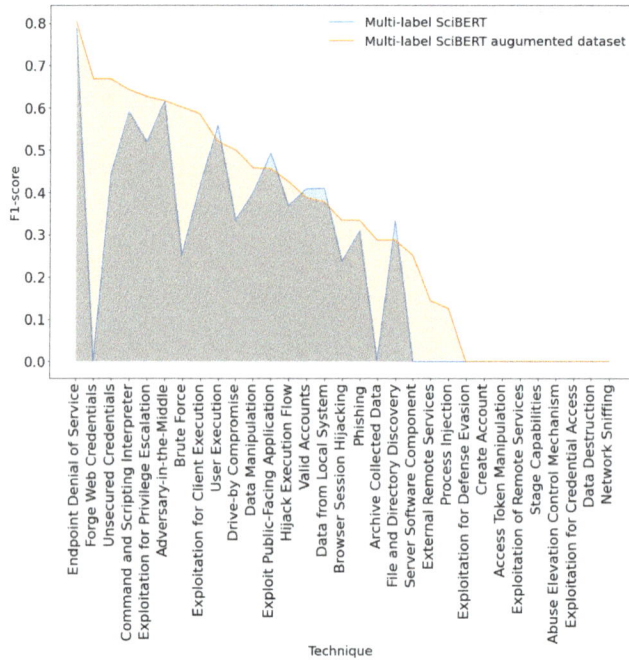

Figure 8. Comparing F1-score per technique between SciBERT model trained on initial and augmented dataset.

The added gain of the multi-label SciBERT model trained on the augmented dataset resides in its ability to maximize the F1-score for techniques where the initial model performed poorly. One such example is *Forge Web Credentials*. The initial model obtained an F1-score of 0% since both recall and precision were 0%. However, the improved version of the model obtained an F1-score of 66.66%, with a recall of 50% and precision of 100% after data augmentation; similarly, data augmentation tuned the model to predict the *Forge Web Credentials* technique with 100% precision. Overall, the number of techniques with which the model had difficulty in learning has decreased substantially.

Figure 9 shows the correlation between the CVE distribution and the F1-score obtained for the SciBERT models, both using the initial dataset and the one trained after augmentation. The techniques are displayed on both graphs in the same order to indicate how the CVE distribution changed after performing the process of data augmentation and how the adjustments in CVE distribution impacted the F1-score. We observe that not only the techniques initially associated with a small number of CVEs benefited from the augmentation method, but also the techniques associated with a high distribution of samples—for example, the F1-score for the *Command and Scripting Interpreter* technique increased from the initial 58.92% to 64.12%.

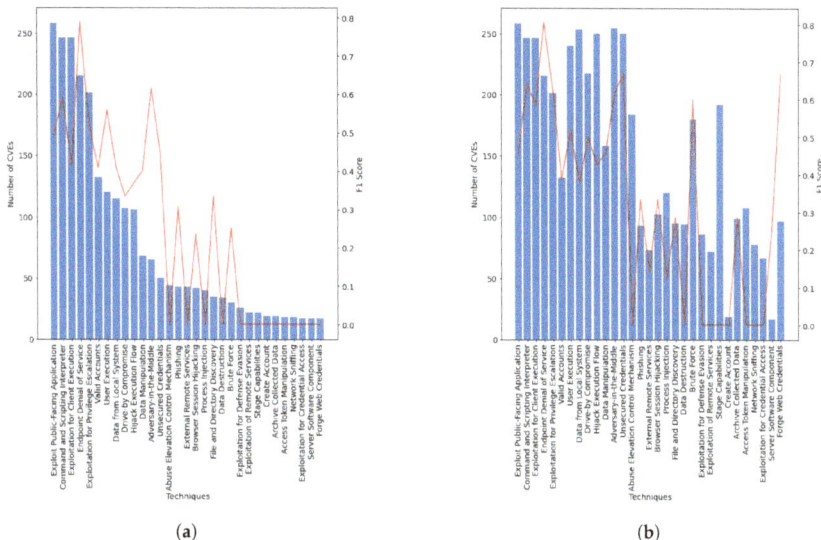

(a)   (b)

Figure 9. Comparing the F1-score over the CVE distribution for the SciBERT model. (**a**) Without augmentation; (**b**) With augmentation.

Table 3. Precision, Recall and F1-Scores for the best model.

Technique	Weighted P	Weighted R	Weighted F1-Score
Endpoint Denial of Service	77.58%	83.33%	80.35%
Forge Web Credentials	100.00%	50.00%	66.66%
Unsecured Credentials	60.00%	75.00%	66.66%
Command and Scripting Interpreter	60.00%	68.85%	64.12%
Exploitation for Privilege Escalation	56.45%	70.00%	62.50%
Adversary-in-the-Middle	80.00%	50.00%	61.53%
Brute Force	100.00%	42.85%	60.00%
Exploitation for Client Execution	50.87%	50.81%	58.49%
User Execution	58.33%	46.67%	51.85%
Drive-by Compromise	64.70%	40.74%	50.00%
Data Manipulation	44.44%	47.05%	45.71%
Exploit Public-Facing Application	48.27%	43.07%	45.52%
Hijack Execution Flow	50.00%	37.03%	42.55%
Valid Accounts	41.37%	36.36%	38.70%
Data from Local System	41.66%	34.48%	37.73%
Browser Session Hijacking	42.85%	27.27%	33.33%
Phishing	42.85%	27.27%	33.33%
Archive Collected Data	50.00%	20.00%	28.57%
File and Directory Discovery	40.00%	22.22%	28.57%
Server Software Component	50.00%	16.66%	25.00%
External Remote Services	50.00%	8.33%	14.28%
Process Injection	25.00%	8.33%	12.50%
Exploitation for Defense Evasion (26)	0.00%	0.00%	0.00%
Create Account (19)	0.00%	0.00%	0.00%
Access Token Manipulation(18)	0.00%	0.00%	0.00%
Exploitation of Remote Services (22)	0.00%	0.00%	0.00%
Stage Capabilities (22)	0.00%	0.00%	0.00%
Abuse Elevation Control Mechanism (44)	0.00%	0.00%	0.00%
Exploitation for Credential Access (17)	0.00%	0.00%	0.00%
Data Destruction (34)	0.00%	0.00%	0.00%
Network Sniffing (18)	0.00%	0.00%	0.00%

4. Discussion

4.1. In-Depth Analysis of the Best Model

Table 3 introduces a complete overview of the results recorded for the best model, the multi-label SciBERT trained on the augmented dataset. The F1-score per technique from the MITRE ATT&CK Enterprise Matrix ranges from 80.35% for *Endpoint Denial of Service* to 0.00%; the last techniques at the end of Table 3 marked with italics and including the corresponding number of training samples in parenthesis. Even though the model scores on a global scale an F1-score of 47.84%, the model fails to capture any knowledge about nine out of the thirty-one techniques, though fewer instances than the other evaluated models. We can associate this inability of the model to recognise the distinct features of these techniques with the extremely reduced number of samples for each technique, even after performing data augmentation. The existing samples in the dataset do not contain enough relevant characteristics for these techniques; as such, the model cannot differentiate them.

Nevertheless, the model successfully captures the essence of other techniques, obtaining a precision of 100.00% for *Forge Web Credentials* and *Brute Force*. For almost all techniques, precision exceeds recall, thus indicating that the general tendency of the model is to omit a label, rather than misplace a technique that cannot be mapped to a particular CVE.

Overall, given the complexity of the multi-label problem and the severe imbalance of the training set, the model obtains promising performance for a subset of techniques, while managing to maximize its overall F1-score.

4.2. Error Analysis

This subsection revolves around understanding the roots of the multi-label SciBERT model limitation. After a methodological investigation that aims to identify the cause of the model's errors, the observed performance deficiencies are further discussed.

Table 4 presents different CVEs whose predicted techniques differ partially or completely from the labeled ones. For most errors in the dataset with multiple techniques tagged, the model succeeds in labeling a subset of correct techniques. This observation stands true for errors 1, 2, and 3 from Table 4. While analyzing error #1, the model extracts the most obvious technique, pointed out by language markers such as *password unencrypted, global file*, but fails to make the deduction that, in order for a user to access the file system, a valid account must be used. In contrast, the model successfully identifies the *Valid Accounts* technique for error #2. In general, techniques that are not clearly textually encapsulated and whose understanding requires prerequisite knowledge are overlooked by the model.

Figure 10 studies the model's choice of labels for CVE #2 from Table 4 using Lime [18], the model successfully recognizes the predominant label (i.e., *Valid Accounts*). Moreover, the model correctly identifies the most important concept, the word *authenticated*, which points in the direction of *Valid Accounts*. We can observe that there are techniques that are not ambiguous for the model and for which the labeling process is straightforward; such an example is *Valid Accounts*. The model extracts only the relevant features for the label and the technique is correctly identified. For the *Exploitation for Client Execution*, the model identifies patterns that suggest that the CVE should be mapped to the given technique, as well as patterns that suggest the contrary. Being capable to identify features that are correlated to both situations confuses the model. This problem results from the fact that the meaning behind multiple techniques is overlapping and, as a result, relevant features for a given technique cannot be differentiated.

Table 4. Comparing predictions with the true values for the best model.

#	CVE Text	True Techniques	Predicted Techniques
1	Jenkins Publish stores password unencrypted in its global configuration file on the Jenkins controller where it can be viewed by users with access to the Jenkins controller file system.	Unsecured Credentials, Valid Accounts	Unsecured Credentials
2	Due to improper input validation in InfraBox, logs can be modified by an authenticated user.	Valid Accounts, Exploitation for Client Execution	Valid Accounts
3	In Django 2.2 MultiPartParser, UploadedFile, and FieldFile allowed directory traversal via uploaded files with suitably crafted file name	File and Directory Discovery, Command and Scripting Interpreter	File and Directory Discovery, Exploit Public-Facing Application
4	Whale browser for iOS before 1.14.0 has an inconsistent user interface issue that allows an attacker to obfuscate the address bar which may lead to address bar spoofing.	Browser Session Hijacking	User Execution
5	isula-build before 0.9.5-6 can cause a program crash, when building container images, part of the functions for processing external data do not remove spaces when processing data.	Exploitation for Client Execution	Endpoint Denial of Service

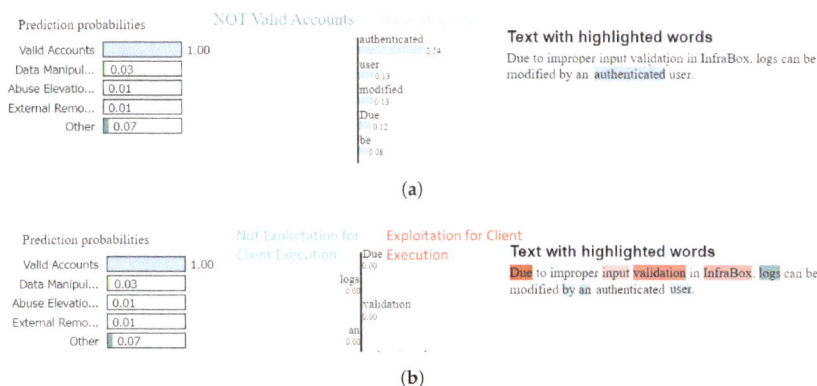

(a)

(b)

Figure 10. Comparison of word mappings for each technique corresponding to CVE #2 from Table 4. (**a**) Mapping Valid Accounts; (**b**) Mapping Exploitation for Client Execution.

An interesting aspect is revealed in error #3, namely that the model correctly tags *File and Directory Discovery*, but also associates the CVE with *Exploit Public Facing Application*, instead of *Command and Scripting Interpreter*. Both techniques in the MITRE ATT&CK Enterprise Matrix could be equally correctly mapped on the given text description. This is an important observation and points out the established CVE labeling methodology; this highlights a fault in the data collection procedure, rather than the model's capacity to learn the multi-labeling problem. Example #4 presents a similar case, since the predicted technique *Endpoint Denial of Service* is a correct label for the CVE, although it does not appear among the true labels.

(a)

(b)

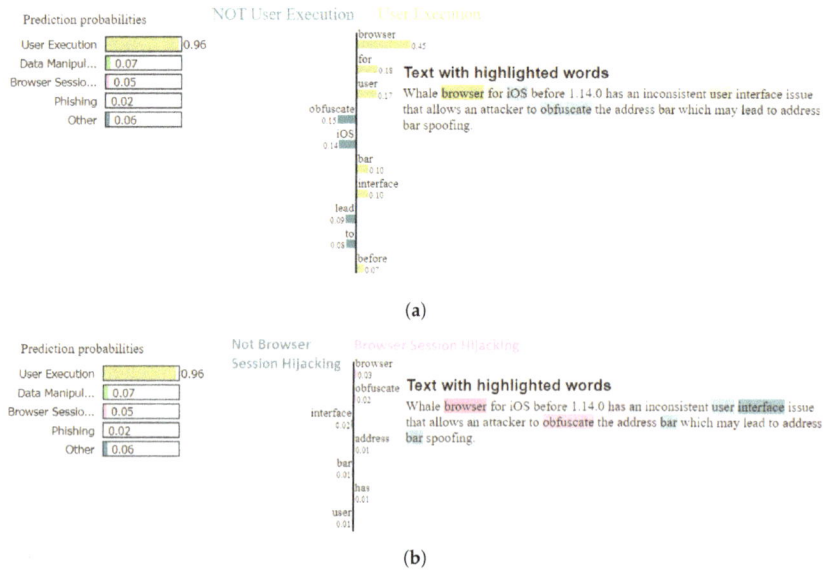

Figure 11. Comparison of word mappings for each technique corresponding to CVE #4 from Table 4. (a) Mapping User Execution technique; (b) Mapping Browser Session hijacking.

Error #4 is analyzed in detail in Figure 11 to observe insights on how the model associates the features. The word *browser* is highlighted for both the predicted and the correct label. However, the difference resides in the relevance percentage associated with the word for each label, namely 0.45 for *User Execution* and 0.03 for *Browser Session Hijacking*. While the word *browser* is recognized as being relevant for both labels, the label with the higher percentage is selected. This finding can be associated with the discrepancy between training examples—240 for *User Execution*, while *Browser Session Hijacking* has only 102. Thus, the class imbalance affects the model's capability to recognize the real correlation between features and techniques, and leads the model to a biased decision.

The model extracts a correct technique for error #5 in Table 4, although it was not among the true labels. As Figure 12 shows, the CVE text description indicates the *Endpoint Denial of Service* technique, since the word *crash* is present and the relevance of the word for the *Endpoint Denial of Service* technique is 0.93. Figure 12 also suggests that the word *crash* is the only word that has a high impact on the model's decision to label the CVE as *Endpoint Denial of Service*.

Two observations can be made based on Figure 12. One is that the model successfully captures a technique overlooked by the reviewer. The technique labeling process is error-prone due to the ambiguity of the CVE text description and also the complexity of the labeling processing given the wide range of available techniques. Second, the model assigns a higher relevance to features that suggest *Endpoint Denial of Service* even though key features for the *Exploitation for Client Execution* are identified (i.e., *program* and *functions*).

Table 5 presents the most relevant words when performing feature extraction for each technique. More than 50% of the techniques have the same most relevant feature in common with other techniques in the MITRE ATT&CK Enterprise Matrix. For example, *Exploitation for Privilege Escalation, Data from Local System, Data Destruction, Browser Session Hijacking, Archive Collected Data,* and *Create Account* are all mapped to the same feature. Having the same most relevant extracted feature implies a strong intersection between techniques. This further emphasizes that the separation between labels is fuzzy. The opinion and consensus among reviewers were used to separate ambiguous examples, making use of previous experience and context obtained from other resources. This is inherited by the model since the labels from the training set reflect the reviewers' perspective. In this context, more information would be valuable to counter the bias encapsulated in the training set by offering more background information to the model.

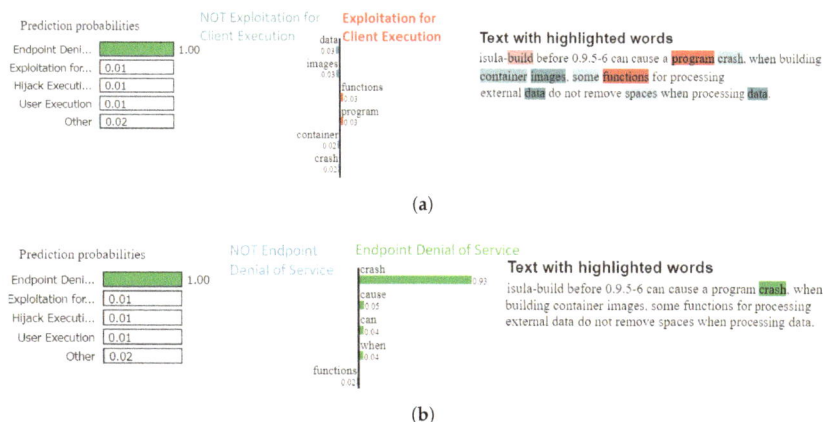

(a)

(b)

Figure 12. Comparison of word mappings for each technique corresponding to CVE #5 from Table 4. (**a**) Mapping Exploitation for Client Execution; (**b**) Mapping Endpoint Denial of Service.

Table 5. The most important words extracted per technique.

Technique	CVEs
Exploitation for Privilege Escalation	arbitrary
Data from Local System	arbitrary
Data Destruction	arbitrary
Browser Session Hijacking	arbitrary
Archive Collected Data	arbitrary
Create Account	arbitrary
Forge Web Credentials	bypass
Unsecured Credentials	bypass
External Remote Services	bypass
Adversary-in-the-Middle	trigger
Phishing	trigger
Stage Capabilities	trigger
Exploitation for Credential Access	wordpress
Brute Force	wordpress
Abuse Elevation Control Mechanism	xml
Endpoint Denial of Service	parameter
Network Sniffing	parameter
User Execution	remote
Drive-by Compromise	remote
Server Software Component	service
Data Manipulation	service
Exploit Public-Facing Application	version
Command and Scripting Interpreter	pointer
Exploitation for Client Execution	attack
Valid Accounts	system
Hijack Execution Flow	cause
Process Injection	privilege
File and Directory Discovery	execute
Exploitation for Defense Evasion	use
Exploitation of Remote Services	possibly
Access Token Manipulation	header

4.3. Limitations

We have identified a number of limitations for our model, which have a toll on the model's performance; these limitations are detailed further. First, the process of manually labeling a CVE is inevitably affected by the subjective perspective of the reviewer. Even though multiple attempts to limit this undesired outcome were taken (i.e., following a clear methodology and establishing general guidelines for the reviewers), the annotators were unable to fully eliminate the inconsistency in the dataset labels.

Second, the quality of the information in the CVE text descriptions must also be taken into consideration when discussing the general limitations of the proposed model. Inconsistencies among the CVE descriptions (incomplete, outdated, or even erroneous details) are highly prevalent [45], thus narrowing the attainable performance of the model.

Third, there is no clear delimitation between certain techniques. Multiple techniques have overlapping meanings and follow the same attack pattern (e.g., *Exploitation for Defence Evasion* and *Abuse Elevation Control Mechanism*). Due to this, a CVE might have multiple possible correct labels, depending on the methodology used to mark the CVE since techniques are closely interconnected and the difference between relating techniques is generally subtle.

Lastly, the rather small dataset and the severe imbalance between the number of CVEs associated with a technique has a toll on the capacity of the model to accumulate enough knowledge to correctly label future samples. Having a larger knowledge base for training the model would help provide samples so that the model perceives also sensitive nuances in CVE text descriptions.

5. Conclusions

In this paper, we emphasized the need for an automatic linkage between the CVE list and MITRE ATT&CK Enterprise Matrix techniques. The problem was transposed into a multi-label task for Natural Language Processing for which we introduce a novel labeled CVE corpus that was augmented using adversarial attacks to limit the severe impact of imbalance between labels. Our baseline includes several classic machine learning models and BERT-based architectures, and the best performing model (i.e., Multi-label SciBERT) was evaluated within a series of experiments from multiple perspectives to extract a complete overview of the data augmentation impact. Comparing the obtained metrics against classical machine learning models accentuates the significant benefits brought by our solution to labeling CVEs with corresponding techniques.

Despite our model obtaining promising results in terms of well-represented techniques, the inherent limitations imposed by the training set tops up the maximum achievable performance. Future work will focus on improving the robustness of the labeled CVE corpus. On one hand, we will focus on enforcing homogeneity among labeling methodology; on the other, we will address the severe imbalance between labels and also its reduced size. Possible new strategies might consider Few-Shot Learning methods [46] for task generalization considering few samples. Semi-supervised learning [47] could also be a possible research direction, given the reduced number of labeled CVEs and the significant number of unlabeled samples that exist in the CVE list. Another aspect that is worth exploring is whether or not gathering extra information from additional sources (e.g., *Common Weakness Enumeration* CWE [48]) can address the incompleteness and inconsistency of the textual CVE description.

Author Contributions: Conceptualization, O.G., A.N., M.D. and R.R.; methodology, O.G. and M.D.; software, A.N. and O.G.; validation, O.G., A.N. and M.D.; formal analysis, O.G., A.N. and M.D.; investigation, A.N., O.G. and M.D.; resources, O.G. and A.N.; data curation, A.N.; writing—original draft preparation, O.G. and A.N.; writing—review and editing, M.D. and R.R.; visualization, A.N.; supervision, M.D. and R.R.; project administration, R.R.; funding acquisition, R.R. All authors have read and agreed to the published version of the manuscript.

Funding: This work was supported by a grant of the Romanian National Authority for Scientific Research and Innovation, CNCS—UEFISCDI, project number 2PTE2020, YGGDRASIL—"Automated System for Early Detection of Cyber Security Vulnerabilities".

Institutional Review Board Statement: The study was conducted in accordance with the Declaration of Helsinki, and approved by the Ethics Committee of the Faculty of Automated Control and Computers, University Politehnica of Bucharest.

Informed Consent Statement: Not applicable.

Data Availability Statement: The dataset is freely available on Tagtog at https://www.tagtog.com/readerbench/MitreMatrix/ (accessed on 8 August 2022), whereas the code is available on Github at https://github.com/readerbench/CVE2ATT-CK (accessed on 8 August 2022).

Acknowledgments: We would also like to show our gratitude to Ioana Nedelcu, Ciprian Stanila, and Ioana Branescu for their contributions to building the labeled CVE corpus.

Conflicts of Interest: The authors declare no conflict of interest.

Abbreviations

The following abbreviations are used in this manuscript:

ATT&CK	Adversarial Tactics, Techniques, and Common Knowledge
BERT	Bidirectional Encoder Representations from Transformers
CAPEC	Common Attack Pattern Enumeration and Classification
CNN	Convolutional Neural Network
CVE	Common Vulnerabilities and Exposures
CVET	Common Vulnerabilities and Exposures Transformer
CWE	Common Weakness Enumeration
EDA	Easy Data Augmentation
ML	Machine Learning
NLP	Natural Language Processing
SciBERT	Scientific Bidirectional Encoder Representations from Transformers
SecBERT	Security Bidirectional Encoder Representations from Transformers
SVM	Support Vector Machine
TF-IDF	Term Frequency-Inverse Document Frequency
TRAM	Threat Report ATT&CK Mapping

References

1. Li, Y.; Liu, Q. A comprehensive review study of cyber-attacks and cyber security; Emerging trends and recent developments. *Energy Rep.* **2021**, *7*, 8176–8186. [CrossRef]
2. Dayalan, M. Cyber Risks, the Growing Threat. *IJNRD-Int. J. Nov. Res. Dev.* **2017**, *2*, 4–6. [CrossRef]
3. Smith, Z.M.; Lostri, E. *The Hidden Costs of Cybercrime*; Technical Report; McAfee: San Jose, CA, USA, 2020.
4. Fichtenkamm, M.; Burch, G.F.; Burch, J. Cybersecurity in a COVID-19 World: Insights on How Decisions Are Made. Available online: https://www.isaca.org/resources/isaca-journal/issues/2022/volume-2/cybersecurity-in-a-covid-19-world (accessed on 8 August 2022).
5. Cremer, F.; Sheehan, B.; Fortmann, M.; Kia, A.N.; Mullins, M.; Murphy, F.; Materne, S. Cyber risk and cybersecurity: A systematic review of data availability. *Geneva Pap. Risk Insur. Issues Pract.* **2022**, *47*, 698–736. [CrossRef] [PubMed]
6. Martin, R.; Christey, S.; Baker, D. *A Progress Report on the CVE Initiative*; Technical Report; The MITRE Corporation: Bedford, MA, USA, 2002.
7. Sönmez, F.Ö. Classifying Common Vulnerabilities and Exposures Database Using Text Mining and Graph Theoretical Analysis. In *Machine Intelligence and Big Data Analytics for Cybersecurity Applications*; Springer: Berlin/Heidelberg, Germany, 2021; pp. 313–338. [CrossRef]
8. Strom, B.E.; Applebaum, A.; Miller, D.P.; Nickels, K.C.; Pennington, A.G.; Thomas, C.B. *MITRE ATT&CK™: Design and Philosophy*; Technical Report; The MITRE Corporation: Bedford, MA, USA, 2018.
9. Hemberg, E.; Kelly, J.; Shlapentokh-Rothman, M.; Reinstadler, B.; Xu, K.; Rutar, N.; O'Reilly, U.M. Linking threat tactics, techniques, and patterns with defensive weaknesses, vulnerabilities and affected platform configurations for cyber hunting. *arXiv* **2021**, arXiv:2010.00533.
10. NVD. NVD Dashboard. Available online: https://nvd.nist.gov/general/nvd-dashboard (accessed on 8 August 2022).
11. The Center for Threat-Informed Defense. *Mapping MITRE ATT&CK® to CVEs for Impact*; The Center for Threat-Informed Defense: Bedford, MA, USA, 2021.

12. Baker, J. CVE + MITRE ATT&CK to Understand Vulnerability Impact. Available online: https://medium.com/mitre-engenuity/cve-mitre-att-ck-to-understand-vulnerability-impact-c40165111bf7 (accessed on 8 August 2022).
13. Roe, S. Using Mitre ATT&CK with threat intelligence to improve Vulnerability Management. Available online: https://outpost24.com/blog/Using-mitre-attack-with-threat-intelligence-to-improve-vulnerability-management (accessed on 24 August 2022).
14. Ampel, B.; Samtani, S.; Ullman, S.; Chen, H. Linking Common Vulnerabilities and Exposures to the MITRE ATT&CK Framework: A Self-Distillation Approach. *arXiv* **2021**, arXiv:2108.01696.
15. Kuppa, A.; Aouad, L.; Le-Khac, N.A. Linking CVE's to MITRE ATT&CK Techniques. In Proceedings of the 16th International Conference on Availability, Reliability and Security, Vienna, Austria, 17–20 August 2021; pp. 1–12.
16. Github. Threat Report ATT&CK Mapping (TRAM). Available online: https://github.com/center-for-threat-informed-defense/tram/ (accessed on 8 August 2022).
17. Yoder, S. Automating Mapping to ATT&CK: The Threat Report ATT&CK Mapper (TRAM) Tool. Available online: https://medium.com/mitre-attack/automating-mapping-to-attack-tram-1bb1b44bda76 (accessed on 8 August 2022).
18. Ribeiro, M.T.; Singh, S.; Guestrin, C. Model-agnostic interpretability of machine learning. *arXiv* **2016**, arXiv:1606.05386.
19. Tagtog. CVE2ATT&CK Dataset. Available online: https://www.tagtog.com/readerbench/MitreMatrix/ (accessed on 8 August 2022).
20. Github. CVE2ATT&CK Repository. Available online: https://github.com/readerbench/CVE2ATT-CK (accessed on 8 August 2022).
21. Vulnerability Database. Available online: https://vuldb.com/ (accessed on 24 August 2022).
22. Exploit Database-Exploits for Penetration Testers, Researchers, and Ethical Hackers. Available online: https://www.exploit-db.com/ (accessed on 24 August 2022).
23. TagTog. API Documentation v1. Available online: https://github.com/tagtog/tagtog-doc/blob/master/API-projects-v1.md (accessed on 8 August 2022).
24. Japkowicz, N.; Stephen, S. The Class Imbalance Problem: A Systematic Study. *Intell. Data Anal.* **2002**, *6*, 429–449. [CrossRef]
25. TextAttack. Documentation Webpage. Available online: https://textattack.readthedocs.io/en/latest/index.html (accessed on 8 August 2022).
26. Morris, J.; Lifland, E.; Yoo, J.Y.; Grigsby, J.; Jin, D.; Qi, Y. TextAttack: A Framework for Adversarial Attacks, Data Augmentation, and Adversarial Training in NLP. In Proceedings of the 2020 Conference on Empirical Methods in Natural Language Processing: System Demonstrations, Online, 16–20 November 2020; Association for Computational Linguistics; pp. 119–126. [CrossRef]
27. TextAttack. Augmentation Recipes. Available online: https://textattack.readthedocs.io/en/latest/3recipes/augmenter_recipes.html (accessed on 8 August 2022).
28. Wei, J.; Zou, K. EDA: Easy Data Augmentation Techniques for Boosting Performance on Text Classification Tasks. In Proceedings of the 2019 Conference on Empirical Methods in Natural Language Processing and the 9th International Joint Conference on Natural Language Processing (EMNLP-IJCNLP), Hong Kong, China, 3–7 November 2019; Association for Computational Linguistics; pp. 6382–6388. [CrossRef]
29. Alazaidah, R.; Ahmad, F.K. Trending Challenges in Multi Label Classification. *Int. J. Adv. Comput. Sci. Appl.* **2016**, *7*. [CrossRef]
30. spaCy. spaCy 101: Everything You Need to Know. Available online: https://spacy.io/usage/spacy-101 (accessed on 8 August 2022).
31. Tsoumakas, G.; Katakis, I.; Vlahavas, I. Mining multi-label data. In *Data Mining and Knowledge Discovery Handbook*; Springer: Berlin/Heidelberg, Germany, 2009; pp. 667–685.
32. Rifkin, R.; Klautau, A. In Defense of One-Vs-All Classification. *J. Mach. Learn. Res.* **2004**, *5*, 101–141.
33. Tsoumakas, G.; Vlahavas, I. Random k-labelsets: An ensemble method for multilabel classification. In Proceedings of the European Conference on Machine Learning, Warsaw, Poland, 17–21 September 2007; Springer: Berlin/Heidelberg, Germany, 2007; pp. 406–417.
34. Rish, I. An Empirical Study of the Naïve Bayes Classifier. *IJCAI 2001 Work. Empir Methods Artif. Intell.* **2001**, *3*, 41–46.
35. Cervantes, J.; Garcia-Lamont, F.; Rodríguez-Mazahua, L.; Lopez, A. A comprehensive survey on support vector machine classification: Applications, challenges and trends. *Neurocomputing* **2020**, *408*, 189–215. [CrossRef]
36. Scikit. Grid Search. Available online: https://scikit-learn.org/stable/modules/generated/sklearn.model_selection.GridSearchCV.html (accessed on 8 August 2022).
37. LeCun, Y.; Haffner, P.; Bottou, L.; Bengio, Y. Object recognition with gradient-based learning. In *Shape, Contour and Grouping in Computer Vision*; Springer: Berlin/Heidelberg, Germany, 1999; pp. 319–345.
38. Yih, W.T.; He, X.; Meek, C. Semantic parsing for single-relation question answering. In Proceedings of the 52nd Annual Meeting of the Association for Computational Linguistics, Baltimore, MD, USA, 23–25 June 2014; Association for Computational Linguistics; Volume 2: Short Papers; pp. 643–648.
39. Kalchbrenner, N.; Grefenstette, E.; Blunsom, P. A Convolutional Neural Network for Modelling Sentences. In Proceedings of the 52nd Annual Meeting of the Association for Computational Linguistics, Baltimore, MD, USA, 23–25 June 2014; Association for Computational Linguistics; Volume 1: Long Papers; pp. 655–665. [CrossRef]
40. Github. Word Representation for Cyber Security Vulnerability Domain. Available online: https://github.com/unsw-cse-soc/Vul_Word2Vec (accessed on 8 August 2022).

41. Devlin, J.; Chang, M.W.; Lee, K.; Toutanova, K. BERT: Pre-training of Deep Bidirectional Transformers for Language Understanding. In Proceedings of the 2019 Conference of the North American Chapter of the Association for Computational Linguistics, Minneapolis, MN, USA, 2–7 June 2019; Association for Computational Linguistics; Volume 1: Long and Short Papers; pp. 4171–4186.
42. Beltagy, I.; Lo, K.; Cohan, A. SciBERT: A Pretrained Language Model for Scientific Text. In Proceedings of the 2019 Conference on Empirical Methods in Natural Language Processing and the 9th International Joint Conference on Natural Language Processing (EMNLP-IJCNLP), Hong Kong, China, 3–7 November 2019; pp. 3615–3620.
43. Huggingface. SecBERT Model. Available online: https://huggingface.co/jackaduma/SecBERT (accessed on 8 August 2022).
44. Pytorch. BCE with Logit Loss. Available online: https://pytorch.org/docs/stable/generated/torch.nn.BCEWithLogitsLoss.html (accessed on 8 August 2022).
45. Dong, Y.; Guo, W.; Chen, Y.; Xing, X.; Zhang, Y.; Wang, G. Towards the Detection of Inconsistencies in Public Security Vulnerability Reports. In Proceedings of the 28th USENIX Security Symposium (USENIX Security 19), Santa Clara, CA, USA, 14–16 August 2019; USENIX Association; pp. 869–885.
46. Wang, Y.; Yao, Q.; Kwok, J.T.; Ni, L.M. Generalizing from a few examples: A survey on few-shot learning. *ACM Comput. Surv. (csur)* **2020**, *53*, 1–34. [CrossRef]
47. Kasieczka, G.; Nachman, B.; Shih, D.; Amram, O.; Andreassen, A.; Benkendorder, K.; Bortolato, B.; Broojimans, G.; Canelli, F.; Collins, J.; et al. The LHC olympics 2020: A community challenge for anomaly detection in high energy physics. *Rep. Prog. Phys.* **2021**, *84*, 124201. [CrossRef] [PubMed]
48. MITRE. Common Weakness Enumeration Webpage. Available online: https://cwe.mitre.org/ (accessed on 8 August 2022).

algorithms

MDPI

Article

Sustainable Risk Identification Using Formal Ontologies †

Avi Shaked [1] and Oded Margalit [2,*]

[1] Department of Computer Science, University of Oxford, Oxford OX1 3QD, UK
[2] Department of Computer Science, Ben Gurion University of the Negev, Be'er Sheva 84105, Israel
*   Correspondence: odedm@cs.bgu.ac.il
†   This Paper Is an Extended Version of Our Paper Published in the Proceedings of the 17th Annual System of Systems Engineering Conference (SOSE 2022) (Rochester, NY, USA, 7–11 June 2022).

Abstract: The cyber threat landscape is highly dynamic, posing a significant risk to the operations of systems and organisations. An organisation should, therefore, continuously monitor for new threats and properly contextualise them to identify and manage the resulting risks. Risk identification is typically performed manually, relying on the integration of information from various systems as well as subject matter expert knowledge. This manual risk identification hinders the systematic consideration of new, emerging threats. This paper describes a novel method to promote automated cyber risk identification: OnToRisk. This artificial intelligence method integrates information from various sources using formal ontology definitions, and then relies on these definitions to robustly frame cybersecurity threats and provide risk-related insights. We describe a successful case study implementation of the method to frame the threat from a newly disclosed vulnerability and identify its induced organisational risk. The case study is representative of common and widespread real-life challenges, and, therefore, showcases the feasibility of using OnToRisk to sustainably identify new risks. Further applications may contribute to establishing OnToRisk as a comprehensive, disciplined mechanism for risk identification.

Keywords: formal ontology; risk identification; cybersecurity; vulnerability

Citation: Shaked, A.; Margalit, O. Sustainable Risk Identification Using Formal Ontologies. *Algorithms* **2022**, *15*, 316. https://doi.org/10.3390/a15090316

Academic Editors: Francesco Bergadano and Giorgio Giacinto

Received: 18 August 2022
Accepted: 31 August 2022
Published: 2 September 2022

Publisher's Note: MDPI stays neutral with regard to jurisdictional claims in published maps and institutional affiliations.

1. Introduction

Risk identification is the process which lays the foundations for establishing the cybersecurity posture of systems, organisations and services. Risk management is a collection of "coordinated activities to direct and control an organisation with regard to risk" [1]. Risk identification provides the infrastructure for all other risk management activities [2].

A risk is a potential for something to go wrong, eventually causing harm or loss [3]. Accordingly, cyber risk is an operational risk which is associated with activities in cyberspace that may cause damage to organisational assets [4].

The goal of risk identification is to "find, recognize and describe risks that may prevent an organization achieving its objectives" [5]. Refsdal et al. identify that risk comprises three elements: asset, vulnerability and threat [3]. In agreement, Strupczewski's meta model of cyber-risk concept includes the same three elements [4]. A vulnerability merely indicates an exploitable system property; a risk is distinguished from a vulnerability by having the potential to harm or reduce the value of an asset. The identification of pertinent assets—such as sensitive information and services—and their business value is therefore an essential risk identification element [6]. Risk identification requires knowing the business environment and the organisational assets in addition to the vulnerabilities [7].

Provided risks are properly identified, they can be then analysed, evaluated for impact and, if necessary, mitigated using appropriate security controls. Otherwise, unidentified risks may go untreated, and misidentified risks may be improperly treated; potentially resulting in considerable damage once they materialise [8].

Continuous organisational changes introduce a major threat to performing risk identification [7]. The dynamics of business environments include changes to processes, products and services, as well as introduction of new information systems and related features. Irrespective of organisational changes, the cyber threat landscape is autonomously evolving. As an example, new software vulnerabilities are published on a daily basis, providing ample opportunities for attackers to exploit them [9]. Moreover, attacker capabilities—tactics, technologies and procedures (TTPs)—continue to improve [10]; sometimes to a military grade level [11]. To address the dynamics of cybersecurity, it is essential to have dynamic and adaptable cyber risk management, with risk identification outputs being revisited often to re-evaluate and establish an up-to-date organisational cybersecurity posture [6,7]. For this purpose, risk register mechanisms, such as those recommended by The European Union Agency for Cybersecurity (ENISA), contain the date of latest assessment as part of the risk register record and are expected to be properly maintained [12].

Relevant, up-to-date and timely information is crucial to robust risk identification [5]. Prevalent risk identification approaches rely on manual analysis by human experts [2]. These include brainstorming, interviews, checklists, statistics and techniques for historical data collection [3]. Risk identification also relies on integration of information from various sources [3,13]. Previous automation attempts with respect to cyber risk activities focused mostly on automated identification of threats and vulnerabilities (for example, [14,15]). Specifically, attributing the actual risk to organisational assets remains a manual analysis effort. The manual nature of risk identification approaches hinders their dynamic application in a sustainable form to meet the challenges of the evolving cybersecurity threat landscape [6].

This paper, which extends [16], proposes the use of a formal ontology to promote rigorous and continuous risk identification. A formal ontology is a well-defined, computer-based representation of concepts and their relations [17]. Formal ontology should not be confused directly with the philosophical term, which is concerned with the understanding of reality. However, formal ontology relates to the philosophical term, by capturing the ontology of a particular domain using a formal, well-structured model. We use the term "ontology" henceforth to relate to formal ontology.

Ontologies are a form of semantic technology. They provide the infrastructure for intelligent applications [18]. Ontologies belong to the content theory branch of Artificial Intelligence (AI) [19], and they are central for building intelligent computational agents [20]. Ontologies can minimise ambiguity and misunderstanding between stakeholders as well as lay the foundations for high-level reasoning and decision making [18,21]. An organisation-specific ontology can be used to facilitate interoperability between domains [22], and, even more specifically, between business and information technology concerns, with which organisational cybersecurity is typically associated [23].

Ontologies can be used to support risk management. Examples of such applications include management of human and ecological health risks [24] and safety risk management in construction [25]. Previous uses of ontologies for cybersecurity risk management did not consider the critical business impact of such risks [26–28]. An ontology-based system was demonstrated for the calculation of cybersecurity risk metrics, but it does not include inferred identification of risks and does not provide actionable risk-related information [29]. An automated security risk identification method to address engineering design issues exists, but it involves only identification of high-level consequence categories [30]. As far as we know, there is no ontology-based method to identify emerging cybersecurity risks which can be employed continuously by organisations, let alone one which allows an organisation to contextualise the risks with respect to the organisational operations.

This paper details and exemplifies a new method—OnToRisk—which uses formal ontology mechanisms to automate cybersecurity risks identification, based on integration of formal definitions and situational information from pertinent sources. OnToRisk is an AI method which employs aspects of knowledge representation to introduce robust information models; and of reasoning to provide actionable insights about situations represented by the models. The information models can include security intelligence related concepts—namely threat, vulnerability, asset and risk—as well as any other technical and organisational concepts that are relevant to provide situational awareness.

We describe a case study of using OnToRisk to identify risks emerging from a newly published software vulnerability, in an undisclosed, international enterprise in the finance sector (henceforth, "the enterprise"). While specific, the case study is representative of a general, desirable practice in every organisation which uses software components. A software vulnerability is "an instance of a flaw, caused by a mistake in the design, development, or configuration of software, such that it can be exploited to violate some explicit or implicit security policy" [31]. While previous work by Wang and Guo used a formal ontology to analyse vulnerabilities from the technical perspective of vulnerability management [21]; our case study uses a formal ontology to capture concepts and relations to analyse cybersecurity vulnerabilities from the organisational operations risk perspective.

The paper continues as follows. Section 2 presents the new, ontology-based risk identification method OnToRisk and overviews the vulnerability-induced risks identification case study. Section 3 details the case study results of using OnToRisk for vulnerability-induced risks identification. Section 4 reflects on the new risk identification method and the case study, as well as discusses further uses, benefits and research potential of the ontology-based method.

2. Materials and Methods

OnToRisk uses formal ontology mechanisms for rigorous, information-based and definitions-based risks identification. The OnToRisk method includes the following activities:

1. formally define concepts associated with a specific risk type, as well as their relations, by authoring an ontology;
2. formally define the risk type in the ontology, using the predefined concepts and relations. This definition of a risk type aims to promote the automatic identification of its instantiations;
3. capture the organisational situation by instantiating the existing ontology definitions. This is achieved by incorporating "individual" definitions into the ontology;
4. apply automated, ontology-based reasoners to the ontology to derive new, inferred insights about the situation.

Activity #3 is meant to be automated as much as possible, e.g., by importing—while translating—existing organisational information from information systems into the formal ontology. Activity #4 is the activity in which new risk-related insights should emerge, automatically, based on the integration of explicit definitions and explicit situations. Ideally, these activities should be performed continuously, reflecting an up-to-date organisational security posture.

We validate OnToRisk using a case study methodology. The selection in a single-case study approach is aligned with the rationale identified by Yin; that the case study is a representative, typical case [32]. The OnToRisk method is applied in a case study of an enterprise seeking to identify risks emerging from the disclosure of a new vulnerability, which is found in a prevalent software component. The widely representative and applicable case study was inspired by real events, following the late 2021 disclosure of a vulnerability in Log4j [33,34].

Risk management is considered a business-related activity in an enterprise. Accordingly, the enterprise established and maintains a system of policies, as well as a hierarchical framework for communicating and assessing operational risks, with cybersecurity risks being included as part of the overall risk management organisational system. The risk-related concepts were identified based on careful reading of official documents and directives, analysis of some of the enterprise's information systems, and on conversations with domain experts. The latter included risk managers and an incident response leader.

First, as the OnToRisk method outlines, relevant concepts and their relations were defined as a formal ontology. Protégé was the tool used for authoring the ontology [35]. The ontology itself is in the standard Web Ontology Language (OWL) format. Relevant concepts are depicted in OWL using "classes"; and relations between concepts are formally expressed in OWL using "object properties". In defining object properties, the source node class is referred to as "Domain," and the target node class is referred to as "Range".

A relevant risk definition was then added to the ontology, using some of the predefined classes and object properties. Next, a situation was captured. The situation was designed using natural language, and then translated into the ontology, as an instantiation of the formalised classes and object properties. Finally, a reasoner (HermiT within Protégé) is used to reason about the situation, i.e., process the explicit situation definitions and present inferred information based on these. The inference was verified to yield the results that are expected based on manual analysis of the situation.

The work, including the ontology and the resulting insight with respect to the enterprise's operations and infrastructure, was presented to domain experts as well as high-level management for both obtaining feedback and promoting the organisational risk management practices.

3. Results

We now describe the results of applying OnToRisk to the case study (of identifying risks to the enterprise as they emerge from the disclosure of a new vulnerability in a software component). Appendix A provides the full definitions, described in Sections 3.1–3.3, in the form of a formal ontology. Appendix B provides the inferred assertions, described in Section 3.4, in the form of a formal ontology.

3.1. Concepts and Relations (Meta Levels Definitions)

Figure 1 shows the concepts and relations, representing the result of performing activity #1 of OnToRisk in the case study. Concepts (classes) appear as graph nodes and relations (object properties) appear as edges between nodes. The concepts are:

1. Application, representing a software application by the enterprise;
2. Component, representing any software component;
3. Business Function, representing any function that relates to the enterprise's business operation;
4. Sensitive Information, representing any sensitive information item owned by the enterprise;
5. Vulnerability, representing any vulnerability of software components;
6. Risk, representing the enterprise's risk definitions;
7. Cybersecurity Risk, representing a specific subclass of risk definitions relating to cybersecurity issues;
8. Vulnerability-Induced Risk, representing any risk to the business emerging from the existence of a vulnerability. Being a risk definition relating to a cybersecurity issue, it is a specific subclass of Cybersecurity Risk.

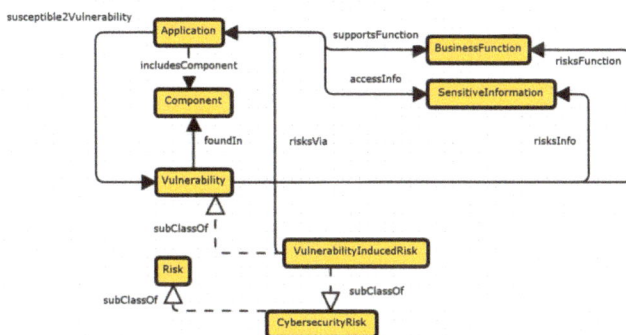

Figure 1. The vulnerability-induced risk case represented as a formal graph of meta-level concepts and relations. This graph is generated by applying the CoModIDE plugin for Protégé [36] to the formal ontology. The graph shows: the classes as nodes; the object properties between classes (from domain to range) as solid, annotated arrows; and subtype (subclass) relations as dashed arrows.

The hierarchical structure of the risk concepts (concepts #6, #7 and #8 above) reflects the hierarchical risk definition architecture which is practiced within the enterprise; with Risk being a layer-1 risk definition, Cybersecurity risk being a layer-2 risk definition, and Vulnerability-induced Risk being introduced as a layer-3 risk definition. This conforms with the prominent business risk typology used in the financial sector with which the enterprise is associated [4].

Figure 1 also shows the object properties—expressing relations between concepts—as graph edges in solid line between the class nodes. The object properties are:

1. accessInfo—represents an ability of an application (Domain) to access a sensitive information item (Range);
2. supportsFunction—represents that an application (Domain) supports a business function (Range);
3. includesComponent—represents a software application composition, linking the application (Domain) with its components (Range);
4. foundIn—represents a vulnerability (Domain) found in a software component (Range);
5. susceptible2Vulnerability—marks an application (Domain) as being susceptible to a vulnerability (Range) due to one of its software components. This object property is formally defined as a composite property using other object properties:

$$susceptible2Vulnerability \equiv inverse(foundIn) \circ includesComponent \qquad (1)$$

6. risksInfo—indicates that a vulnerability (Domain) may risk sensitive information (Range). This object property is formally defined as a composite property using other object properties:

$$risksInfo \equiv accessInfo \circ inverse(includesComponent) \circ foundIn \qquad (2)$$

7. risksFunction—indicates that a vulnerability (Domain) may risk a business function (Range). This object property is formally defined as a composite property using other object properties:

$$risksFunction \equiv supportsFunction \circ inverse(includesComponent) \circ foundIn \qquad (3)$$

8. risksVia—identifies the application (Range) through which a specific Vulnerability-Induced Risk (Domain) can be realised. This object property is formally defined as a composite property using other object properties:

$$\begin{aligned} risksVia \equiv inverse(accessInfo) \circ risksInfo \,| \\ inverse(supportsFunction) \circ risksFunction \end{aligned} \qquad (4)$$

Reflecting on the derived ontology, we note that it is realistic and practical to acquire relevant information, which can be used for instantiating a situation using the ontology meta-level definitions. The enterprise operates an information system which records all the enterprise applications, along with attributes. Some of these attributes are the category of information that can be accessed by the application; and the application's business criticality score, which is established based on supported business functions. Extracting software components used by an application—a Software Bill Of Material (SBOM)—is a feature provided by various software composition analysis tools (by analysing either the source code or the final software artifacts). Information about software components vulnerabilities is found online in vulnerability repositories, such as [37].

3.2. Risk Definition

Following the activity #2 guideline of the OnToRisk method, a Vulnerability-Induced Risk concept is formally defined using the established concepts and relations:

$$\text{vulnerabilityInducedRisk} \equiv \text{vulnerability and}$$
$$((\text{risksFunction some BusinessFunction}) \text{ or} \qquad (5)$$
$$(\text{risksInfo some SensitiveInformation}))$$

This formally defines the specific risk as a vulnerability which risks either a business function and/or sensitive information. Ideally, this definition could instantiate new information elements (of the VulnerabilityInducedRisk type). However, due to limitation in both the OWL ontology standard and the Protégé ontology authoring tool, instantiation of new elements is not possible by inference, and instead this tags a Vulnerability typed individual element as a VulnerabilityInducedRisk. Accordingly, VulnerabilityInducedRisk is also considered as a subclass of Vulnerability (in addition to being a subclass of Cybersecurity Risk); this is shown in Figure 1. This is merely a technical adaptation, which has no effect on the results as it can be easily interpreted to the ideal case, and we discuss this shortly. The formal definition of the set of risks (*R*) in this implementation is:

$$R \equiv \{v \in V \mid (\exists x \in BF \,\&\, (v, x) \in RF) \text{ or } (\exists y \in SI \,\&\, (v, y) \in RI)\} \qquad (6)$$

with:

V—the set of Vulnerability class (i.e., concept) instantiations
BF—the set of Business Function class instantiations
RF—the set of risksFunction object properties instantiations
SI—the set of Sensitive Information class instantiations
RI—the set of risksInfo object properties instantiations
i.e., the set of risks is a subset of all vulnerabilities that have either a risksFunction object property (stating the vulnerability risks an existing business function) or a riskInfo object property (stating the vulnerability risks existing sensitive information).

The formal definition itself is more than a technical definition. This is the first concrete layer-3 risk definition, which extends the existing conceptual and abstract layer-2 enterprise risk definition (Cybersecurity risk). This fairly simple, formal ontology-based definition of a "vulnerability-induced risk" rigorously expresses a concrete type of risk. This specific risk type is of high importance to enterprise stakeholders, including its high-level management, and had not been declared until our OnToRisk implementation named it explicitly.

3.3. Situation

The case study situation details a risk assessment scenario which considers a newly disclosed vulnerability. It is based on real-life situations—specifically, the discovery and public disclosure of the vulnerability known as "Log4shell" [33,34]. The case study is designed as an alternative, what-if scenario of detecting risks associated with the vulnerability using OnToRisk.

According to OnToRisk activity #3, the situation is captured as a collection of instantiations of the ontological concepts and relations (derived in activity #1 and reported in Section 3.1).

The baseline situation captures the organisational situation with respect to its operational applications and their business context. Four applications exist:

1. App1, which does not include Log4j as one of its software components;
2. App2, which includes Log4j as one of its software components;
3. App3, which includes Log4j as one of its software components and has access to the sensitive information item named ClientIDsList;
4. App4, which includes Log4j as one of its software components and supports the business function named OpenAccount.

Now, consider the publication of a new Common Vulnerabilities and Exposures (CVE) record, related to the Log4j component. This results in a new situation, captured in formal ontology form by adding the newly disclosed vulnerability into our ontology, as an instantiation of the "Vulnerability" concept. We name this entity "Log4shell." Additionally, the vulnerability is associated with the affected software component—Log4j—by adding a "foundIn" object property from the Log4shell individual to the Log4j individual.

3.4. Ontology-Based Inferrence

According to OnToRisk activity #4, we use the ontology-based reasoner to make inferences about the developing situation, and, ultimately, identify the emerging risks. The resulting inferred assertions that extend the explicitly declared assertions appear in Appendix B.

The reasoner provides the following new inferences:

1. The "susceptible2Vulnerability" object property is attributed to App2, App3 and App4. This suggests that each of these applications is susceptible to the vulnerability.
2. The Log4shell vulnerability is categorized—automatically—as a VulnerabilityInducedRisk. This indicates that this specific vulnerability introduces new risk/s to the enterprise, as Figure 2 shows. This is the automatic identification of new risks.
3. The object property "risksInfo ClientIDsList" emerges with respect to the Log4shell vulnerability (Figure 2). This suggests that ClientIDsList, which is one of the enterprise's sensitive information items, is at risk.
4. The object property "risksFunction OpenAccount" emerges with respect to the Log4shell vulnerability (Figure 2). This suggests that OpenAccount—one of the enterprise's business functions—is at risk.
5. Two new risksVia object property assertions emerge, with respect to the Log4shell vulnerability (Figure 2). Each of these suggests a possible attack surface through which the risk can realise. In the specific case, App3 is the attack surface for the risk on ClientIDsList and App4 is the attack surface for the risk on the OpenAccount. While this is not captured explicitly in the inferred assertions, the reasoner explanation mechanism provides this traceability, as Figure 3 shows.

Figure 2. The Log4shell ontology-based assertions in Protege. Manually stated (explicit) assertions appear in bold font, while automatically inferred assertions appear in regular font.

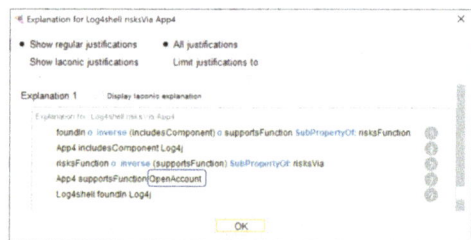

(a)   (b)

Figure 3. The reasoner explanation for asserting the risksVia properties. (**a**) for App3, as a result of the risksInfo with respect to ClientIDsList; (**b**) for App4, as a result of the risksFunction with respect to OpenAccount.

The automatically derived inferences are aligned with a manual analysis of the situation. While the manual analysis can be considered straightforward, performing such an analysis is time consuming, and this is exactly the effort that OnToRisk is designed to make redundant. The vulnerability in App2 does not present a new risk to the enterprise, from an operational perspective. Still, using the "susceptible2Vulnerability" property which now characterises the application, the potential of App2 being affected by the vulnerability can be communicated with the App2 application owner. The application owner can then choose whether to further analyse the vulnerability impact on the application and/or solve any vulnerability-related issues in a future version. The vulnerabilities in App3 and in App4, however, should be of interest to the enterprise management, as they introduce business risks. Continuously applying the reasoner to the enterprise situation allows pertinent managers to be notified immediately of such risks as they emerge; and the enterprise management can then promptly act to solve them, by identifying and empowering the appropriate personnel—such as application owners, risk managers, security officers and information officers—to do so.

4. Discussion

The dynamic cybersecurity threat landscape requires risk identification to be performed continuously to achieve up-to-date situational awareness. This paper proposes a new, formal ontology-based method—OnToRisk—for promoting automated risk identification. The method relies on the use of AI—through its formal ontology branch—for information-based, systematic and continuous risk identification. The method employs formal ontology definitions of domain concepts and relations, as well of the associated risk, to analyse organisational situations and automatically provide actionable insights.

The OnToRisk method was successfully applied to identify risks emerging from a vulnerability disclosure, which is a widely applicable challenge in enterprises. As a given enterprise situation has changed to reflect existence of a new vulnerability, a reasoning mechanism—applied to the situation—automatically yielded a list of potentially affected applications as well as of the potential business impact. In practice, typical software applications may include hundreds of re-used lower-level components, which may lead to a significant effort in their manual analysis. The automated approach of OnToRisk decouples the risk identification effort from the quantity of software components. Moreover, new risks are identified, along with their potential business impact and the respective attack surface. A reasoning mechanism can act continuously on the information. These provide a strong basis for sustainable risk management, which is essential to creating a valid cybersecurity situational awareness.

Our method provides a step forward with respect to a previously identified need for a conceptual framework to drive the rapid and automated integration of Cyber Threats Intelligence (CTI) [10]. Specifically, our method conforms with the requirement that both internal and external information be factored into the automated integration process; and it provides a rigorous infrastructure for such integration. The case study demonstrates the integration of internal, enterprise-owned information—about applications composition as well as about their business context—with external vulnerability information. Currently, some of the data was integrated manually, by importing data—exported from various information systems—into the ontology. In the case study, information about enterprise applications was adopted from the enterprise's information system which is used to catalogue applications and their metadata. A likely technical future effort is to develop mechanisms to automate the integration of data into the ontology, using both internal data sources (such as application inventory information systems) and external data sources (such as CVE repositories).

Furthermore, with OnToRisk being a technology-agnostic and vendor-neutral method, the formal representation of a domain of interest may lead to identification of gaps in information, which in turn may justify the introduction of new technology and/or tools into the enterprise. Specifically, the case study's formal ontology relies on associating each application with its SBOM. However, at the time of performing the case study, the enterprise has only employed SBOM tools to ingest open-source software packages and did not apply the relevant technology to produce the SBOM of its own applications. Our case study highlights the need to incorporate the technology and tools to extract SBOM from the enterprise applications that are in production in order to support risk assessment with respect to vulnerability-induced risks.

In the technical implementation of the case study, vulnerability-induced risks are represented by "tagging" vulnerabilities as vulnerability-induced risk, i.e., the risks are a subset of the vulnerabilities (as captured formally in Equation (6)). This is due to limitations in the OWL standard and the standard Protégé implementation that prevents inferring the existence of new individuals. We chose to adhere to the standard implementation to demonstrate the feasibility and practicality of OnToRisk. Ideally, however, the risk identification implementation can be easily improved when developing an ontology-based application or information system by using a proprietary mechanism to yield new individuals. Such individuals can be derived formally as the tuple (vulnerability, impacted element, attack surface), i.e.:

$$
\begin{aligned}
(v,i,a) \equiv \{(v,i,a)| \\
(\ni v \in V \ \& \ \ni i \in BF \ \& \ \ni a \in A \ \& \ (a,i) \in SF \ \& \ (v,i) \in RF \ \& \ (v,\,a) \in RV) \ or \\
(\ni v \in V \ \& \ \ni i \in SI \ \& \ \ni a \in A \ \& \ (a,i) \in AI \ \& \ (v,i) \in RI \ \& \ (v,\,a) \in RV)\}
\end{aligned}
\tag{7}
$$

with:

V—the set of Vulnerability class (i.e., concept) instantiations
BF—the set of Business Function class instantiations
A—the set of Application class instantiations
SF—the set of supportsFunction object properties instantiations
RF—the set of risksFunction object properties instantiations
SI—the set of Sensitive Information class instantiations
AI—the set of accessInfo object properties instantiations
RI—the set of risksInfo object properties instantiations
RV—the set of risksVia object properties instantiations

OnToRisk currently provides the identification of potential risks. Identified risks should be further analysed. In the case study implementation, for example, an application marked as susceptible to a vulnerability due to the identity of one of its components may not present an actual risk, e.g., in a case where an application uses the component in a version which is not susceptible to the vulnerability or if the context of use or security controls prevent the exploitation of the disclosed vulnerability. Future research can establish the use of other ontology elements, such as data properties—in addition to classes and object properties—for improving the risk identification and its automation. Expanding the ontology with additional elements may also contribute to the prioritisation of risks (e.g., by introducing impact levels) and to the inclusion of additional CTI information.

With OnToRisk currently validated for the specific case study of a vulnerability-induced risk, additional research can utilise the method to identify other types of cybersecurity risks, such as those emerging from a compromised supply chain or from existence of Common Weaknesses Enumeration (CWE) in applications and application development.

Whereas a previous method by Eckhart et al. employs automated risk identification for improving engineering artifacts [30], OnToRisk provides automated risk identification for better organisational situational awareness. OnToRisk provides a more concrete view of business consequences, compared with the high-level consequence categories of the engineering-focused method proposed by Eckhart et al. OnToRisk relies on continuous integration of information within operational context, as opposed to initiated engineering design verification, which is the domain of the method by Eckhart et al. Both methods share a formal ontology approach as well as the goal of relieving personnel from tedious risk identification so that it can concentrate on other aspects of risk management. Therefore, future research may seek to integrate the two methods and deliver an ontology-based risk identification method for the full system lifecycle.

5. Conclusions

In this paper we describe a new method—OnToRisk—which promotes the automatic identification of risks. The method is validated using a widely applicable, realistic and representative case study implementation of identifying risks emerging from software vulnerabilities.

Future research may demonstrate the use of the proposed method to support the automated identification of risks of additional types. Furthermore, elaborating the ontology definitions and the ontology-based reasoning can improve the output of the method, providing a more accurate and prioritised risk identification.

Author Contributions: Conceptualization, A.S.; methodology, A.S.; software, A.S.; validation, A.S.; resources, O.M.; writing—original draft preparation, A.S.; writing—review and editing, O.M.; visualization, A.S.; project administration, O.M. All authors have read and agreed to the published version of the manuscript.

Funding: This research received no external funding.

Institutional Review Board Statement: Not Applicable.

Informed Consent Statement: Not Applicable.

Data Availability Statement: Not applicable.

Conflicts of Interest: The authors declare no conflict of interest.

Appendix A. The Case Study Formal Ontology (OWL Format)

This appendix provides the full ontology of the reported case study. The results are fully reproducible by copying the ontology into a text file and opening it with the Protégé ontology authoring tool.

```
<?xml version="1.0"?>
<Ontology xmlns="http://www.w3.org/2002/07/owl#">
  <Prefix name="owl" IRI="http://www.w3.org/2002/07/owl#"/>
  <Declaration>
    <Class IRI="owlapi:ontology578765402008551#Risk"/>
  </Declaration>
  <Declaration>
    <Class IRI="owlapi:ontology578765402008553#CybersecurityRisk"/>
  </Declaration>
  <Declaration>
    <Class IRI="owlapi:ontology578765402008555#VulnerabilityInducedRisk"/>
  </Declaration>
  <Declaration>
    <Class IRI="owlapi:ontology578765402008557#Application"/>
  </Declaration>
  <Declaration>
    <Class IRI="owlapi:ontology578765402008559#Component"/>
  </Declaration>
  <Declaration>
    <Class IRI="owlapi:ontology578765402008561#BusinessFunction"/>
  </Declaration>
  <Declaration>
    <Class IRI="owlapi:ontology578765402008563#SensitiveInformation"/>
  </Declaration>
  <Declaration>
    <Class IRI="owlapi:ontology578765402008565#Vulnerability"/>
  </Declaration>
```

```
<Declaration>
  <ObjectProperty IRI="owlapi:ontology578765402008567#includesComponent"/>
</Declaration>
<Declaration>
  <ObjectProperty IRI="owlapi:ontology578765402008569#foundIn"/>
</Declaration>
<Declaration>
  <ObjectProperty IRI="owlapi:ontology578765402008571#risksFunction"/>
</Declaration>
<Declaration>
  <ObjectProperty IRI="owlapi:ontology578765402008573#risksInfo"/>
</Declaration>
<Declaration>
  <ObjectProperty IRI="owlapi:ontology578765402008577#supportsFunction"/>
</Declaration>
<Declaration>
  <ObjectProperty IRI="owlapi:ontology578765402008581#accessInfo"/>
</Declaration>
<Declaration>
  <ObjectProperty IRI="owlapi:ontology578765402008588#risksVia"/>
</Declaration>
<Declaration>
  <ObjectProperty IRI="owlapi:ontology578765402008590#susceptible2Vulnerability"/>
</Declaration>
<Declaration>
  <NamedIndividual IRI="http://www.co-ode.org/ontologies/ont.owl#App1"/>
</Declaration>
<Declaration>
  <NamedIndividual IRI="http://www.co-ode.org/ontologies/ont.owl#App2"/>
</Declaration>
<Declaration>
  <NamedIndividual IRI="http://www.co-ode.org/ontologies/ont.owl#App3"/>
</Declaration>
<Declaration>
  <NamedIndividual IRI="http://www.co-ode.org/ontologies/ont.owl#App4"/>
</Declaration>
<Declaration>
  <NamedIndividual IRI="http://www.co-ode.org/ontologies/ont.owl#ClientIDsList"/>
</Declaration>
<Declaration>
  <NamedIndividual IRI="http://www.co-ode.org/ontologies/ont.owl#Log4j"/>
</Declaration>
<Declaration>
  <NamedIndividual IRI="http://www.co-ode.org/ontologies/ont.owl#Log4shell"/>
</Declaration>
<Declaration>
  <NamedIndividual IRI="http://www.co-ode.org/ontologies/ont.owl#OpenAccount"/>
</Declaration>
<EquivalentClasses>
  <Class IRI="owlapi:ontology578765402008555#VulnerabilityInducedRisk"/>
  <ObjectIntersectionOf>
    <Class IRI="owlapi:ontology578765402008565#Vulnerability"/>
    <ObjectUnionOf>
      <ObjectSomeValuesFrom>
```

```xml
            <ObjectProperty IRI="owlapi:ontology578765402008571#risksFunction"/>
            <Class IRI="owlapi:ontology578765402008561#BusinessFunction"/>
        </ObjectSomeValuesFrom>
        <ObjectSomeValuesFrom>
            <ObjectProperty IRI="owlapi:ontology578765402008573#risksInfo"/>
            <Class IRI="owlapi:ontology578765402008563#SensitiveInformation"/>
        </ObjectSomeValuesFrom>
      </ObjectUnionOf>
    </ObjectIntersectionOf>
  </EquivalentClasses>
  <SubClassOf>
    <Class IRI="owlapi:ontology578765402008553#CybersecurityRisk"/>
    <Class IRI="owlapi:ontology578765402008551#Risk"/>
  </SubClassOf>
  <SubClassOf>
    <Class IRI="owlapi:ontology578765402008555#VulnerabilityInducedRisk"/>
    <Class IRI="owlapi:ontology578765402008553#CybersecurityRisk"/>
  </SubClassOf>
  <SubClassOf>
    <Class IRI="owlapi:ontology578765402008555#VulnerabilityInducedRisk"/>
    <Class IRI="owlapi:ontology578765402008565#Vulnerability"/>
  </SubClassOf>
  <ClassAssertion>
    <Class IRI="owlapi:ontology578765402008557#Application"/>
    <NamedIndividual IRI="http://www.co-ode.org/ontologies/ont.owl#App1"/>
  </ClassAssertion>
  <ClassAssertion>
    <Class IRI="owlapi:ontology578765402008557#Application"/>
    <NamedIndividual IRI="http://www.co-ode.org/ontologies/ont.owl#App2"/>
  </ClassAssertion>
  <ClassAssertion>
    <Class IRI="owlapi:ontology578765402008557#Application"/>
    <NamedIndividual IRI="http://www.co-ode.org/ontologies/ont.owl#App3"/>
  </ClassAssertion>
  <ClassAssertion>
    <Class IRI="owlapi:ontology578765402008557#Application"/>
    <NamedIndividual IRI="http://www.co-ode.org/ontologies/ont.owl#App4"/>
  </ClassAssertion>
  <ClassAssertion>
    <Class IRI="owlapi:ontology578765402008563#SensitiveInformation"/>
    <NamedIndividual IRI="http://www.co-ode.org/ontologies/ont.owl#ClientIDsList"/>
  </ClassAssertion>
  <ClassAssertion>
    <Class IRI="owlapi:ontology578765402008559#Component"/>
    <NamedIndividual IRI="http://www.co-ode.org/ontologies/ont.owl#Log4j"/>
  </ClassAssertion>
  <ClassAssertion>
    <Class IRI="owlapi:ontology578765402008565#Vulnerability"/>
    <NamedIndividual IRI="http://www.co-ode.org/ontologies/ont.owl#Log4shell"/>
  </ClassAssertion>
  <ClassAssertion>
    <Class IRI="owlapi:ontology578765402008561#BusinessFunction"/>
    <NamedIndividual IRI="http://www.co-ode.org/ontologies/ont.owl#OpenAccount"/>
  </ClassAssertion>
```

```xml
<ObjectPropertyAssertion>
  <ObjectProperty IRI="owlapi:ontology578765402008567#includesComponent"/>
  <NamedIndividual IRI="http://www.co-ode.org/ontologies/ont.owl#App2"/>
  <NamedIndividual IRI="http://www.co-ode.org/ontologies/ont.owl#Log4j"/>
</ObjectPropertyAssertion>
<ObjectPropertyAssertion>
  <ObjectProperty IRI="owlapi:ontology578765402008567#includesComponent"/>
  <NamedIndividual IRI="http://www.co-ode.org/ontologies/ont.owl#App3"/>
  <NamedIndividual IRI="http://www.co-ode.org/ontologies/ont.owl#Log4j"/>
</ObjectPropertyAssertion>
<ObjectPropertyAssertion>
  <ObjectProperty IRI="owlapi:ontology578765402008581#accessInfo"/>
  <NamedIndividual IRI="http://www.co-ode.org/ontologies/ont.owl#App3"/>
  <NamedIndividual IRI="http://www.co-ode.org/ontologies/ont.owl#ClientIDsList"/>
</ObjectPropertyAssertion>
<ObjectPropertyAssertion>
  <ObjectProperty IRI="owlapi:ontology578765402008567#includesComponent"/>
  <NamedIndividual IRI="http://www.co-ode.org/ontologies/ont.owl#App4"/>
  <NamedIndividual IRI="http://www.co-ode.org/ontologies/ont.owl#Log4j"/>
</ObjectPropertyAssertion>
<ObjectPropertyAssertion>
  <ObjectProperty IRI="owlapi:ontology578765402008577#supportsFunction"/>
  <NamedIndividual IRI="http://www.co-ode.org/ontologies/ont.owl#App4"/>
  <NamedIndividual IRI="http://www.co-ode.org/ontologies/ont.owl#OpenAccount"/>
</ObjectPropertyAssertion>
<ObjectPropertyAssertion>
  <ObjectProperty IRI="owlapi:ontology578765402008569#foundIn"/>
  <NamedIndividual IRI="http://www.co-ode.org/ontologies/ont.owl#Log4shell"/>
  <NamedIndividual IRI="http://www.co-ode.org/ontologies/ont.owl#Log4j"/>
</ObjectPropertyAssertion>
<ObjectPropertyDomain>
  <ObjectProperty IRI="owlapi:ontology578765402008567#includesComponent"/>
  <Class IRI="owlapi:ontology578765402008557#Application"/>
</ObjectPropertyDomain>
<ObjectPropertyDomain>
  <ObjectProperty IRI="owlapi:ontology578765402008569#foundIn"/>
  <Class IRI="owlapi:ontology578765402008565#Vulnerability"/>
</ObjectPropertyDomain>
<ObjectPropertyDomain>
  <ObjectProperty IRI="owlapi:ontology578765402008571#risksFunction"/>
  <Class IRI="owlapi:ontology578765402008565#Vulnerability"/>
</ObjectPropertyDomain>
<ObjectPropertyDomain>
  <ObjectProperty IRI="owlapi:ontology578765402008573#risksInfo"/>
  <Class IRI="owlapi:ontology578765402008565#Vulnerability"/>
</ObjectPropertyDomain>
<ObjectPropertyDomain>
  <ObjectProperty IRI="owlapi:ontology578765402008577#supportsFunction"/>
  <Class IRI="owlapi:ontology578765402008557#Application"/>
</ObjectPropertyDomain>
<ObjectPropertyDomain>
  <ObjectProperty IRI="owlapi:ontology578765402008581#accessInfo"/>
  <Class IRI="owlapi:ontology578765402008557#Application"/>
</ObjectPropertyDomain>
```

```xml
<ObjectPropertyDomain>
  <ObjectProperty IRI="owlapi:ontology578765402008588#risksVia"/>
  <Class IRI="owlapi:ontology578765402008555#VulnerabilityInducedRisk"/>
</ObjectPropertyDomain>
<ObjectPropertyDomain>
  <ObjectProperty IRI="owlapi:ontology578765402008590#susceptible2Vulnerability"/>
  <Class IRI="owlapi:ontology578765402008557#Application"/>
</ObjectPropertyDomain>
<ObjectPropertyRange>
  <ObjectProperty IRI="owlapi:ontology578765402008567#includesComponent"/>
  <Class IRI="owlapi:ontology578765402008559#Component"/>
</ObjectPropertyRange>
<ObjectPropertyRange>
  <ObjectProperty IRI="owlapi:ontology578765402008569#foundIn"/>
  <Class IRI="owlapi:ontology578765402008559#Component"/>
</ObjectPropertyRange>
<ObjectPropertyRange>
  <ObjectProperty IRI="owlapi:ontology578765402008571#risksFunction"/>
  <Class IRI="owlapi:ontology578765402008561#BusinessFunction"/>
</ObjectPropertyRange>
<ObjectPropertyRange>
  <ObjectProperty IRI="owlapi:ontology578765402008573#risksInfo"/>
  <Class IRI="owlapi:ontology578765402008563#SensitiveInformation"/>
</ObjectPropertyRange>
<ObjectPropertyRange>
  <ObjectProperty IRI="owlapi:ontology578765402008577#supportsFunction"/>
  <Class IRI="owlapi:ontology578765402008561#BusinessFunction"/>
</ObjectPropertyRange>
<ObjectPropertyRange>
  <ObjectProperty IRI="owlapi:ontology578765402008581#accessInfo"/>
  <Class IRI="owlapi:ontology578765402008563#SensitiveInformation"/>
</ObjectPropertyRange>
<ObjectPropertyRange>
  <ObjectProperty IRI="owlapi:ontology578765402008588#risksVia"/>
  <Class IRI="owlapi:ontology578765402008557#Application"/>
</ObjectPropertyRange>
<ObjectPropertyRange>
  <ObjectProperty IRI="owlapi:ontology578765402008590#susceptible2Vulnerability"/>
  <Class IRI="owlapi:ontology578765402008565#Vulnerability"/>
</ObjectPropertyRange>
<SubObjectPropertyOf>
  <ObjectPropertyChain>
    <ObjectProperty IRI="owlapi:ontology578765402008567#includesComponent"/>
    <ObjectInverseOf>
      <ObjectProperty IRI="owlapi:ontology578765402008569#foundIn"/>
    </ObjectInverseOf>
  </ObjectPropertyChain>
  <ObjectProperty IRI="owlapi:ontology578765402008590#susceptible2Vulnerability"/>
</SubObjectPropertyOf>
<SubObjectPropertyOf>
  <ObjectPropertyChain>
    <ObjectProperty IRI="owlapi:ontology578765402008569#foundIn"/>
    <ObjectInverseOf>
      <ObjectProperty IRI="owlapi:ontology578765402008567#includesComponent"/>
```

```
      </ObjectInverseOf>
      <ObjectProperty IRI="owlapi:ontology578765402008577#supportsFunction"/>
    </ObjectPropertyChain>
    <ObjectProperty IRI="owlapi:ontology578765402008571#risksFunction"/>
  </SubObjectPropertyOf>
  <SubObjectPropertyOf>
    <ObjectPropertyChain>
      <ObjectProperty IRI="owlapi:ontology578765402008569#foundIn"/>
      <ObjectInverseOf>
        <ObjectProperty IRI="owlapi:ontology578765402008567#includesComponent"/>
      </ObjectInverseOf>
      <ObjectProperty IRI="owlapi:ontology578765402008581#accessInfo"/>
    </ObjectPropertyChain>
    <ObjectProperty IRI="owlapi:ontology578765402008573#risksInfo"/>
  </SubObjectPropertyOf>
  <SubObjectPropertyOf>
    <ObjectPropertyChain>
      <ObjectProperty IRI="owlapi:ontology578765402008571#risksFunction"/>
      <ObjectInverseOf>
        <ObjectProperty IRI="owlapi:ontology578765402008577#supportsFunction"/>
      </ObjectInverseOf>
    </ObjectPropertyChain>
    <ObjectProperty IRI="owlapi:ontology578765402008588#risksVia"/>
  </SubObjectPropertyOf>
  <SubObjectPropertyOf>
    <ObjectPropertyChain>
      <ObjectProperty IRI="owlapi:ontology578765402008573#risksInfo"/>
      <ObjectInverseOf>
        <ObjectProperty IRI="owlapi:ontology578765402008581#accessInfo"/>
      </ObjectInverseOf>
    </ObjectPropertyChain>
    <ObjectProperty IRI="owlapi:ontology578765402008588#risksVia"/>
  </SubObjectPropertyOf>
</Ontology>
```

Appendix B. Inferred Assertions by the Reasoner (OWL Format)

```
    <ClassAssertion>
      <Class IRI="owlapi:ontology578765402008551#Risk"/>
      <NamedIndividual IRI="http://www.co-ode.org/ontologies/ont.owl#Log4shell"/>
    </ClassAssertion>
    <ClassAssertion>
      <Class IRI="owlapi:ontology578765402008553#CybersecurityRisk"/>
      <NamedIndividual IRI="http://www.co-ode.org/ontologies/ont.owl#Log4shell"/>
    </ClassAssertion>
    <ClassAssertion>
      <Class IRI="owlapi:ontology578765402008555#VulnerabilityInducedRisk"/>
      <NamedIndividual IRI="http://www.co-ode.org/ontologies/ont.owl#Log4shell"/>
    </ClassAssertion>
    <ObjectPropertyAssertion>
      <ObjectProperty IRI="owlapi:ontology578765402008590#susceptible2Vulnerability"/>
      <NamedIndividual IRI="http://www.co-ode.org/ontologies/ont.owl#App2"/>
      <NamedIndividual IRI="http://www.co-ode.org/ontologies/ont.owl#Log4shell"/>
    </ObjectPropertyAssertion>
    <ObjectPropertyAssertion>
```

```
            <ObjectProperty IRI="owlapi:ontology578765402008590#susceptible2Vulnerability"/>
            <NamedIndividual IRI="http://www.co-ode.org/ontologies/ont.owl#App3"/>
            <NamedIndividual IRI="http://www.co-ode.org/ontologies/ont.owl#Log4shell"/>
        </ObjectPropertyAssertion>
        <ObjectPropertyAssertion>
            <ObjectProperty IRI="owlapi:ontology578765402008590#susceptible2Vulnerability"/>
            <NamedIndividual IRI="http://www.co-ode.org/ontologies/ont.owl#App4"/>
            <NamedIndividual IRI="http://www.co-ode.org/ontologies/ont.owl#Log4shell"/>
        </ObjectPropertyAssertion>
        <ObjectPropertyAssertion>
            <ObjectProperty IRI="owlapi:ontology578765402008571#risksFunction"/>
            <NamedIndividual IRI="http://www.co-ode.org/ontologies/ont.owl#Log4shell"/>
            <NamedIndividual IRI="http://www.co-ode.org/ontologies/ont.owl#OpenAccount"/>
        </ObjectPropertyAssertion>
        <ObjectPropertyAssertion>
            <ObjectProperty IRI="owlapi:ontology578765402008573#risksInfo"/>
            <NamedIndividual IRI="http://www.co-ode.org/ontologies/ont.owl#Log4shell"/>
            <NamedIndividual IRI="http://www.co-ode.org/ontologies/ont.owl#ClientIDsList"/>
        </ObjectPropertyAssertion>
        <ObjectPropertyAssertion>
            <ObjectProperty IRI="owlapi:ontology578765402008588#risksVia"/>
            <NamedIndividual IRI="http://www.co-ode.org/ontologies/ont.owl#Log4shell"/>
            <NamedIndividual IRI="http://www.co-ode.org/ontologies/ont.owl#App3"/>
        </ObjectPropertyAssertion>
        <ObjectPropertyAssertion>
            <ObjectProperty IRI="owlapi:ontology578765402008588#risksVia"/>
            <NamedIndividual IRI="http://www.co-ode.org/ontologies/ont.owl#Log4shell"/>
            <NamedIndividual IRI="http://www.co-ode.org/ontologies/ont.owl#App4"/>
        </ObjectPropertyAssertion>
    </Ontology>
```

References

1. *ISO 31073:2022*; Risk Management—Vocabulary. International Standardization Organization (ISO): Geneva, Switzerland, 2022.
2. Atkinson, C.; Cuske, C.; Dickopp, T. Concepts for an Ontology-Centric Technology Risk Management Architecture in the Banking Industry. In Proceedings of the 2006 10th IEEE International Enterprise Distributed Object Computing Conference Workshops (EDOCW'06), Hong Kong, China, 16–20 October 2006; p. 21. [CrossRef]
3. Refsdal, A.; Solhaug, B.; Stølen, K. *Cyber-Risk Management*; SpringerBriefs in Computer Science; Springer International Publishing: Cham, Switzerland, 2015; ISBN 978-3-319-23569-1.
4. Strupczewski, G. Defining Cyber Risk. *Saf. Sci.* **2021**, *135*, 105143. [CrossRef]
5. *ISO 31000:2018*; Risk Management—Guidelines. International Standardization Organization (ISO): Geneva, Switzerland, 2018.
6. Eling, M.; Schnell, W. What Do We Know about Cyber Risk and Cyber Risk Insurance? *J. Risk Financ.* **2016**, *17*, 474–491. [CrossRef]
7. Kosub, T. Components and Challenges of Integrated Cyber Risk Management. *Z. Für Die Gesamte Versicher.* **2015**, *104*, 615–634. [CrossRef]
8. Jackson, G. Contingency for Cost Control in Project Management: A Case Study. *Constr. Econ. Build.* **2003**, *3*, 1–12. [CrossRef]
9. Radanliev, P.; Charles, D.; Roure, D.; Nicolescu, R.; Huth, M.; Mantilla, R.; Cannady, S.; Burnap, P. Computers in Industry Future Developments in Cyber Risk Assessment for the Internet of Things. *Comput. Ind.* **2018**, *102*, 14–22. [CrossRef]
10. Shin, B.; Lowry, P.B. A Review and Theoretical Explanation of the 'Cyberthreat-Intelligence (CTI) Capability' That Needs to Be Fostered in Information Security Practitioners and How This Can Be Accomplished. *Comput. Secur.* **2020**, *92*, 101761. [CrossRef]
11. Kotsias, J.; Ahmad, A.; Scheepers, R. Adopting and Integrating Cyber-Threat Intelligence in a Commercial Organisation. *Eur. J. Inf. Syst.* **2022**, 1–17. [CrossRef]
12. Risk Registers (ENISA). Available online: https://www.enisa.europa.eu/topics/threat-risk-management/risk-management/current-risk/bcm-resilience/bc-plan/supporting-documents/risk-registers (accessed on 24 August 2022).
13. Chen, Y.; Boehm, B.; Sheppard, L. Value Driven Security Threat Modeling Based on Attack Path Analysis. In Proceedings of the Annual Hawaii International Conference on System Sciences, Waikoloa, HI, USA, 3–6 January 2007.
14. Zhao, J.; Yan, Q.; Li, J.; Shao, M.; He, Z.; Li, B. TIMiner: Automatically Extracting and Analyzing Categorized Cyber Threat Intelligence from Social Data. *Comput. Secur.* **2020**, *95*, 101867. [CrossRef]

15. Schauer, S.; Polemi, N.; Mouratidis, H. MITIGATE: A Dynamic Supply Chain Cyber Risk Assessment Methodology. *J. Transp. Secur.* **2019**, *12*, 1–35. [CrossRef]
16. Shaked, A.; Margalit, O. OnToRisk–A Formal Ontology Approach to Automate Cyber Security Risk Identification. In Proceedings of the 2022 17th Annual System of Systems Engineering Conference (SOSE), Rochester, NY, USA, 7–11 June 2022; pp. 74–79.
17. Gruber, T.R. Towards Principles for Design of Ontologies Used for Knowledge Sharing. *Int. J. Hum.-Comput. Stud.* **1995**, *43*, 907–928. [CrossRef]
18. Benjamins, V.R.; Davies, J.; Baeza-Yates, R.; Mika, P.; Zaragoza, H.; Greaves, M.; Gomez-Perez, J.M.; Contreras, J.; Domingue, J.; Fensel, D. Near-Term Prospects for Semantic Technologies. *IEEE Intell. Syst.* **2008**, *23*, 76–88. [CrossRef]
19. Chandrasekaran, B.; Josephson, J.R.; Benjamins, V.R. What Aro Ontologies, and Why Do We Need Them? *IEEE Intell. Syst. Appl.* **1999**, *14*, 20–26. [CrossRef]
20. Poole, D.L.; Mackworth, A.K. *Artificial Intelligence*; Cambridge University Press: Cambridge, UK, 2017; ISBN 9781107195394.
21. Wang, J.A.; Guo, M. Security Data Mining in an Ontology for Vulnerability Management. In Proceedings of the 2009 International Joint Conference on Bioinformatics, Systems Biology and Intelligent Computing (IJCBS 2009), Shanghai, China, 3–5 August 2009; pp. 597–603.
22. Gailly, F.; Alkhaldi, N.; Casteleyn, S.; Verbeke, W. Recommendation-Based Conceptual Modeling and Ontology Evolution Framework (CMOE+). *Bus. Inf. Syst. Eng.* **2017**, *59*, 235–250. [CrossRef]
23. Thomas, O.; Michael Fellmann, M.A. Semantic Process Modeling—Design and Implementation of an Ontology-Based Representation of Business Processes. *Bus. Inf. Syst. Eng.* **2009**, *1*, 438–451. [CrossRef]
24. Meng, X.; Wang, F.; Xie, Y.; Song, G.; Ma, S.; Hu, S.; Bai, J.; Yang, Y. An Ontology-Driven Approach for Integrating Intelligence to Manage Human and Ecological Health Risks in the Geospatial Sensor Web. *Sensors* **2018**, *18*, 3619. [CrossRef]
25. Shen, Y.; Xu, M.; Lin, Y.; Cui, C.; Shi, X.; Liu, Y. Safety Risk Management of Prefabricated Building Construction Based on Ontology Technology in the BIM Environment. *Buildings* **2022**, *12*, 765. [CrossRef]
26. Välja, M.; Heiding, F.; Franke, U.; Lagerström, R. Automating Threat Modeling Using an Ontology Framework: Validated with Data from Critical Infrastructures. *Cybersecurity* **2020**, *3*, 19. [CrossRef]
27. Aranovich, R.; Wu, M.; Yu, D.; Katsy, K.; Ahmadnia, B.; Bishop, M.; Filkov, V.; Sagae, K. Beyond NVD: Cybersecurity Meets the Semantic Web. In *ACM International Conference Proceeding Series*; Association for Computing Machinery: New York, NY, USA, 2021; pp. 59–69.
28. Mozzaquatro, B.; Agostinho, C.; Goncalves, D.; Martins, J.; Jardim-Goncalves, R. An Ontology-Based Cybersecurity Framework for the Internet of Things. *Sensors* **2018**, *18*, 3053. [CrossRef]
29. Vega-Barbas, M.; Villagrá, V.A.; Monje, F.; Riesco, R.; Larriva-Novo, X.; Berrocal, J. Ontology-Based System for Dynamic Risk Management in Administrative Domains. *Appl. Sci.* **2019**, *9*, 4547. [CrossRef]
30. Eckhart, M.; Ekelhart, A.; Weippl, E. Automated Security Risk Identification Using AutomationML-Based Engineering Data. *IEEE Trans. Dependable Secur. Comput.* **2022**, *19*, 1655–1672. [CrossRef]
31. Ghaffarian, S.M.; Shahriari, H.R. Software Vulnerability Analysis and Discovery Using Machine-Learning and Data-Mining Techniques: A Survey. *ACM Comput. Surv.* **2017**, *50*, 1–36. [CrossRef]
32. Yin, R.K. *Case Study Research: Design and Methods*; SAGE: New York, NY, USA, 2009; ISBN 9781412960991.
33. Adkins, H. *Review of the December 2021 Log4j Event*; Cybersecurity and Infrastructure Security Agency: Rosslyn, VA, USA, 2022.
34. Tuttle, H. 2022 Cyber Landscape. *Risk Manag.* **2022**, *69*, 18–23.
35. Protégé. Available online: http://protege.stanford.edu (accessed on 14 March 2022).
36. The CoModIDE Plugin for Protégé Repository. Available online: https://github.com/comodide/CoModIDE (accessed on 21 March 2022).
37. MITRE CVE Website. Available online: https://cve.mitre.org/ (accessed on 17 August 2022).

MDPI

St. Alban-Anlage 66

4052 Basel

Switzerland

Tel. +41 61 683 77 34

Fax +41 61 302 89 18

www.mdpi.com

Algorithms Editorial Office

E-mail: algorithms@mdpi.com

www.mdpi.com/journal/algorithms

Milton Keynes UK
Ingram Content Group UK Ltd.
UKHW050728190124
436268UK00004B/92

9 783036 582641